PLANETS, STARS, AND GALAXIES

PLANETS, STARS, AND GALAXIES

FOURTH EDITION

STUART J. INGLIS
CHABOT COLLEGE
LIVERMORE, CALIFORNIA

JOHN WILEY & SONS, INC.
NEW YORK LONDON SYDNEY TORONTO

Text design by Suzanne G. Bennett.

Cover illustration by Ed Cassel.

Library of Congress Cataloging in Publication Data

Inglis, Stuart J
Planets, stars, and galaxies.

Includes bibliographies and index.
1. Astronomy. I. Title.
QB43.2.I54 1976 520 75-31542
ISBN 0-471-42737-3
ISBN 0-471-42738-1 pbk.

Printed in the United States of America
10 9 8 7 6 5 4 3 2 1

TO
JENNIFER,
ADRIENNE,
AND JEFF

PREFACE

This fourth and brief edition is designed for colleges on the quarter system, or for instructors who want to use this book as part of a physical science course.

Since this edition is shorter than the previous ones, material of course has been omitted; the major consideration in this revision is the nonscience student. Extensive use of astronomical terms is not only frightening but rather useless to the nonscience student. I have, therefore, kept the astronomical terms at a reasonable level. The term meteorite, for example, is used to the exclusion of the term meteoroid. Why use two terms when one will suffice?

Nearly every chapter has been extensively rewritten. Each includes the latest possible observations. Today, however, it is inevitable that any edition will be the antecedent of some new and interesting observations. And so be it. Consequently, it falls to the instructor to introduce these new observations and to relate them to the material in the book.

The bibliography has been up-dated and includes many recent articles from *Sky and Telescope*, *Scientific American*, and *Natural History*. It is intended to be used by both the student and the instructor. The interested student will find much valuable material in the articles referred to.

With each edition of a book the author is able to correct some of the inevitable errors that creep in, but then new ones appear. I am grateful to those who took time to write and call my attention to errors. In particular I thank Professors Ernest H. Cherrington, Jr. of Hood College, Maryland and Emilia P. Belserene of Herbert H. Lehman College of the City University of New York.

With great pleasure I dedicate this book to my two daughters and son. Each has heard me say: "But I must work on the book now." Each has been patient with me. My wife continues to help me in every way. I am grateful for their consideration and help. I also thank Henrietta Picaud for helping me with the gruelling job of proof reading. Her care and cheerfulness are greatly appreciated.

STUART J. INGLIS

Livermore, California

TO THE STUDENT

To just read a chapter produces some learning. But to first skim and then to study the chapter carefully produces a deeper learning. This study often includes rereading and working over important words and concepts.

Each chapter has several aids to your study. At the beginning of the chapters are section headings and learning objectives. The section headings give you an idea of the chapter's content and organization. The learning objectives have been written to indicate some of the more important aspects of each chapter. Although these are not all inclusive, they can be used effectively as a study guide.

At the end of most chapters are vocabularies and questions. The vocabularies give you an idea of the more important words. The questions are used not only to review but occasionally to extend concepts covered in the chapter.

CONTENTS

PLANETS, STARS, AND GALAXIES

LEARNING OBJECTIVES

BE ABLE TO:

1. EXPLAIN HOW ASTRONOMY WAS A PRACTICAL SCIENCE BEFORE GALILEO.
2. GIVE THE MAIN REASONS FOR THE DEVELOPMENT OF THE CALENDAR.
3. DESCRIBE THE JULIAN CALENDAR.
4. DESCRIBE THE GREGORIAN CALENDAR.
5. DESCRIBE HOW GALILEO CHANGED THE STUDY OF ASTRONOMY.
6. COMPARE A PURE SCIENCE WITH AN APPLIED SCIENCE.

Chapter One
MAN AND THE HEAVENS

Man is curious about the heavens. Other animals see the stars; apparently some birds navigate by the stars during their long migrations. Yet man alone is curious about the heavens, and the world about him.

Without this curiosity, scientists, for example, would not have devised better microscopes to investigate the tiny microbes they had discovered. Since these microbes are, in themselves, interesting creatures, the early studies were not done with any thought of helping or improving the lot of mankind. Those studies were conducted for their own sake — because of man's inherent curiosity.

Yet the early studies of microbes paid huge dividends when, a century and a half later, it was discovered that many diseases are caused by those tiny microscopic creatures. None of the early scientists could have foreseen the intimate relationship between the objects of their study and the health of mankind.

Astronomers study celestial objects, but not because it will directly benefit mankind. They study the universe because the universe is, of itself, interesting. The studies of the universe have not only had an immense influence on man's concept of himself and his world, but they have added to the understanding of physics, lately even to chemistry, and most recently to the possible understanding of the origin of life on Earth. Organic compounds that may have led to more complex organic molecules are being found not only in meteorites that were formed in the solar system, but also in gaseous clouds between the stars in our galaxy.

If, the astronomer argues, the sun has planets about it, one with life on it, then why shouldn't other stars have planets and some of them have life? Would that life be like ours? Why not? What observations will help us answer these questions?

Although searching for life on other worlds is not the main occupation of astronomers, the recognition of such life will come only as a result of a more thorough understanding of the celestial objects, many of which we see as stars in the nighttime sky. Mankind wants to understand his environment, and that environment extends beyond our own backyard or the nearest polluted river. It extends to the microscopic beings and to the distant galaxies. If our horizons of interest shrink completely to our mundane lives, then our philosophical outlook of life shrinks accordingly, and man becomes smaller.

In the beginning of astronomy, the celestial objects were thought to be gods who intimately affected the people's lives. It was noticed, for example, that the seasons, which meant so much to the agrarian peoples of 3000 B.C., could be

Chapter Opening Photo
Stonehenge, an astronomical observatory in southern England, built perhaps 3800 years ago. (George Gerster/Rapho-Photo Researchers.)

Figure 1-1
The stars of the constellation Orion setting behind buildings of Lick Observatory. This photograph is a time exposure taken with a stationary camera. (Lick Observatory photograph.)

correlated with the motions of the sun, and later with the risings and settings of certain stars (Figure 1-1). Accordingly the sun, the stars, and the moon, too, played an important role in ancient astronomy.

As the sun stays in the sky for a longer period each day the weather becomes warmer. It is time to plant the crops. To the people who lived in ancient Egypt the lengthening day meant something more: the Nile, the source of their livelihood, would overflow its banks and supply water and fresh soil to the ground so that their crops could grow.

To be able to predict when the Nile would overflow became a vital concern of the ancient Egyptians. Their priests noticed that when a very bright star, the one we call Sirius, rose above the horizon concurrently with the sun, they could expect the river to overflow within a matter of days. The importance of the Nile in their lives led the priests to study more carefully the motions of the sun.

They found that the sun seemed to travel not only across the sky, but through the field of stars as well. If the sun rose with Sirius one morning, it would rise a little later than Sirius the next. After about 91 days Sirius would be high in the sky when the sun appeared on the eastern horizon. After about 182 days Sirius would be setting in the west when the sun was just rising above the horizon in the east. After 365 days, however, the sun would again rise with Sirius, the Nile would overflow, and a new year would begin. Thus the calendar was born.

The month was born of the motions of the moon. There is a new moon every 29½ days, which made it difficult to establish a year with an integral number of full months. This could not and cannot be done simply because 29½ does not divide into 365 an integral number of times.

When Sirius and the sun rise concurrently they appear at different points on the eastern horizon, Sirius more to the southeast than the sun. Huge temples were built with long narrow corridors directed to the exact spot where Sirius would appear. These dark corridors eliminated most of the light of the dawn and enabled the priests to see Sirius more clearly. Through prolonged observation, coupled with this improved method, they came to realize that Sirius rose concurrently with the sun not once every 365 days but every 365¼ days.

If the priests set up a calendar with only 365 days for every year, in 4 years the sun would rise with Sirius one day later than their calendar predicted; in 8 years it would be 2 days late. After 100 years the Nile would overflow its banks 25 days later than the calendar date which had been set 100 years previously.

The priests realized that in order to correct this error in the length of the year they would have to add one day every fourth year and thus make their predictions more accurate. The calendar itself, however, was not changed until the time of Julius Caesar when the Romans took over this knowledge and officially adopted leap year.

In the course of the ensuing centuries it was found that the simple leap year overcorrected the calendar since the year is actually 365.2422 days long or a trifle less than 365¼ days. This inaccuracy of the Julian Calendar, as it is called, was corrected in the sixteenth century by Pope Gregory XIII. The Gregorian Calendar, the one we use today, drops leap year at the close of each century excepting every fourth century. The century year is not a leap year unless it is a multiple of 400. The year 2000 will be a leap year, the years 1700, 1800, and 1900 were not. By this means 3 days out of every 400 years are dropped and the calendar is made accurate for centuries to come.

The Egyptians had other troubles, too. After many years of observation they found that Sirius could no longer be used for predicting the overflow of the Nile. The year stayed the same length, but something had happened that made

Figure 1-2
Diurnal (daily) motion of the stars about the North Celestial Pole. (Lick Observatory photograph.)

Sirius and the other stars appear to move very slowly in the sky. As a result, the priests had to find other stars that rose concurrently with the sun when the Nile was to overflow and new temples were built to observe these stars.

The Egyptians were well aware of the four cardinal points of the compass (North, South, East, and West). They noticed that each night the stars seemed to travel through the sky in circles centered on a common point in the north (see Figure 1-2). For many years there was a star near that point, but this star did not stay there through the centuries; it moved just as Sirius seemed to move in the sky. After hundreds of years another star was at that special place in the sky, about which the stars seem to rotate.

If we look out into the sky at night we can locate this spot, the north celestial pole, because there is a fairly bright star near it. This star is now called the North Star, or Polaris. Just as 3000 years ago a different star marked the north pole of the sky, 3000 years from now Polaris will not mark the north celestial pole; in fact there will be no star so close to that point. Polaris, however, will again be the North Star 26,000 years from now. Even before the time of Christ the ancient astronomers knew about this apparent motion of the stars, a motion we now call precession, and whose cycle lasts not 24 hours nor 365¼ days, but 26,000 years.

The brighter stars were named by the ancient peoples, chiefly the Arabs and the Greeks. Such names as Aldebaran, Betelgeuse, Deneb, and Vega are Arabic in origin. Others, such as Antares, Canopus, Procyon, and Pollux, were given to us by the Greeks. A few, such as Capella, were derived from the Latin.

These ancient peoples watched the sky closely and told stories about the figures they imagined to be represented by the configurations of stars. These figures, our constellations, divide our sky as counties divide a state. Examples are the Big Dipper (correctly called Ursa Major, the Big Bear); Orion, the hunter; Gemini, the twins (from which we derive our expression "by Jiminy"); Scorpius; and the archer, Sagittarius. Most of the constellations seen from the Northern Hemisphere are the subject of much myth and folklore.

The myth and folklore, generated by the ancient Greeks and Arabs telling stories night after night under the clear desert sky, led to the idea that the stars are somehow related to our daily lives. And thus astrology was born. Astrology depends on the motions of the planets and on the ability to judge the time of birth of an individual. Consequently, as astrology became more popular, it stimulated interest in increasing the accuracy of astronomical observations.

Astronomy has become more precise, the fulfillments of its predictions today would be classed as miracles by the ancients. The accuracy and predicting ability of astrology, however, have not become noticeably better than they were 2000 years ago.

The fainter stars were not named in antiquity but are now named according to the catalog in which they appear. For example, B.D. +30°3639 is a star in the catalog compiled in Bonn called the *Bonner Durchmusterung*. Nonstellar objects are also named after the catalog in which they are listed. M 31 is in Messier's catalog, and N.G.C. 6523 is listed in the New General Catalog. In addition to catalog numbers some of the more famous nonstellar objects have names that derive from the appearance of the object (for example, the Crab nebula, the Owl nebula) or from the constellation in which they appear (for example, the Orion nebula).

Many people take pride in our particular modern civilization with its many miraculous inventions, but should we not take pride also in the other great civilizations that mankind has produced? Many people in different parts of the world have known a great deal about the very complicated motions of the sun, moon, stars, and planets without the aid of our technology.

In 2254 B.C. the early Chinese civilizations had a calendar that was essentially correct. In A.D. 1279, at the time of Kubla Khan, Chinese astronomers built the first observatory in the world to be equipped with instruments. The Mayas, of what is now Central America, had a calendar that accounted for not only the motions of the sun but also the motions of the moon, Venus, and Mars. This calendar was revised in the year A.D. 1091. The ancient Hindus had developed a calendar that included periods of time up to 4,320,000,000 years. This, interestingly enough, is about the age of the Earth.

Over 2000 years ago the Phoenicians were using the stars to guide their ships on long voyages across the Mediterranean Sea, through the Pillars of Hercules, and up to the British Isles. The Polynesians made even more astounding sea voyages. Unlike the Phoenicians, they had no coast lines which could be used as landmarks. Sometimes the Polynesians sailed between islands more than a thousand miles apart.

This seems amazing to us who tend to look at any but our modern technical society as primitive. We have newspapers, radios, and printed calendars to tell us the day of the month. As a result the average person today is neither aware of nor concerned with the movements of celestial objects. The sun and stars are not our timepieces; the clock and the calendar serve us better. But both our calendar and the rate of the clock are determined by the motion of the Earth as it turns on its axis once a day and as it makes its complete trip around the sun each year. Modern man, living in large metropolitan areas ablaze with lights and shrouded with the gases and dust of civilization, scarcely sees the stars on the *celestial sphere* over his head. He hardly glances at the beauty of the celestial sphere, the myriads of stars, and the graceful motions of the planets. He is aware only of the sun and occasionally of the moon, but he is not aware that both the sun and the moon move on the celestial sphere — the dome of the heavens.

Astronomy was a practical science in its early conception, when it was needed to navigate and to develop a calendar. In 1609, however, the approach to astronomy changed. It was then that Galileo first looked through his homemade telescope and saw the mountains on the moon, as well as four other moons revolving about Jupiter. He noticed that Mars and Saturn are not like the stars, for when they are magnified they appear as disks, while the stars remain but points of light. He saw that Venus is different from the other planets in that it goes through phases like the moon. Galileo was fascinated by these phenomena, and his interest in them for their own sake helped turn astronomy into a pure science.

Galileo was concerned with things so distant, so vast, and so mysterious that he followed the urge to learn about our universe even though what he learned might have no direct effect upon human life. He and the other astronomers who have followed have seen the unseen; they are revealing mysteries that seem almost beyond the comprehension of mankind.

Yet studying the vastness of space and time may have some effect, not on our daily life perhaps, but upon our thinking and concept of life. Little by little we realize that the Earth, once thought to be the very center of the entire universe, is really just a bit of matter revolving around one of billions of stars that form, together with vast volumes of cosmic dust and gas, a huge galaxy. Comprehending this, we must next confront the fact that this huge galaxy of stars, which we call the Milky Way system, is only one of billions of such galaxies, each with its billions of stars. We are lost in the universe so long as we speak in terms of miles and not light years; so long as we look with our eyes and not with huge telescopes; so long as we think of mankind as having control of vast amounts of energy when we control the nucleus of the atom, and fail to compare this energy with that released by an exploding star.

Perhaps this concept of the universe will, after all, have some effect on our daily lives: that Earth, with powerful mankind scurrying over its surface, is less than insignificant in size and influence when measured on a universal scale.

Perhaps, as humans travel out into space and look back, we will finally realize that the world's inhabitants are common occupants of the spacecraft we have called The Earth. We are all neighbors, for to travel to the far ends of the Earth is like going next door when compared with a trip to the moon. Perhaps, when we realize what close neighbors we are, we will learn to get along better one with another.

We hope we will learn how to care for our Earth better as we come to realize that it is really a cage whirling through space, a cage with distinct and emphatic limitations. If space travel will help us realize that we must not only live together in this cage, but within the limitations set by Nature when She made the Earth, then space travel will have made a truly significant contribution to society.

QUESTIONS AND PROBLEMS

1. The star trails in Figure 1-2 were photographed with a stationary camera set on time exposure. Recognizing that the star trails would make one complete circle in 24 hours, estimate the length of the time used for this photograph.

2. According to the Julian calendar, one day must be added every 4 years (leap year), so that the sun continues to be in the Vernal Equinox on March 21 (or thereabouts). This is the first day of spring. If that one day were not added every 4 years, would the first day of spring regress from March 21 to the 20th, 19th, etc., over the years? Or would it advance to the 22nd, 23rd, etc.?

3. The Gregorian calendar recognizes that the year is not quite 365.25 days but 365.2422 days. Therefore adding one day every 4 years is too much. From the year A.D. 323, the last year the first day of spring fell on March 21 during the use of the Julian calendar, until the year 1582, when our present calendar was adopted, too many days were added because one day was added every 4 years. (a) What decimal fraction of a day was added in excess each year? (b) How many excess days were added for the entire 1259 years, from A.D. 323 until 1582?

4. Great Britain did not adopt the Gregorian calendar (for political and religious reasons) until 1752. By then the calendar had shifted by 11 days. The first day of spring had shifted 11 days from March 21. (a) Did it occur on March 10 or April 1? (b) Did the government have to add 11 days to the year of 1752 or delete them? (c) How do you suspect this affected the relationship between landlord and renter? Between employer and employee?

5. Locate Polaris, and by watching the stars during the night verify the statement that each travels in a circle concentric with the North Celestial Pole.

6. With a little effort, a great deal of enjoyment can be gained by learning to recognize the more important constellations.

7. As a topic of discussion consider the question: Is astronomy a practical science, or is it a pure science?

FOR FURTHER READING

Bernhard, Herbert J. et al., *New Handbook of the Heavens,* McGraw-Hill paperback, New York, 1964.

Hawkins, Gerald S., *Stonehenge Decoded,* Dell Publishing Co. paperback, New York, 1965.

Menzel, Donald H., *A Field Guide to the Stars and Planets,* Houghton Mifflin Co., Boston, 1964.

Ronan, Colin, *The Astronomers,* Hill and Wang, New York, 1964.

Thiel, Rudolf, *And There Was Light,* Mentor, New York, 1960.

Emlen, Stephen T., "The Celestial Guidance System of a Migrating Bird," *Sky and Telescope,* p. 4, July 1969.

Emlen, Stephen T., "The Stellar-Orientation System of a Migratory Bird," *Scientific American,* p. 102, August 1975.

Graubard, Mark, "Under the Spell of the Zodiac," *Natural History,* p. 10, May 1969.

Kals, William S., "Polynesian Navigation," *Sky and Telescope,* p. 358, June 1967.

Price, D. J. de Solla, "Unworldly Mechanics," *Natural History,* p. 9, March 1962.

Price, D. J. de Solla, "Astronomy's Past Preserved at Jaipur," *Natural History,* p. 48, June–July 1964.

Ronan, Colin A., "Astronomy and Music," *Sky and Telescope,* p. 145, September 1975.

"The Sun," *Natural History,* p. 38, April 1963.

"Visits to Stonehenge and Herstmonceux," *Sky and Telescope,* p. 197, October 1970.

CHAPTER CONTENTS

LEARNING OBJECTIVES

BE ABLE TO:

1. WRITE SIMPLE NUMBERS IN THE EXPONENTIAL SYSTEM.
2. DESCRIBE ROEMER'S OBSERVATION OF THE SPEED OF LIGHT.
3. GIVE EXAMPLES OF REFRACTION AND REFLECTION.
4. DESCRIBE THE REFRACTING TELESCOPE: THE OBJECTIVE LENS, FOCAL PLANE, FOCAL LENGTH, IMAGE, EYEPIECE.
5. DESCRIBE THE REFLECTING TELESCOPE: CONCAVE MIRROR, PRIME FOCUS, NEWTONIAN FOCUS.
6. DESCRIBE THE SPECTROGRAPH: SLIT, COLLIMATOR, PRISM.
7. DESCRIBE HOW EACH KIND OF SPECTRUM IS FORMED.
8. EXPLAIN WHAT IS MEANT BY MEASURING THE WAVELENGTH OF LIGHT.
9. EXPLAIN HOW RADIAL VELOCITY IS DETERMINED.
10. EXPLAIN WHAT HAPPENS TO THE ELECTRON IN A HYDROGEN ATOM AS THE ATOM FIRST ABSORBS A PHOTON AND THEN EMITS ONE.
11. DESCRIBE HOW LINES IN THE BALMER SERIES ARE FORMED.
12. GIVE THE SEVEN PARTS OF THE RADIANT ENERGY SPECTRUM.
13. DESCRIBE HOW A RADIO TELESCOPE RESEMBLES AN OPTICAL TELESCOPE.

Chapter Two
BASIC TOOLS AND METHODS

The physicist, the geologist, the biologist, and the chemist are all able to study their subjects in the laboratory or at least to go into the field and make direct observations by many different methods. The astronomer, on the other hand, studies objects that are too big and too far away. With the exception of the moon and meteorites (bits of material that move in orbits about the sun much as the Earth does and that occasionally fall onto the Earth), astronomical objects can be studied mainly by the light that we receive from them.

Since the astronomer is dependent upon light we must first consider how it is possible to learn so much about the planets, stars, and galaxies from it alone. To understand this we will need to know some of the characteristics of light. These will be taken up one at a time.

2.1
SOME CHARACTERISTICS OF LIGHT

A. LIGHT TRAVELS IN A STRAIGHT LINE An important characteristic of light is its property to travel in a straight line. We observe this straight-line travel in the laboratory and on the Earth's surface, and if we assume that light travels in a straight line in free space, beyond the Earth's atmosphere and beyond the solar system, then a star's position in the sky can be determined by noting the direction from which the light comes; that is, by pointing the telescope to the star.

B. WAVE CHARACTERISTICS OF LIGHT Many experiments lead us to believe that light has wavelike characteristics; in some respects it behaves like a wave caused by a pebble dropped into a smooth pond. A series of these waves on a surface of water is called a wave train (Figure 2-1). Each wave in that train is made up of one crest and one trough. The length of this crest and trough is called the *wavelength*. Waves on the surface of a pond might have a length of 1 centimeter [1 inch equals 2.54 centimeters, or 1 centimeter (cm) equals about 0.4 inch]. The wavelength of light is much shorter. The Greek letter λ (lambda) has come to designate the wavelength.

Although it is difficult to measure, the wavelengths of light have been measured with great precision. The wavelength of violet light is about 4.0 ×

Chapter Opening Photo
The dome housing the 4-meter Mayall telescope on Kitt Peak, Arizona. (Kitt Peak National Observatory photograph.)

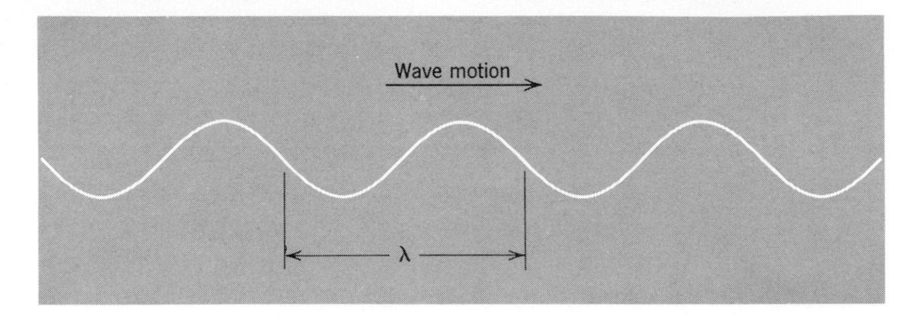

Figure 2-1
One wavelength, λ, is one complete crest and trough.

10^{-5} cm* (about 1.6×10^{-5} in.); the wavelength of red light is about 7.0×10^{-5} cm (about 2.8×10^{-5} in.). The wavelengths of all the other colors lie between these two extremes. Since the wavelengths of light are so short a unit of measure, the *angstrom* (abbreviated Å) is generally used. One angstrom equals 10^{-8} cm, so the wavelength of violet light is about 4000 Å; the wavelength of red light is about 7000 Å.

2.2 THE VELOCITY OF LIGHT

Light travels at a very high velocity; a velocity so high that it becomes of interest to learn how it was first measured. Before it was measured, no one knew whether the velocity of light was finite or infinite. If it were finite, light would require a certain length of time to travel from one point to another. If the velocity were infinite, however, light would need no time at all to travel from one point to another.

The first measurement was made as early as 1675 by an astronomer named Roemer, who at the time was studying the motions of the satellites of Jupiter. Each satellite goes around Jupiter in a set length of time, called a *period*. The

*When dealing with very large and very small numbers, it is much easier to use the exponential system for keeping track of the number of zeros. This system is used throughout this book:

$$1 \times 10^{-10} = 0.000,000,000,1$$
$$4.0 \times 10^{-5} = 0.000,04$$
$$2.1 \times 10^{-2} = 0.021$$
$$10^{-1} = 0.1$$
$$10^{0} = 1.0$$
$$3 \times 10^{1} = 30$$
$$6.1 \times 10^{2} = 610$$
$$10^{3} = 1000$$
$$4.0 \times 10^{6} = 4,000,000$$
$$5.5 \times 10^{9} = 5,500,000,000$$

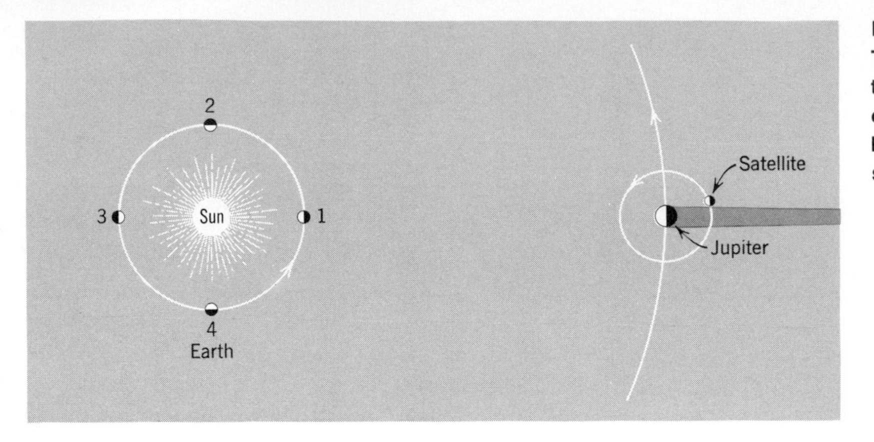

Figure 2-2
The relative positions of the Earth, Jupiter, and one of Jupiter's satellites used by Roemer to measure the speed of light.

period of each satellite had been determined and Roemer was trying to make a more accurate determination. He timed one of the satellites from the moment it emerged from the shadow of Jupiter until it emerged once again, the satellite having made one complete revolution during this time.

Roemer realized that he would be able to determine a period more accurately if he timed many periods and took their average. This involved him in observations over a considerable length of time, in the course of which a curious phenomenon came to his attention. For the first six months the satellite seemed to emerge from Jupiter's shadow somewhat later than he had anticipated. Then for the second six months it emerged a little earlier until it was back on schedule again.

Roemer explained this by a hypothesis that combined the motion of the Earth around the sun with the assumption that light must have a finite velocity (Figure 2-2). When the Earth was in position 1, he obtained a good value for the length of the period. But as the Earth moved on to position 2 and then to position 3, the satellite appeared to emerge a little later because the light that it reflected and by which alone it could be observed had to travel farther each time to reach the Earth as the Earth receded from Jupiter. (Jupiter moves so much more slowly than the Earth that we may consider it stationary for the purposes of this explanation.)

Since the Earth was moving away from Jupiter the fastest when at position 2, the period was the longest there. The period observed with the Earth at position 3 would have been equal to that observed near position 1, for the Earth is neither moving away from nor toward Jupiter in either position. Acually the planet is not visible from position 3, for the sun is in the way. So mathematics based on the observed changes in the period as the Earth moves from 1 to 2 and on toward 3 must be used to predict the values at position 3.

At position 3 the satellite emerges 16 minutes 37 seconds (Roemer's less accurate value was 22 minutes) later than the schedule established with the Earth at position 1. It takes this much time for light to travel the full diameter (2 radii) of the Earth's orbit. This is nearly 1000 seconds to travel a distance of 186,000,000 miles (the radius of the Earth's orbit is close to 93,000,000 miles — the distance of the Earth from the sun) or a velocity of light close to 186,000 mi/sec.

By this observation Roemer not only demonstrated that light has a finite velocity, but acceptance of this observation in the late 1670s carried with it the acceptance of the notion that the Earth revolves around the sun and not the sun around the Earth.

From this discussion it is evident that when we observe any celestial object, we observe not the object itself as it is at the moment of observation, but as it was at the moment the light we observe left it. That is, what we see is the object itself at a moment in the past. The interval between that moment and the present is measured by the length of time it has taken the light to travel from the object to the Earth.

2.3 REFLECTION, REFRACTION, AND THE TELESCOPE

Light, then, travels at a very high velocity and generally in a straight line. The two most obvious exceptions to the generalization that it travels in a straight line are reflection and refraction.

Reflection occurs when light is reflected from a mirror or some other object; *refraction* occurs when light travels from one substance, such as air, into some other substance, such as glass or water. This can be seen when a pencil is placed in a glass of water. The pencil appears to bend at the surface of the water. That the pencil itself is not bent can be shown by removing it. It is the light bending as it emerges from the water that creates the illusion that the pencil is bent. Refraction results from a change of velocity when light passes from one medium into another.

Only because we can change the direction of light's travel are we able to make telescopes. These two ways of changing the direction of travel enable us to build two basically different types of telescopes.

A. THE REFRACTING TELESCOPE The most common refracting telescopes are built according to the simplified drawing in Figure 2-3. The main lens of a telescope is called the *objective lens*. It is the purpose of the objective lens to gather the parallel rays that go to make up a parallel beam of light and to focus them at one point.

The light from a single star (*a* or *b*) enters our figure from the right. Since the star is so far away the light travels in an essentially parallel beam. However, the beam of light from star *a* is not parallel to the beam from star *b*; therefore these two beams will be brought to a focus at two different points. All such points from many stars define what is called the *focal plane*. If the objects observed with

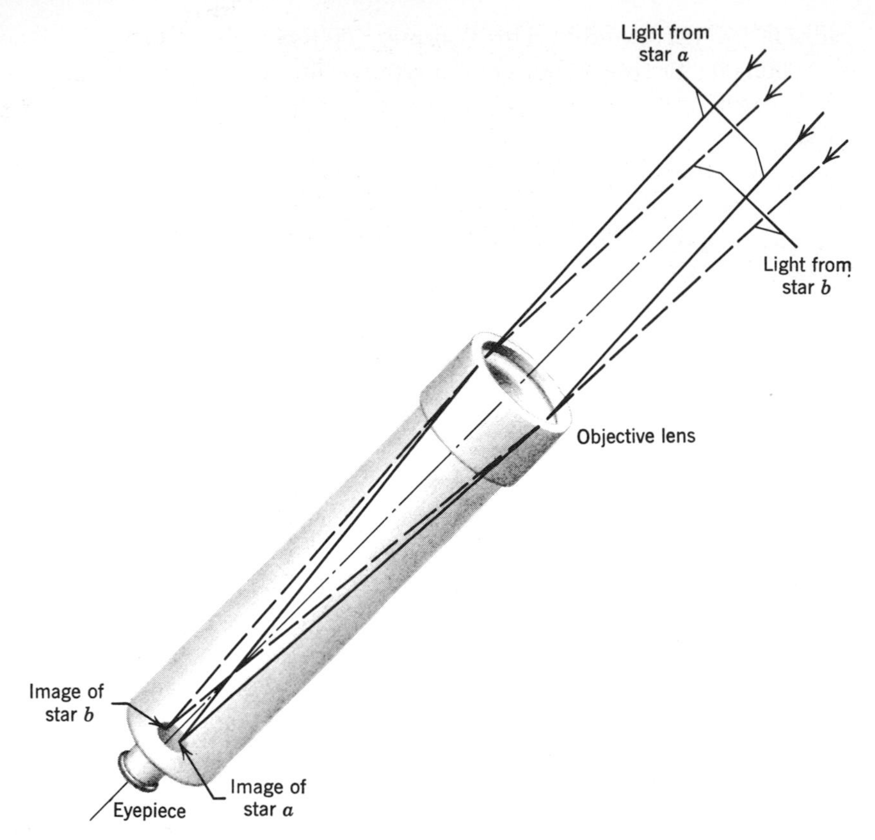

Figure 2-3
Schematic diagram of a refracting telescope.

the telescope are a long ways away, such as all astronomical objects, then the distance between the focal plane and the lens is called the *focal length* of that lens. If objects closer to the telescope are viewed, their images are located at a greater distance from the lens than the focal length.

It is possible to place a photographic plate at the focal plane and take a picture of a star field or a planet. The telescope is then a very large and specialized camera.

If the telescope is pointed toward a planet, the light from any one point on the planet's surface will be traveling in a parallel beam and all the light from this given point will be brought to a sharp focus. This is true for every point on the surface of the planet. All these focused points considered together are called an *image* of the planet. The image is real in the sense that a piece of paper could be placed at the focal plane and the image could be seen on the paper. Again, a photographic plate can be placed at the focal plane and a photograph taken of the planet; or an eyepiece, which is essentially a magnifying glass, can be placed just beyond the focal plane and the image, thus magnified, may be seen by the eye.

The larger the objective lens, the more light it can gather and thus the fainter the object that can be seen or photographed. It is because of this that astronomers

Figure 2-4
The 36-in. refracting telescope of the Lick Observatory, Mount Hamilton, California. (Lick Observatory photograph.)

have built larger and larger telescopes and have seen correspondingly fainter stars. The three biggest refracting telescopes in the world today are the one at Yerkes Observatory in Wisconsin which has an objective that is 40 in. in diameter; the one at Lick Observatory in California with an objective 36 in. in diameter (Figure 2-4); and a 33-in. refractor at the Meudon Observatory in Paris, France.

B. THE REFLECTING TELESCOPE We have noted that the direction of light can be changed by reflection as well as by refraction. It was Sir Isaac Newton who, in developing the first reflecting telescope, put this principle of optics to use. A schematic drawing of a reflecting telescope is shown in Figure 2-5. Light coming in a parallel beam from a distant star strikes the concave mirror and is converged and focused to a point, F. This point is called the *prime focus*. In order to look into an eyepiece placed near the prime focus of this telescope, however, the head of the observer must obstruct the incoming light and thus reduce the amount of light. So Newton placed a small flat mirror in the beam of light converged by the concave mirror and reflected that converging beam to a point, F', called the *Newtonian focus*. The Newtonian focus is outside the telescope tube and therefore readily available to the astronomer. It is true that even this small mirror prevents some light from reaching the objective, but since it is comparatively small little light is lost.

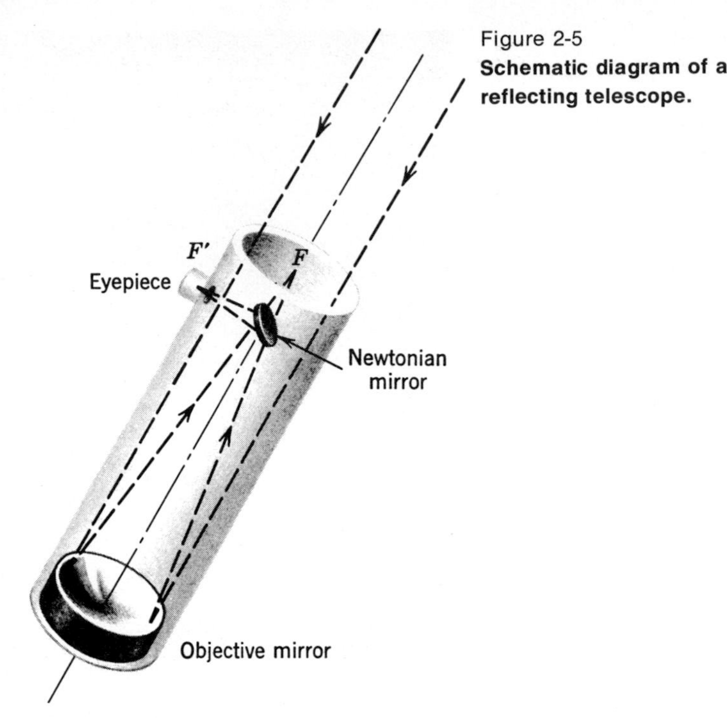

Figure 2-5
Schematic diagram of a reflecting telescope.

There are many large reflecting telescopes in the world. The largest are so large that, instead of a small flat mirror at the prime focus, they have a little cage where the observer rides. The largest are: the 236 in. telescope in the Caucasus Mountains in Russia, the 200-in. of the Hale Observatories on Mount Palomar in California, and two nearly identical 158-in. (about 4 meters) telescopes, one on Kitt Peak in Arizona (Figure 2-6) and the other at Cerro Tololo in Chile. Large reflecting telescopes are even housed in observatories on the sometimes snowy summit of a dormant volcano in Hawaii (Figure 2-7).

2.4 THE SPECTROGRAPH

When sunlight shines through a raindrop, the refraction of the light passing from air into water and back into the air again causes the original white light to spread out into a rainbow. White light entering a glass prism may be refracted so that a rainbow, or *spectrum,* is formed. In studying the spectrum of the sun, a German astronomer, Fraunhofer, found that if he put the image of the sun on a narrow slit in front of the prism the spectrum contained a number of dark lines. A narrow slit can be made by placing the cutting edge of two razor blades very close together. This will allow only a very narrow beam of light to pass through (Figure 2-8). The narrow beam will diverge until it passes through a first lens, called the *collimator.* It then becomes a parallel beam of light. The parallel beam

Figure 2-6
N. U. Mayall standing beside the 4-meter Mayall reflecting telescope at the Kitt Peak National Observatory. (Kitt Peak National Observatory photograph.)

is sent through a prism, which disperses the light into its component colors. It then travels through a second lens by which it is focused on a photographic plate. Thus the spectrum is photographed. The entire instrument is called a *spectrograph*.

Each dark line that Fraunhofer found is an image of the narrow slit. If he had placed the original beam of light on a small circular hole he would have found small dark circular spots on the spectrum rather than dark lines. If the slit is made wider, the lines become wider. Fraunhofer did not know the cause of these dark lines but he did letter them *A*, *B*, *C*, etc., according to their position in the spectrum. This lettering is still used. It remained for another German, Kirchhoff, to discover what in the sun caused the dark lines.

2.5 THREE KINDS OF SPECTRA

What Kirchhoff discovered was that there are really three kinds of spectra, each formed under different conditions.

A. CONTINUOUS SPECTRUM A solid object like a red-hot toaster filament will give a *continuous spectrum*, that is, a continuous array of colors from violet through blue, green, yellow, and orange to red. An incandescent gas under pressure or an incandescent liquid will also emit a continuous spectrum.

Figure 2-7
The Mauna Kea Observatory on the Island of Hawaii. Mauna Loa, a shield volcano, is in the background. (Courtesy D. Morrison, Mauna Kea Observatory, University of Hawaii.)

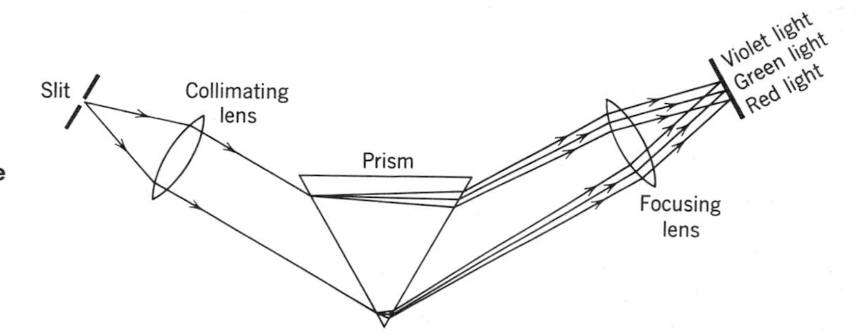

Figure 2-8
A schematic diagram of a spectrograph. The spectrograph is attached to the telescope so that the slit is in the focal plane of the objective.

B. BRIGHT-LINE SPECTRUM On the other hand, a gas under low pressure, that is, at much less than atmospheric pressure, when caused to emit light (as in a neon sign) will not give a continuous spectrum but a series of bright lines on a dark background. Such a spectrum is called a *bright-line spectrum* and is seen in Figure 2-9*a*. Here again each line is an image of the slit. If a circular hole had been used, a series of differently colored round spots would have been seen. It is easier, however, to work with thin lines.

Each gas has its own distinctive spectrum composed of a certain number of lines arranged in a fixed sequence. The spectrum of hydrogen (Figure 2-11) is quite simple, whereas that caused by iron when vaporized (Figure 2-9*a*) has many lines and is quite complex. It may seem difficult to vaporize iron, but all that need be done is to cause electricity to spark between two pieces of iron; the spark vaporizes a bit of the metal and thus gives a bright-line spectrum.

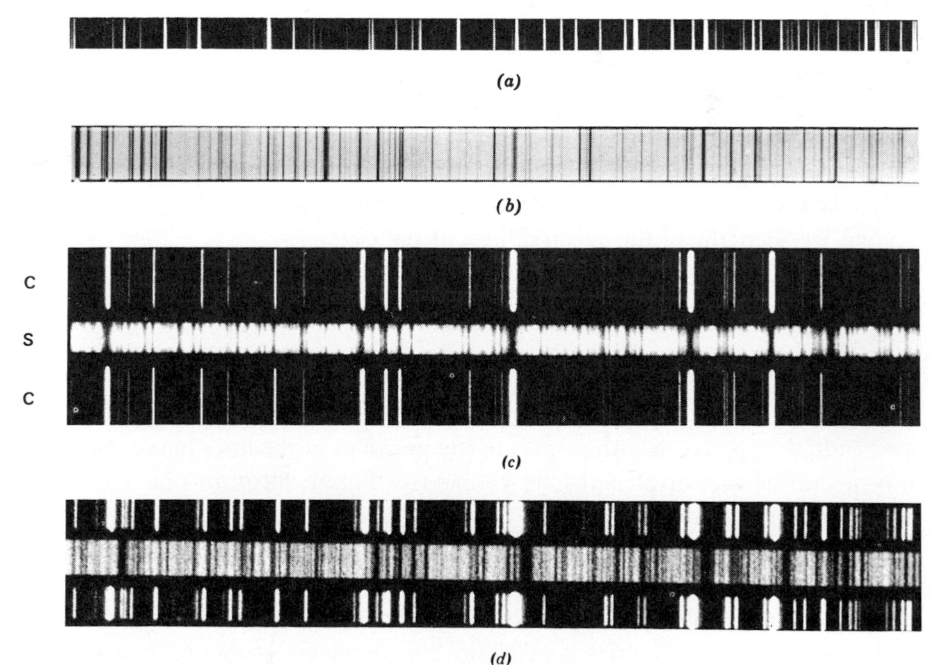

Figure 2-9
***(a)* A bright-line spectrum of iron. *(b)* A dark-line stellar spectrum. *(c)* A stellar dark-line spectrum *S* has a comparison (bright-line) spectrum *C* above and below it. The iron lines in the stellar spectrum line up with the iron lines in the comparison spectra. *(d)* The iron lines in this stellar spectrum are each shifted a bit to the right, to the longer wavelength, from the lines in the iron comparison spectra. This star is receding from us. (Photographs from the Hale Observatories.)**

C. DARK-LINE SPECTRUM If the light from a source that emits a continuous spectrum shines through a large container full of gas that is cooler than the source and under low pressure, the continuous spectrum from the source is crossed by a series of dark lines. These dark lines are in the same position as the bright lines for the same gas when incandescent. This type of spectrum is known as a *dark-line* or *absorption spectrum* (Figure 2-9*b*).

It should be noted that these lines can originate only with a gas under low pressure. Even the smallest dust particles between the observer and the source of a continuous spectrum will not yield absorption lines.

The work of Fraunhofer and Kirchhoff enables us to determine what chemical elements are in the atmosphere of a star. An atmosphere acts like a relatively cool gas that absorbs the light from a hotter source which emits a continuous spectrum. Thus the sun's atmosphere caused the dark lines that Fraunhofer first saw.

To determine what chemicals go to make up a star's atmosphere we can take a bright-line source such as a spark between two pieces of iron, and let its light travel through the spectrograph parallel to the star's light. This gives the comparison spectra shown in Figure 2-9*c* and *d*. The star's spectrum (an absorption spectrum) is in the middle while the two comparison spectra (both bright-line spectra) are on the top and bottom. We know that the lines in the comparison spectra are caused by iron, and we can see that there are lines in the star's spectrum that are located in nearly the same position. This tells us that the star's atmosphere must contain iron.

Since there are more than 100 elements it would not be practicable to form a comparison spectrum for each one of them. There must be an easier means of identifying the dark lines in a stellar spectrum. The wavelengths of the spectral lines of iron can be measured, and by using them as a reference scale we can measure the wavelengths of all the lines in a star's spectrum. The physicists and chemists, too, are interested in the spectra of elements, and they have measured the wavelength of spectral lines for nearly all of them. The astronomer need only take the wavelengths of the star's spectral lines and match them against the known wavelengths of the spectral lines of the elements.

2.6 MEASURING WAVELENGTHS OF SPECTRAL LINES

"Measuring" the wavelength of a stellar spectral line consists in determining its *position* in the spectrum with respect to the positions of the lines in the comparison spectra whose wavelengths are known (see Figure 2-10). In practice, however, this is not quite such an easy matter. There are many, many lines in the spectrum of almost every element, and the gases that make up the atmospheres of the stars are not under the same conditions as gases in a laboratory. Consequently, there are always differences that complicate the situation immensely.

There are other difficulties, too. Notice the placement of the stellar iron lines

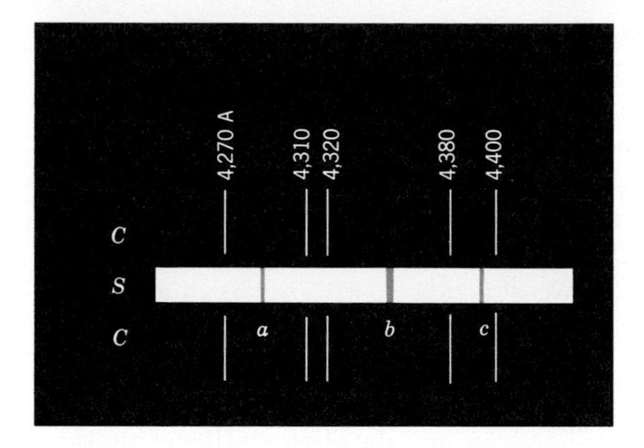

Figure 2-10
Estimate the wavelengths for the three stellar lines *a*, *b*, and *c*.

in Figure 2-9*d* with respect to the iron lines in the comparison spectra. The stellar iron lines are all shifted to the right (the red). It might be said that the lines we have attributed to iron are not iron lines but something else very much like iron. But there is no other element that has spectral lines in this same sequence, so there must be some other explanation for the shift.

2.7 THE DOPPLER EFFECT AND RADIAL VELOCITY

For the explanation of the shifting of spectral lines let us look to sound, which also travels by wave motion. Although the nature of sound waves is quite different from that of light waves the effect we wish to discuss is very similar in both.

The wavelength of light determines its color; the wavelength of sound determines its pitch. A very high sound has a short wavelength and a low-pitched sound has a long wavelength.

Let us consider the sound of the horn of a car as it comes toward us, goes past, and travels away from us. The horn emits a sound made up of a series of wave crests and troughs that results in a characteristic pitch if the car is standing still. Since the velocity of sound depends only on the material (such as air) through which it travels, it is not affected by the speed of the car. Let us further consider the crests only, one at a time. The horn will emit a certain crest which will travel towards our ear with the velocity of sound. If the horn is approaching us, it will move a little closer to our ear before it emits the next crest. Consequently, the next crest will reach us a little more quickly than if it were emitted when the car is standing still. Not only are these two crests closer together but all the crests emitted will be closer together as long as the car is approaching us. Since these crests are closer together when they reach our ear, the pitch of the horn will be higher. Nothing need be said about how far away the car may be, for we are concerned not with its distance but with its velocity toward us.

When the car is directly opposite it will be neither approaching us nor receding from us, and the crests will be spaced the same as if the car were standing still. The pitch of the sound will be the same as if both the car and the listener were at rest.

As the car recedes from us it will move a little farther away after it emits each crest, and the crests will be spaced a little farther apart, that is, the wavelength will be a little longer. In this case fewer crests will reach our ear each second, and the pitch will become lower.

The change in pitch, called the *Doppler effect*, is often effectively mimicked by the entertainer who assumes the role of a radio sportscaster at an automobile race when he imitates the sound of the racing cars as each passes his broadcasting booth.

The Doppler effect applied to light means that the spectrum shifts a bit if there is relative motion between the source and the observer. If the measurements of wavelength of the spectral lines of a stellar spectrum are all a bit too short, we assume that the star and the earth are approaching each other. The entire spectrum is shifted to the short wavelength, which for the visible part of the spectrum is toward the violet.

If the measurements of the wavelength of the stellar spectral lines are all a bit too long, then we declare that that star and the earth are receding from each other. The entire spectrum is shifted toward the longer wavelength, which for light is toward the red.

The amount the wavelength changes depends on both the relative velocity of the source and the observer, and on the wavelength. The faster the two objects are moving away from or toward each other the greater the shift in the wavelength and in direct proportion. A star approaching the Earth at twice the velocity of another will have its spectral lines shifted by twice the amount.

The shift in wavelength also depends on the wavelength of the unshifted spectral line. A line with an unshifted wavelength of 8000 Å will be shifted by twice the amount of a spectral line with an unshifted wavelength of 4000 Å.

With this well in mind, we can interpret the spectrum in Figure 2-9*d*. All the iron lines have been shifted slightly to the longer wavelength (red), and thus the star must be receding from the Earth.

When the Doppler effect for light was discovered in stellar spectra, it was realized what a powerful tool astronomy had acquired, for by means of the *Doppler shift*, as it is called, an astronomer can determine the velocity with which a star is moving away from or toward us. The motion of a star either away from or toward the Earth is called the *radial velocity* because it is along a radius (in the line of sight) with the Earth at the center.

2.8 THE HYDROGEN SPECTRUM

The most important spectral lines for the astronomer are those of hydrogen, since it is the most abundant element in the universe. This element has several series of spectral lines. The first of these series is in the ultraviolet; hence, it is difficult to observe because it is in a part of the spectrum that is absorbed by the atmosphere. (Artificial satellites, however, have made and are making many observations in the ultraviolet.) This series is called the *Lyman series*. The line in this series with the longest wavelength (1216 Å) is called *Lyman-alpha*, the next is *Lyman-beta* (1025 Å), the third is *Lyman-gamma*, etc.

The next series in the hydrogen spectrum is in the visible part of the spectrum (or very close to it) so it is easily observable by astronomers using earth-based telescopes. This series is called the *Balmer series*, and since it is so readily observable, the lines in this series are the most familiar to astronomers (Figure 2-11). The lines with the longest wavelength are called *H-alpha* (wavelength 6563 Å), *H-beta* (4861 Å), then *H-gamma*, etc.

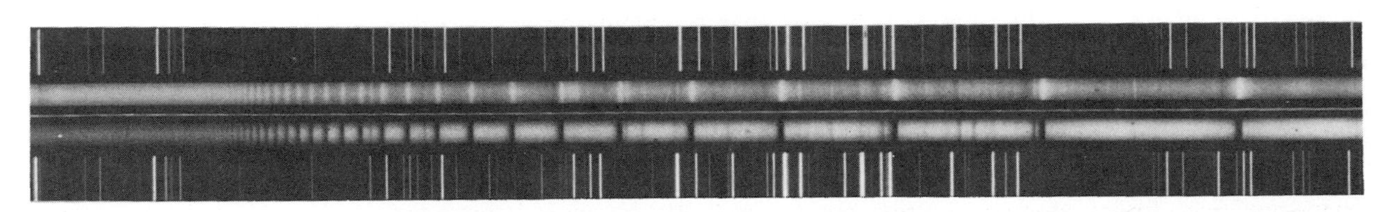

Figure 2-11
Two stellar spectra are contained between the two comparison spectra. The upper stellar spectrum shows the lines of the Balmer series as bright lines. The lower stellar spectrum shows the Balmer lines as dark lines. (Photographs from the Hale Observatories.)

Spectral lines result from changes in the atom. The hydrogen atom is composed of a proton and an electron. The *proton* has a positive electric charge. It is close to 1800 times more massive than the electron and, therefore, forms the nucleus of the atom. The *electron* has a negative charge and thus is held to the proton by an attractive force (unlike electric charges attract). The electron about the nucleus can have only certain — discrete — values of energy. The electron can change its energy from one value to another and in so doing can convert energy from one form to another.

By way of analogy, the energy of a plate in an empty cupboard is limited to certain values. The plate can exist on only one shelf at a time; therefore, the plate's height above the floor can only equal the height of any one of the empty shelves. Its energy depends on its height, and since its height is limited to certain discrete (distinct) values, its energy is also limited to certain discrete values. The higher the plate, the greater the energy. If the plate falls to the floor, the energy given it in placing it on one of the shelves is converted to sound and thermal energy (heat).

An atom has a "ground state" or 1st level of energy; a plate on the floor would be similar to the electron in the ground state of the hydrogen atom. You must give the plate energy to move it up to one of the shelves (or higher energy

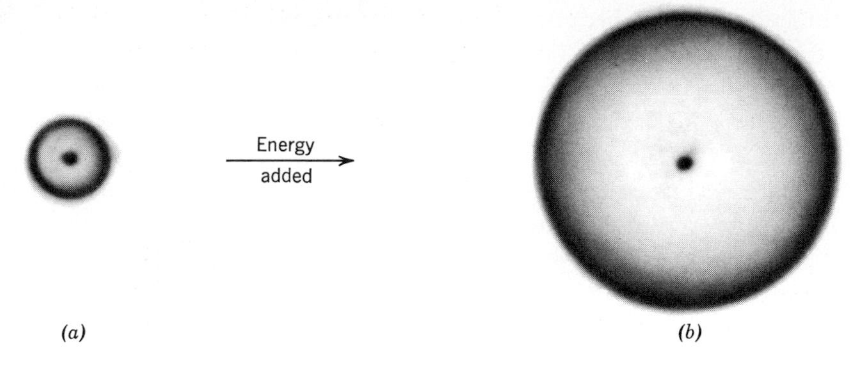

Figure 2-12
***(a)* A hydrogen atom with its electron in the 1st energy level. *(b)* Given the right amount of energy, the electron in that hydrogen atom jumps to the 2nd energy level. Having more energy the atom now has a larger radius. The nucleus, a dot in the center, is unaffected by this change in energy.**

levels). The plate on any shelf has more energy than the plate on the floor; the higher the shelf, the more energy the plate has.

Similarly, energy must be given an atom if the electron is to move up from the 1st energy level (ground state) to the 2nd energy level, the 3rd, or to higher energy levels. Figure 2-12 illustrates this. The fuzzy little circle in Figure 2-12*a* represents a hydrogen atom in the first energy level. By giving it energy the electron jumps to a higher energy level. With this increased energy, the radius of the atom increases. Figure 2-12*b* represents a hydrogen atom in the second energy level.

But a hydrogen atom does not remain in a higher energy level long. Within 10^{-8} seconds (0.000,000,01 sec) the electron falls back down to the 1st energy level. The size of the atom decreases, and the extra energy goes off in the form of a little bit of light called a *photon* (Figure 2-13). If a plate were to fall from a cupboard shelf, its extra energy would go off in the form of sound.

Whenever the electron of a hydrogen atom falls from the 2nd to the 1st energy level, the photon given off always has the same amount of energy. Many photons of this amount of energy entering a spectrograph would appear as a Lyman-alpha spectral line. Photons that result from the electrons of hydrogen atoms falling from the 3rd level to the 2nd result in the H-alpha spectral line, the first line in the Balmer series.

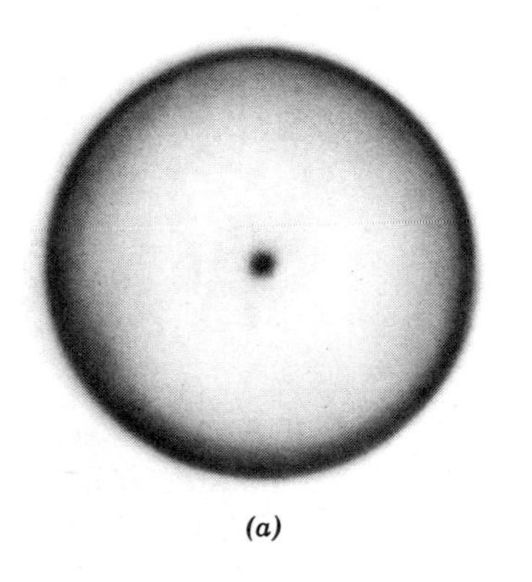

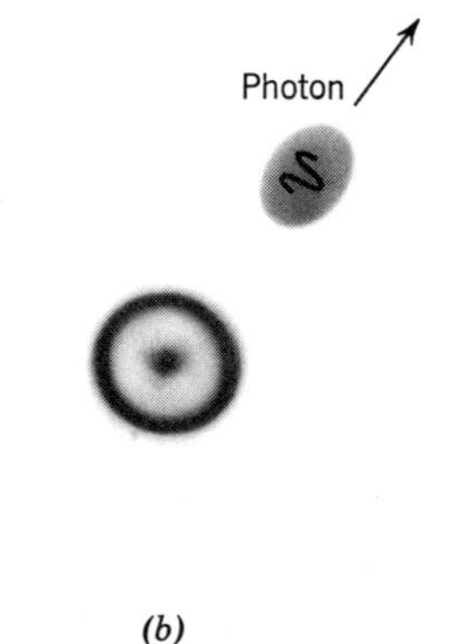

Figure 2-13
***(a)* A hydrogen atom with its electron in the 2nd energy level. *(b)* As the electron falls back to the 1st energy level a photon of light is emitted.**

Absorption spectra are caused by the reverse process. An absorption line is a result of a photon being absorbed by an atom. The electron involved moves from one energy level to a higher one, and the photon disappears completely in the process. Its energy is converted into increased energy of the atom. Similarly, sound energy is converted to another form (thermal energy) when the sound is absorbed by the walls of a room.

2.9 THE SPECTRUM OF RADIANT ENERGY (ELECTROMAGNETIC SPECTRUM)

The visible light that enables us to see the stars at night is not the only energy that the Earth receives from the universe. We receive energy over the entire range of the spectrum of radiant energy: gamma rays, X rays, ultraviolet, visible, infrared, microwaves, and radio waves (in order of decreasing energy; see Figure 2-14). Gamma rays have the shortest wavelength and the most energy; radio waves have the longest wavelength and the least energy. Most of this radiation is absorbed by our atmosphere. Although this absorption does protect us from the harmful effects of the very energetic radiation, it also robs the earth-based astronomer of information. All of the radiation that is absorbed contains information that the astronomer can use to help him understand the universe. Hence, the astronomer wants to make more and more observations from above the atmosphere. To make these observations he uses high altitude balloons (100,000 ft), rockets, artificial satellites, and more distant space probes.

2.10 THE RADIO TELESCOPE

The radiation that does reach the Earth's surface, the visible and part of the radio waves, has told us a good deal about the universe, and continues to do so. To obtain the information from the visible light, we use optical telescopes and spectrographs. To obtain the information from the radio waves, we must use radio telescopes. A radio telescope is a reflecting telescope with a reflecting surface not of glass but of wire mesh or a metal membrane (Figure 2-15). An antenna at the prime focus picks up the signal and feeds it to an amplifier, which in turn feeds the signals into a recording device that records the intensity of the radio waves received at each frequency studied.

The use of radio telescopes has vastly extended our knowledge of the universe. Many of the recently discovered objects were first detected by radio telescopes and then later studied by other means, such as optical telescopes and space probes.

The instruments used by the astronomer to study the universe are limited mainly by his ingenuity, and he has not seemed to be at a loss on this account.

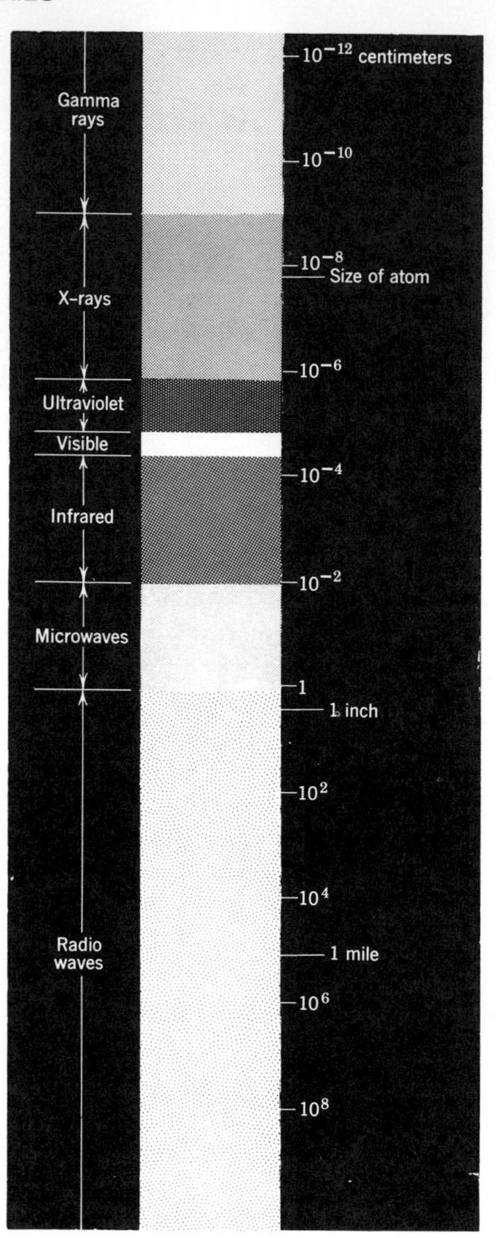

Figure 2-14
The entire spectrum of radiant energy.

Figure 2-15
The 300-ft radio telescope at Green Bank, West Virginia. (National Radio Astronomy Observatory.)

BASIC VOCABULARY FOR SUBSEQUENT READING*

Absorption spectrum
Ångstrom
Balmer series
Bright-line spectrum
Continuous spectrum
Doppler effect
Electromagnetic spectrum
Focal length
Focal plane
Lyman series
Objective
Photon
Prime focus
Radiant energy spectrum
Radial velocity
Spectral line
Spectrograph
Wavelength

QUESTIONS AND PROBLEMS

1. Write the following numbers in the exponential system:
(a) 93,000,000 (c) 0.034
(b) 24,000,000,000,000 (d) 0.000 000 45

2. Write the following wavelengths in angstroms:
(a) 4.861×10^{-5} cm (c) 3.4×10^{-7} cm
(b) 6.0×10^{-3} cm (d) 21 cm

3. Sketch a graph of the changes in the period of Jupiter's satellite as observed by Roemer. This sketch should indicate at which point in the Earth's orbit 1, 2, 3, or 4 (see Figure 2-2), the period is the greatest length, at which the shortest, and at which (other than 1) it is the standard length. The horizontal axis should be the position of the Earth in its orbit; 1, 2, 3, and 4. The vertical axis can be the period; upward can be longer periods, downward can be shorter.

4. If the wavelength, λ, of a spectral line is 4500 Å, what is its frequency (the velocity of light $c = 3 \times 10^{10}$ cm/sec)? (The frequency $f = c/\lambda$.)

5. The radial velocity of a star can be determined numerically by the equation for the Doppler shift:

$$\frac{\Delta\lambda}{\lambda o} = \frac{v}{c}$$

where λo is the laboratory wavelength of the spectral line, $\Delta\lambda$ the amount the line is shifted in the stellar spectrum, c is the velocity of light (3×10^5 kilometers per second) and v is the velocity of the star. Determine the Doppler shift $\Delta\lambda$ for the hydrogen spectral line ($\lambda o = 4860$ Å) for a star whose radial velocity is:
(a) 10 kilometers per second
(b) 100 kilometers per second
(c) 1000 kilometers per second

6. A spectral line (the K-line of calcium) with a laboratory wavelength $\lambda o = 3930$ Å has an observed wavelength of 4930 Å in the spectrum of a star.
(a) Is the star receding from or approaching the Earth?
(b) Determine the radial velocity of the star.

7. In question 6, the K-line was shifted by 1000 Å. How far would the 21-cm line be shifted if radio observations were made of the same object?

*A list of words is given at the end of each chapter to indicate which words are important for further reading. The student should be very familiar with these words.

FOR FURTHER READING

Larmore, L., *Introduction to Photographic Principles,* Dover Publications, New York, 1965.

Mayall, R. N., and M. Mayall, *Skyshooting; Hunting Stars with your Camera.* The Ronald Press Co., New York, 1949.

Miczaika, G. R., and W. M. Sinton, *Tools of the Astronomer,* Harvard University Press, Cambridge, Mass., 1961.

Page, T., and L. W. Page, eds., *Telescopes,* The Macmillan Co., New York, 1968.

Ruechardt, E., *Light, Visible and Invisible,* University of Michigan Press, Ann Arbor, Mich., 1958.

Struve, O., and V. Zebergs, *Astronomy of the 20th Century,* Crowell Collier and Macmillan, New York, 1962, Chapters I, II and VI.

Texereau, J., *How to Make a Telescope,* Interscience Publishers, New York, 1957.

Woodbury, D. O., *The Glass Giant of Palomar,* Dodd, Mead and Co., New York, 1953.

Brown, R. H., "The Stellar Interferometer at Narrabri Observatory," *Sky and Telescope,* p. 64, Aug. 1964.

Findlay, John W., "The National Radio Astronomy Observatory," *Sky and Telescope,* p. 352, Dec. 1974.

Irwin, John B., "Chile's Mountain Observatory Revisited," *Sky and Telescope,* p. 11, Jan. 1974.

Kaufman, P., and R. D'Amato, "A Brazilian Radio Telescope for Millimeter Wavelengths," *Sky and Telescope,* p. 144, March 1973.

Kiepenheuer, K. O., "European Site Survey for a Solar Observatory," *Sky and Telescope,* p. 84, Aug. 1974.

Morrison, D., and J. T. Jefferies, "Hawaii's Mauna Kea Observatory Today," *Sky and Telescope,* p. 361, Dec. 1972.

Philip, A. G. David, "A Visit to the Soviet Union's 6-meter Reflector," *Sky and Telescope,* p. 290, May 1974.

"Arecibo Observatory Today," *Sky and Telescope,* part I, p. 214, April 1972; part II, p. 293, May 1972.

"Astronomy in the Owens River Valley," *Sky and Telescope,* p. 217, October 1974.

"Giant Mirror Blanks Poured for Chile and Australia," *Sky and Telescope,* p. 140, Sept. 1969.

"Giant X-Ray Telescope," *Sky and Telescope,* p. 300, May 1969.

"The Latest Flight of Stratoscope II," *Sky and Telescope,* p. 365, June 1970.

"Photographic Report from Kitt Peak," *Sky and Telescope,* p. 10, January 1973.

"Radio Interferometers with Very Long Base Lines," *Sky and Telescope,* p. 143, September 1967.

CHAPTER CONTENTS

LEARNING OBJECTIVES

BE ABLE TO:

1. DIFFERENTIATE BETWEEN ROTATION AND REVOLUTION.
2. EXPLAIN HOW WE KNOW MERCURY HAS NO ATMOSPHERE.
3. DESCRIBE THE GENERAL FEATURES OF MERCURY'S SURFACE.
4. DESCRIBE TWO METHODS BY WHICH VENUS' ATMOSPHERE HAS BEEN OBSERVED WITH SPACE PROBES.
5. DESCRIBE THE GREENHOUSE EFFECT.
6. DESCRIBE THE EARTH'S INTERIOR AND CONDITIONS NECESSARY FOR ITS EVOLUTION.
7. EXPLAIN WHY THE SKY IS BLUE.
8. DESCRIBE THE SOLAR WIND AND ITS EFFECT ON THE EARTH'S MAGNETIC FIELD.
9. COMPARE THE RELATIVE MERITS OF OBSERVING PLANETS, SUCH AS MARS, VISUALLY AND PHOTOGRAPHICALLY.
10. DESCRIBE THE MAJOR FEATURES ON MARS' SURFACE.
11. GIVE OBSERVATIONS INDICATING THAT WATER MAY HAVE EXISTED ON MARS.
12. COMPARE THE COMPOSITION OF THE ATMOSPHERES OF VENUS, EARTH, AND MARS.
13. DESCRIBE THE ATMOSPHERIC CLOUDS OF MARS.

Chapter Three
THE INNER PLANETS

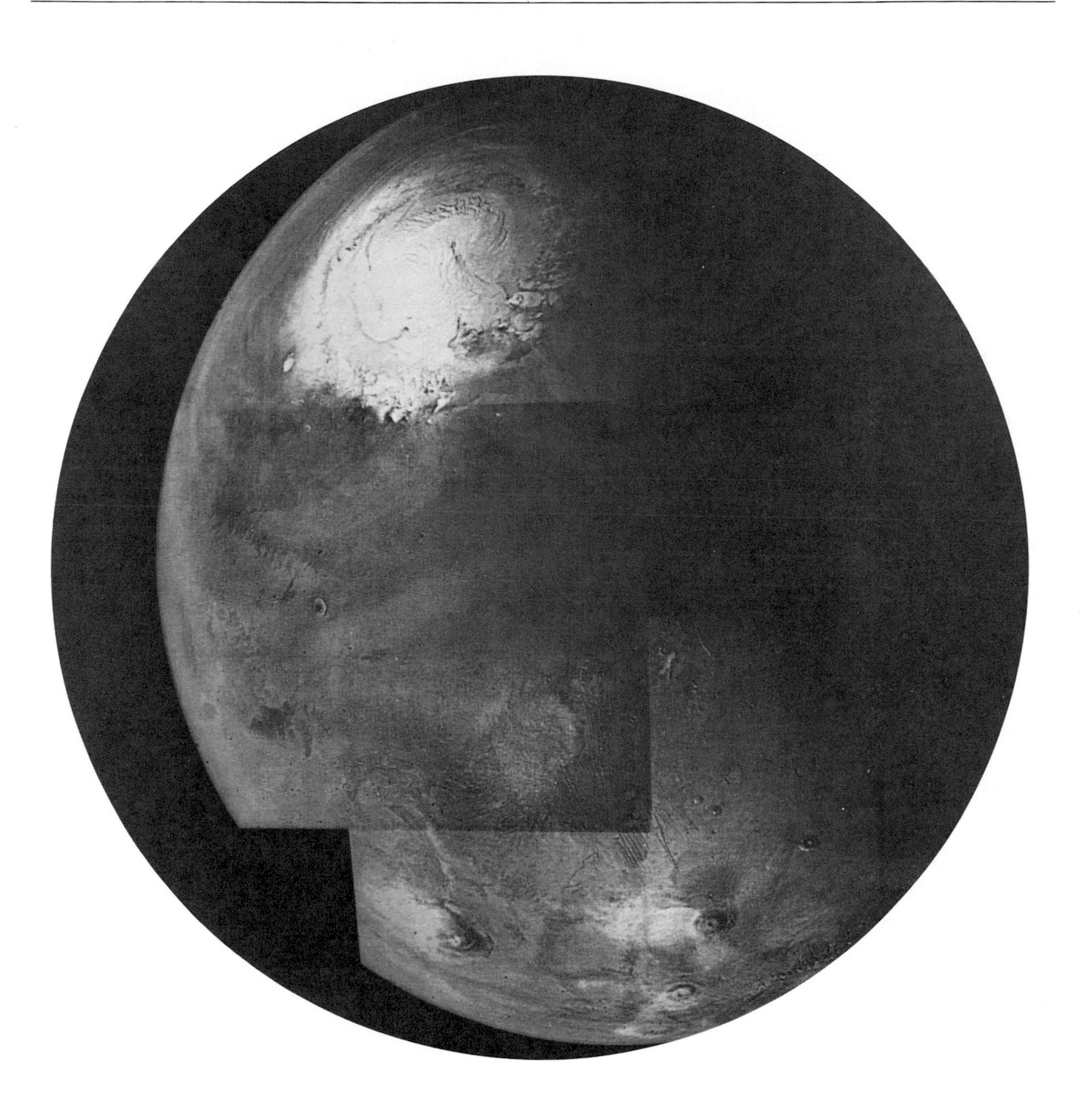

Now that we have some idea of the astronomer's methods, we can better understand the objects of his studies. To the naked eye all the stars (except the sun) and planets appear as points of light. But in the early telescopes those objects closest to the Earth took on a new appearance. The same is true today: planets appear as disks, whereas the stars appear only as points of light. The early observations led to the first great change in man's concept of the universe — the idea of a *solar system*.

The solar system is composed of the sun, its largest and central member; the nine planets that revolve about the sun; satellites, comets, asteroids, meteors, and something we might call dust.

The sun is really a star and we shall consider it along with the other stars. For the present we are concerned with the planets, which in the solar system are second only to the sun in importance and size.

Within the solar system, the nine planets have at least two different motions. Each of them *revolves* about the sun in a given length of time just as the satellites of Jupiter revolve about Jupiter. The planet Earth, for instance, takes one year to revolve about the sun; in fact, it is because of this revolution and the tilt of the Earth's axis that we have the four seasons.

At the same time each planet *rotates* on its own axis like a merry-go-round. The Earth makes one complete rotation on its axis each day. During part of this rotation you, the observer, are facing the sun; during part of it you are facing away from the sun, in the shadow of the Earth.

All planets describe these two basic motions of revolution and rotation, although with differences. But what of their other physical characteristics? Do planets other than the Earth have atmospheres, soil, conditions that would favor life? Conditions on them are largely determined by their distance from the sun because each, like the Earth, receives its energy from the sun. The closer the planet, the more energy it receives.

3.1 MERCURY

Mercury, less than half as far from the sun as the Earth, is the planet closest to the sun. To put it more briefly, it is 0.4 A. U. from the sun; A. U. is the abbreviation for a measure of length called the *astronomical unit* which is used by astronomers because the units of length employed on the Earth are all too short. 1 A. U. is equal to the average radius of the Earth's orbit, about 93,000,000 miles.

Chapter Opening Photo

A mosaic showing part of the northern hemisphere of Mars. The north polar cap is at the top, Nix Olympica at the lower left, three other volcanoes and the equatorial canyon are in the lower right. (NASA photograph.)

With a diameter of 3021 miles and a mass only 0.05 that of the Earth, Mercury is also the smallest planet. It revolves about the sun in 88 days and was long thought to keep the same face toward the sun. However, radar observations since 1965 made by G. Pettengill and R. Dyce using the 1000-ft radio telescope in Puerto Rico indicate that, with respect to the stars, Mercury rotates on its axis with a period of 58.6 days. Radar observations are made by sending a radio signal to an object then receiving and studying the echo. Consequently, during one period of rotation the entire planet receives sunlight; no face is perpetually dark. But since the direction of rotation is the same as the direction of revolution, the period of rotation with respect to the sun is about 175 days. The sun will shine on one point of the surface for a period of time equal to about 87½ days here on Earth.

A. THE ATMOSPHERE Determining if Mercury has an atmosphere may depend in part upon how the word "atmosphere" is defined. If the term is applied to a layer of gas about a planet which is not only permanent in nature but which is dense enough to be easily detectable, then the answer is an unqualified no. If, however, the term atmosphere is taken to include very rarified gases which may or may not be permanently associated with the planet, then Mercury may have an atmosphere.

Mercury, like every other planet, shines by reflected sunlight. Therefore, its spectrum must be the same as the sun's spectrum — unless it has an atmosphere of its own. (Surface materials do absorb some light but not as spectral lines.) If Mercury has an atmosphere, the sunlight would have to pass through it to reflect from the surface, and then pass back through the atmosphere again before reaching our telescopes. Since this atmosphere would act as a cool gas at low pressure, it would add absorption lines to the solar spectrum. The wavelengths of these lines would indicate the chemical nature of the atmosphere. Thus far spectral studies of this nature have not been successful in detecting even a rarified atmosphere.

Whether a planet can retain an atmosphere or not depends upon the mass and radius of the planet, its temperature, and the kind of gas. The mass and radius of the planet determine the escape velocity at the surface. The *escape velocity* is the velocity an object must achieve in order to escape from that planet. The escape velocity at the surface of Mercury is 2.6 miles per second. (By comparison the escape velocity at the surface of the Earth is 7.0 miles per second.)

The velocity with which the molecules of gas travel depends upon the temperature of the gas and upon the kind of gas. Even if an atmosphere is composed of only one kind of gas, its molecules, because of collisions, will be traveling at different velocities. As a result of a collision, one molecule may acquire a greater velocity, another a lower velocity. The number of molecules at each particular velocity depends upon the temperature of the gas. As a result, the fastest molecules will escape from the planet if they acquire the escape velocity.

In March 1974, the space probe Mariner 10 passed within 450 miles of

Mercury's surface. Instruments on board the craft detected a tenuous atmosphere composed principally of helium. These observations revealed no trace of hydrogen, argon, or oxygen.

B. THE SURFACE Until March 1974, little was known of Mercury's surface. Common sense indicated that since Mercury has no appreciable atmosphere, its surface might be marked by craters as the moon, Mars, and to a much lesser degree, the Earth. Mercury had been observed by every available means from the Earth, but its great distance from us and its proximity to the sun guaranteed that its surface would remain a mystery until it was viewed from a passing space probe.

As expected, the surface viewed by the television cameras on board Mariner 10 appears just like that of the moon (Figure 3-1). Even the composite of many smaller photographs appears very much like the moon (Figure 3-2).

Special instruments on board Mariner 10 recorded surface temperatures as high as 950° F (hot enough to melt lead, zinc, tin, and solder) on the hot side and as low as −350° F on the dark side.

One of the surprise findings of the very successful flight of Mariner 10 — it flew past Venus once and Mercury three times — is that Mercury has a permanent magnetic field of its own. That field is not very strong, about 3% of the Earth's, but that it has a magnetic field at all is significant. It had been thought that the two main requirements for a magnetic field were a hot partially molten core and rapid rotation. Mercury may have the hot core, but it certainly

Figure 3-1
The surface of Mercury as viewed by Mariner 10. The prominent sharp crater near the center is 19 miles across and has a central peak. The sun is shining from the right. (NASA photograph.)

Figure 3-2
A mosaic of a portion of the surface of Mercury. On the left side of this photomosaic is part of a large circular mountain-ringed basin some 800 miles in diameter. The floor of this basin has many fractures and ridges on it. (NASA photograph.)

does not rotate rapidly. This finding, therefore, may help astronomers better understand why some planets have magnetic fields and others do not.

3.2 VENUS

The second planet from the sun (0.7 A. U.) is Venus. With a diameter of 7522 miles and a mass 0.89 times that of the Earth, the planet is only a little smaller than the Earth. The optical telescope reveals only a layer of clouds which are opaque to visible light. Occasionally these clouds show markings, although these markings are not permanent (Figure 3-3*a*).

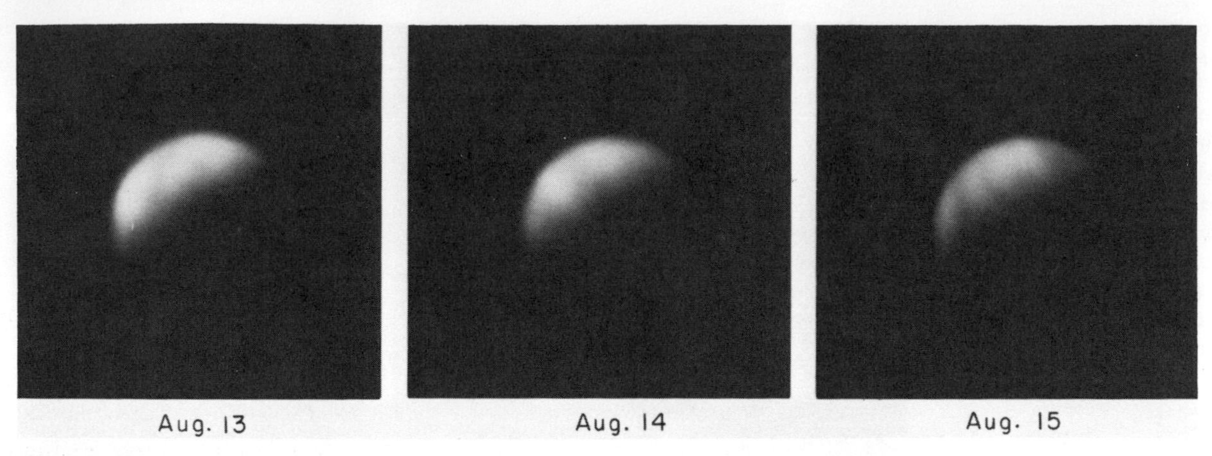

Figure 3-3
(a) Three photographs of Venus taken on consecutive days to show the changes in its cloud cover. (Lick Observatory.)

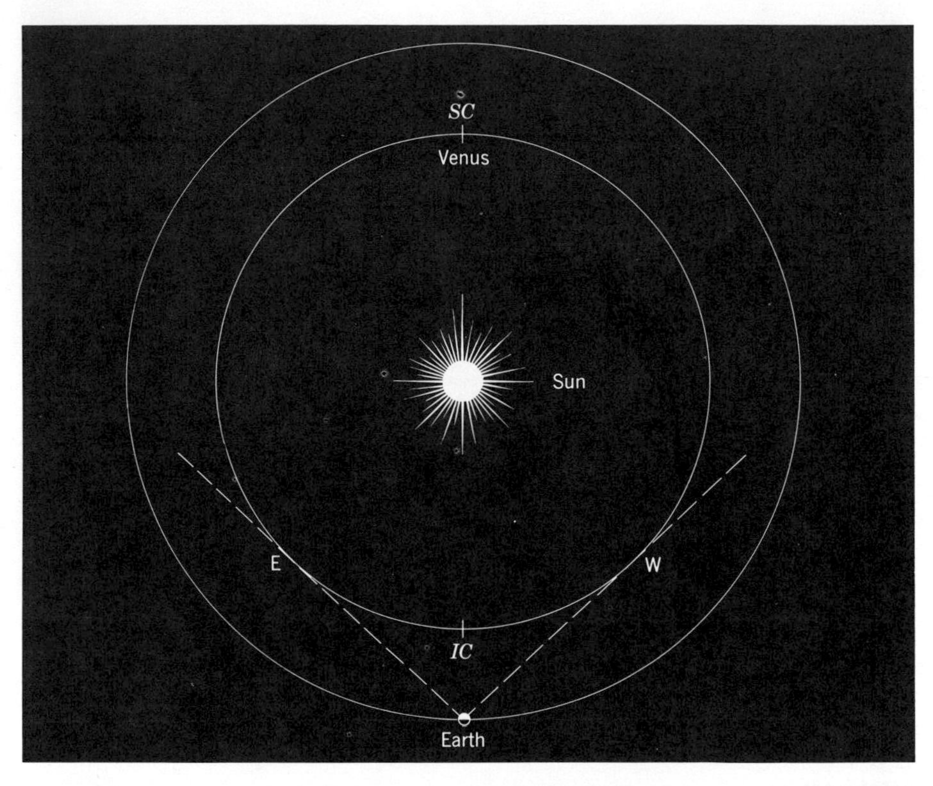

(b) From the Earth we cannot observe either Mercury or Venus in full phase, for at full phase the planets are at *superior conjunction (SC)* behind the sun. We can only see a thin crescent at most when either planet is at *inferior conjunction (IC).* As viewed from the Earth, the planets are farthest from the sun at *eastern elongation (E)* or *western elongation (W),* and are most easily observed at these times. Radar studies can be made near inferior conjunction, however.

A. RADAR AND ROTATION Since the markings of the visible surface of Venus are not permanent, astronomers had to resort to other means to determine the period of rotation. Radio techniques offered the solution.

Radio signals reflect off of most objects — even planets. *Radar* is the term applied to the observation of radio echoes or reflections. The size, shape, surface, and motion of an object all influence the reflection.

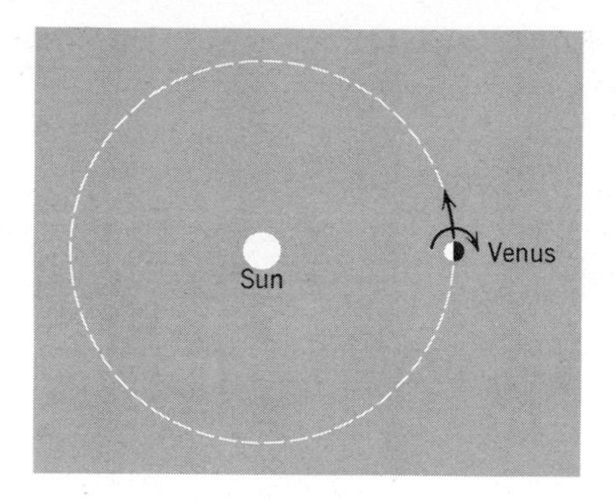

Figure 3-4
Venus revolves about the sun in the same direction as all the planets, but its rotation is in the opposite direction and is thus called retrograde motion.

Radar observations of Venus indicate a period of rotation of 243 days, but in the direction opposite to the direction of revolution. Rotation in the opposite direction is called *retrograde rotation* (Figure 3-4). The period of rotation with respect to the sun is, therefore, about 122 days. Consequently, the sun shines on one location for about 61 of our days.

Retrograde rotation is a bit disconcerting, for it is not the easiest situation to explain. Present theories of the origin and evolution of the solar system (see Chapter 12) attempt to explain only those planets that rotate in the same direction in which they revolve about the sun. At this stage, we simply do not know enough of the past history of the solar system to explain retrograde rotation. We must wait until we have gathered much more information, particularly about the very early history of our solar system. Space exploration appears to be the only method by which we will be able to gather such information.

B. OTHER TECHNIQUES OF OBSERVATION The most exciting observations of Venus have been made by space probes. The United States has sent a number of space probes of the Mariner class; the Russians, a number of the Venera (Venus) class.

Space probes that simply travel close to Venus can take photographs and gather other information during that "flyby." For instance, if the craft goes behind Venus (as seen from the Earth) it can send radio signals to the Earth through the atmosphere of Venus. Before the spacecraft goes behind the planet, its radio signal is uninterrupted; but as it goes behind the planet, its radio signal passes first through the upper atmosphere, and then through the full thickness of the atmosphere before being blocked out by the planet (Figure 3-5). As the spacecraft emerges from behind the planet, its signal again passes through successive layers of the atmosphere. That radio signal is altered by the atmosphere, and the manner in which it is altered depends on the composition and characteristics of the atmosphere. Consequently, a study of that radio signal reveals very useful information about the planet's atmosphere.

Space probes have been parachuted through the atmosphere by the Russians to make soft landings on the surface of Venus. Two such craft have gathered information and returned signals for up to 50 minutes after landing.

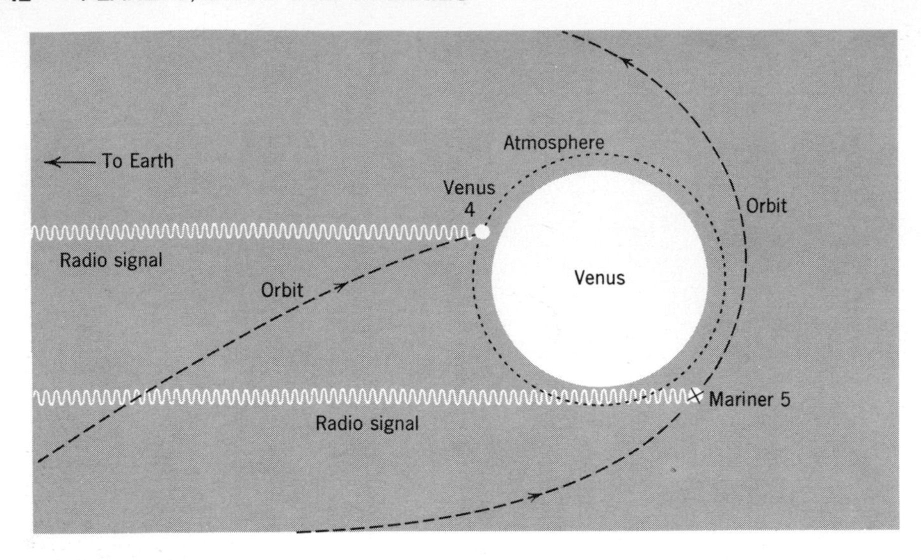

Figure 3-5
The signal sent to Earth from Mariner 5 passed through Venus' atmosphere only 34 hours after Venus 4 had crashed onto the planet's surface.

C. THE SURFACE Radar observations, similar to those made of Mercury, indicate that Venus's surface is as rough as the surfaces of Mercury and the moon, and probably made of similar material. But this finding is not really surprising. Recent observations suggest the existence of craters on Venus.

D. THE ATMOSPHERE AND THE GREENHOUSE EFFECT The space probes have returned some surprising information about the atmosphere of Venus. Information they have radioed back can best be interpreted by assuming that the atmosphere of Venus is nearly pure carbon dioxide. This is a striking difference from the atmosphere of the Earth, which is 78% nitrogen, 21% oxygen, and only 0.03% carbon dioxide. The atmospheric pressure at the surface of Venus is close to 90 times that at the surface of the Earth.

The cloud cover that obscures the surface of Venus was observed in detail in 1974 by Mariner 10 using television cameras sensitive to the ultraviolet. The complex swirling nature of cloud patterns shown in Figure 3-6 apparently results in part from the difference in solar heating between the equator and the poles, and in part by the slow retrograde motion of the planet. The observed cloud cover appears to move rapidly westward relative to the planet's surface, yielding an apparent retrograde rotation of about 4 days. Observations from both Venera 8 and Mariner 10 lead to the conclusion that much of the complex cloud cover is composed of tiny droplets of sulfuric acid.

Observations made by radio waves of length 2 to 3 centimeters indicate that Venus radiates energy as if its surface temperature is near 800° F. Observations made by Mariner 2 and 10 (1962 and 1974) and Venera 7 and 8 (1970 and 1972) confirm that the surface is indeed hot, close to 900° F. The nighttime temperature appears to be nearly as high as the daytime temperature.

That the surface of Venus is so hot is at first puzzling. Why should the surface of Venus be nearly as hot as the surface of Mercury? It is, after all, nearly twice as far from the sun as Mercury. Mercury has no atmosphere,

Figure 3-6
Venus as viewed from Mariner 10 in 1974. (NASA photograph.)

however, and Venus does, so perhaps Venus's atmosphere is the cause of the high temperature.

The atmosphere apparently acts as a heat trap. The term describing this action is the *greenhouse effect*. It is recognized that radiation of one wavelength — the short wavelengths of visible and infrared light — can travel through the glass of a greenhouse and warm up the interior, but that since the radiations emitted by the interior are of longer wavelength, they are blocked by the glass. An automobile parked in the summer sun with all the windows rolled up demonstrates this principle only too well.

From studies here on Earth, it is known that carbon dioxide is a particularly effective producer of the greenhouse effect. So, although the cloudy atmosphere of Venus does prevent most of the sun's radiation from reaching the surface, apparently an ample supply of radiation does reach the surface to warm it up. If heat is then trapped by the atmosphere, the temperature will rise until the planet is giving off as much to outer space as it receives from the sun. At that point the surface temperature becomes stable. Planets radiate heat out into space just as the sun does — by radiant energy. Since the planets are a good deal cooler than the sun, however, they radiate less energy, and that energy has a much longer wavelength. Even though they do not glow, a soldering iron and a clothes iron radiate energy in the infrared region of the spectrum.

Studies of the information radioed back from Venera 7 and 8 have led astronomers to conclude that the atmosphere of Venus can best be described as being composed of three layers. The layer beneath the clouds is apparently clear with no particles suspended in the form of clouds or fine dust. This layer extends up to about 22 miles above the surface where the pressure is about equal to atmospheric pressure on the surface of the Earth, and the temperature is about 70° F.

The second layer includes the clouds and extends from about 22 miles high to about 30 miles above the surface where the pressure is much less and the temperature is close to −36° F. The highest layer appears to be rather hazy.

The amount of water vapor in Venus' atmosphere is less than 1/300 of the total amount of water on Earth. Why so little water in the atmosphere of Venus and so little carbon dioxide in the atmosphere of the Earth? Answers to these very interesting questions will surely come from future studies of both planets.

E. THE MAGNETIC FIELD Information returned by both Mariner 5 and 10 clearly indicates that Venus has only a very weak magnetic field, perhaps 0.03% the strength of the Earth's magnetic field. Whether this is because of its slow rotation or because it does not have a molten core or both is not yet known.

3.3 EARTH

The third planet out from the sun is the Earth with a diameter of 7917 miles. It is, to our present knowledge, unique because it does support life. As our studies of the universe become more precise and penetrating, however, it is logical to expect that we shall find many, many stars with planets about them and many of these may well have life on them. We have advanced to the stage of sophistication where we do not believe that the universe was made for the sole benefit of mankind here on Earth.

We know more about the Earth's surface and its interior than we do about any other planet; consequently, in our study of the solar system we can take advantage of this knowledge of our particular planet. This knowledge will add greatly to our understanding of all the planets and perhaps help us understand how they were all formed.

A. THE INTERIOR Most of the ideas about the Earth's interior are based on information gained from the study of earthquake waves. As these waves travel through the Earth, their speed and direction are changed according to the type of wave and the material through which they are traveling.

These studies show that it is convenient to consider the Earth as four concentric spheres: the inner core, the outer core, the mantle, and the crust (Figure 3-7). The core is composed principally of iron and nickel; the inner core is believed to be solid, the outer core molten. Above the core is the mantle, composed chiefly of the denser rock-forming substances. Atop the mantle is the crust, the rocky surface of the Earth. It will be seen in Chapter Seven that knowledge of the Earth's iron-nickel core has influenced our thinking about the history of the solar system.

The separation of chemicals, as has occurred in the Earth, can take place in a solid astronomical body only if the interior of that body becomes heated to temperatures in the order of 1000° F. Only then can the heavier chemicals, such as iron, slowly sink to the center. The most plausible explanation for such high temperatures is that radioactive substances, such as uranium, trapped inside, act like a furnace, converting nuclear energy into thermal energy.

B. THE ATMOSPHERE The Earth's atmosphere has influenced the surface of the Earth a greal deal. The atmosphere carries the water vapor which forms the rains, and the rains have been the chief agent in erosion and transportation of material from one location on the crust to another. The crust has responded to this relocation of material by sinking where the material was deposited and rising in that location from which the material was taken, changing the patterns of erosion.

The atmosphere actually contains very little water vapor. Nitrogen is the dominant gas in the atmosphere (78%), oxygen is next (21%), and argon is third

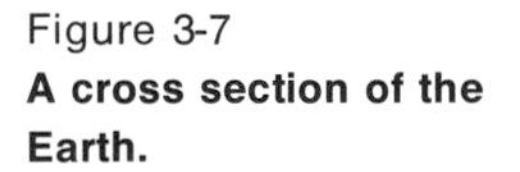
Figure 3-7
A cross section of the Earth.

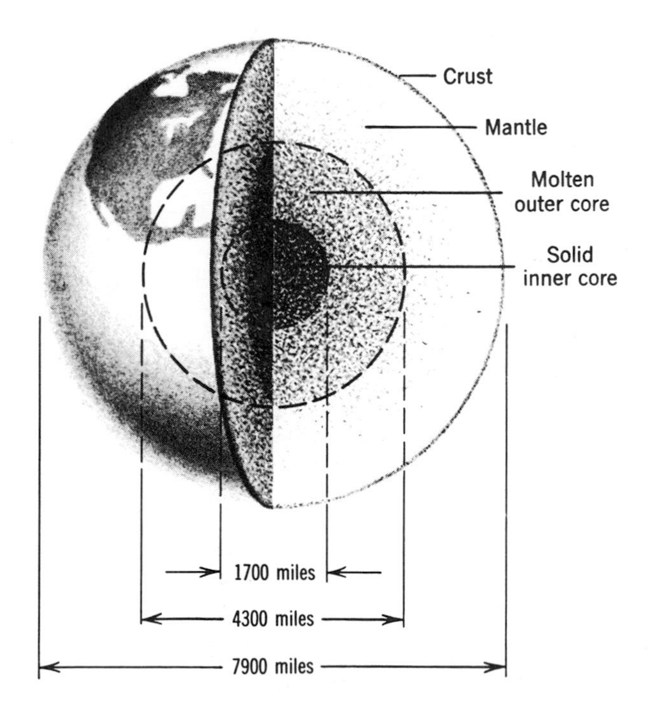

(0.9%), with the rest of the gases comprising the very small percentage which remains.

It is the atmosphere that gives rise to the blue sky which outshines the stars in the daytime. Sunlight that enters the atmosphere is intercepted not only by the gases that give rise to absorption lines but also by tiny dust particles. These dust particles, as well as the gas molecules, act somewhat like reflectors in that they can change the direction of light's travel. The process, which is different from reflection, is called *scattering* and is selective in that molecules and particles scatter the blue light (the shorter wavelengths) more effectively than they scatter red light. Since the blue is scattered more thoroughly, our sky is blue for it consists of this scattered light; the setting sun is red and the harvest moon orange because we see them after their light has had much of the blue taken out. When the sun and the moon are on the horizon their light travels through a longer path in the atmosphere, with the result that more of the blue is scattered out of the direct beam of light.

C. THE EARTH'S MAGNETIC FIELD From above the Earth's atmosphere, rockets and space probes have made observations and measurements of features not dreamt of before. Two of the first discoveries of the space age are the Van Allen radiation belts and the extended magnetic field of the Earth. Radiation counters aboard satellites and space probes sent to both Venus and Mars have revealed that the Earth is surrounded by a large magnetic field, a portion of which traps high-energy electrons (Figure 3-8). These electrons travel in corkscrew paths from the region of one pole to the other, along the magnetic lines of force. In the polar regions, these electrons dip into the upper atmosphere, causing it to glow

Figure 3-8
Van Allen radiation belts. (Reprinted by permission from *Scientific American*.)

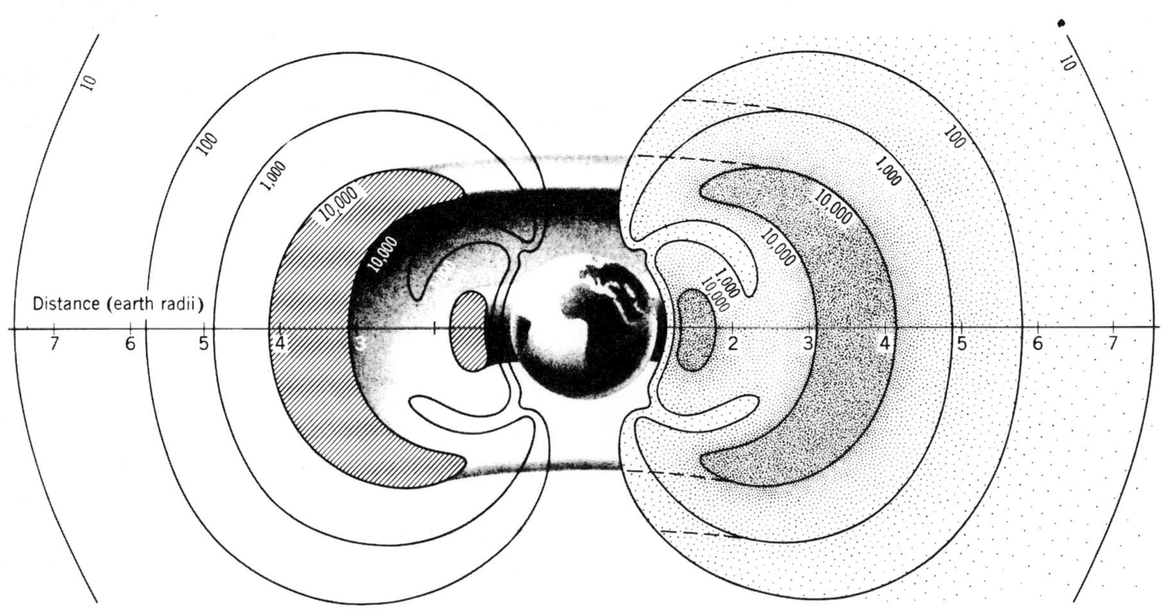

Figure 3-9
Aurora Borealis in Alaska. (Photograph by V. P. Hessler.)

in much the same manner that a gaseous tube glows. This glow is observed as the *aurora*, both the aurora borealis (northern lights) (Figure 3-9) and the aurora australis (southern lights).

The magnetic field of the Earth, however, is not free from external disturbing influences. Space probes have detected the existence of what is called the *solar wind*, a stream of charged particles, mostly electrons and protons, emitted by the sun and flowing past the Earth. Since these particles are charged, they carry a magnetic field with them. This magnetic field reacts with the magnetic field of

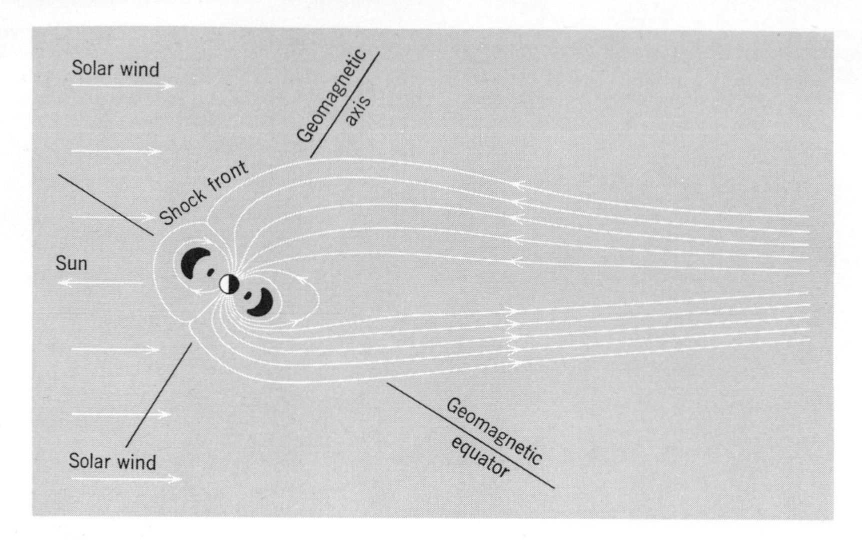

Figure 3-10
The Earth's magnetic tail caused by the solar wind.

the Earth in a way which appears to "blow" the magnetic field away from the Earth (Figure 3-10).

On the sunward side, the Earth's magnetic field extends to a distance of about 10 Earth's radii. Just beyond 10 Earth's radii, a region of magnetic turbulence has been observed, with an outer boundary consisting of a shock wave. The shock wave is the encounter between the particles of the solar wind and the Earth's magnetic field. It resembles a shock wave set up by a rock projecting above the surface of a river, or the wave set up by the bow of a boat as it cuts through the water. Outside this shock wave, the magnetic field is that of the solar system.

On the "leeward" (nighttime) side of the Earth the magnetic field forms a "magnetic tail." The Earth's magnetic field has been detected in this direction as far away as 800 Earth's radii, and it appears that the tail is about 40 Earth's radii in diameter. The magnetic lines of force in the northern part of the magnetic tail point toward the Earth, that is, the north end of a compass needle if placed there would point to the Earth. The magnetic lines of force in the southern part of the tail point away from the Earth. The boundary between these two regions of oppositely directed magnetic fields within the tail shows very little — if any — magnetic field and is called the *neutral surface.* The similarity between the shape of the Earth's magnetic tail, with its encircling region of magnetic turbulence, and the shape of a comet's tail is striking.

3.4 MARS

Mars, 1.5 A. U. from the sun, rotates on its axis in 24 hours, 37 minutes, so its day is just a little longer than ours on Earth. It revolves about the sun once every 687 days, so its year is less than twice as long as ours. Its diameter is 4200 miles, about one half that of the Earth; its mass, however, is only 0.11 times that of the

Earth. Presumably it does not have a large iron-nickel core as does the Earth.

Mars was once called the "red planet of mystery," and mysterious it used to be. Its red color led the ancient Greeks and Romans to call it the god of war. In the 19th century improved telescopes revealed markings on Mars that were called *canali* by Schiaparelli, one of that century's great Martian observers. In Italian, *canali* means channels which need not be artificial waterways. But the word was translated as *canals,* a word that implies that they were built by intelligent life.

The Martian canals were only part of the features of the visible surface that heightened the mysteries. A snow-white polar cap forms and nearly disappears on each pole as the seasons progress in each hemisphere, winter in the north with summer in the south, etc. The changing seasons also bring about changes in the brightness and colors of the bluish-greenish-grayish permanent markings on the planet's surface. Clouds of several sorts are visible. There were ample observations to feed the hypotheses that flourished, but not enough to help decide which one was the best description of Mars. It was a case of just enough observations to prick speculation and not enough to settle anything, so the controversies continued.

A. TECHNIQUES OF OBSERVATION Mars has been studied both extensively and intensively for more than 100 years. In the 1830s it was recognized that its darker markings are permanent features, so maps were made to study these features. Shortly after the middle of the 19th century, two famous Martian observers, Father Secchi in Italy and Camille Flammarion in France, published maps of Mars in color and recognized changes in the coloration.

Telescopic visual observations of Mars are exciting to make, but they are subjective and rely on the observer's ability to jot down or sketch what he sees. Photographic observations, on the other hand, permit an objective and permanent record that can be studied later and by other people. However, the disturbing effect of the Earth's atmosphere, called *bad seeing* by astronomers, causes the image formed by any ground-based telescope to move about on the photographic plate. This causes detail in the Martian features to become smeared out during even a relatively short time exposure. Although considerable reliance was placed on good visual observations, the thousands on thousands of good photographs have proved to be trustworthy, as the two photographs in Figure 3-11 demonstrate.

In 1965 the study of Mars made an abrupt change. The old controversies and mysteries were cleared. Or rather they were shifted from one plane, that of sheer speculation for want of better observations, to another plane, that of interpretation of detailed observations obtained by spacecraft. Mariner 4 passed behind Mars and not only took photographs of the surface, but also made observations of temperature and sent radio signals back to the Earth through the Martian atmosphere. In 1969 the flyby of Mariner 6 and 7 produced more photographs with greater detail. In 1971, however, Mariner 9 was placed in an orbit about

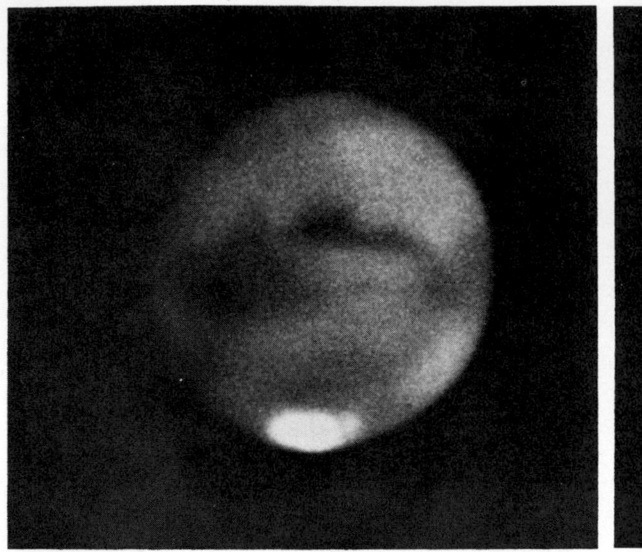

Figure 3-11
(a) Mars photographed by the 24-in. refractor at Flagstaff, Arizona, in 1924. (Lowell Observatory.)

(b) Mars as photographed by Mariner 6 when within only 460,000 miles of that planet in 1969. (NASA photograph.)

Mars; it became the first artificial satellite of a planet other than the Earth. It was equipped with numerous cameras and instruments to gather information about the physical nature of Mars and to relay that information back to Earth.

B. THE SURFACE The permanent dark markings on the surface of Mars have commanded the strongest interest and aroused the most controversy. The Mariner spacecraft, however, ended that long-standing controversy. The Martain surface, like the moon, is pockmarked with craters.

1. Impact craters As with the moon, it is strongly suspected that many of these craters are the result of large chunks of rock falling from outer space onto the surface (Figure 3-12). Before impact, each of these rocks had presumably been revolving in an orbit about the sun. The Earth, too, has some impact craters (see Figure 7-10), and it must have had many more that have been erased by persistent erosion and the catastrophic upheavals of the Earth's crust.

The impact craters on Mars come in all sizes, from the smallest visible in the telescopic cameras aboard Mariner 9 to large ones visible with telescopes from Earth and plotted as deserts by the early Martian mapmakers. One of these, known as Hellas, is about 900 miles in diameter.

2. Volcanic Craters Not all of the craters on the surface of Mars are impact craters, however. There are a number of very large volcanoes all in a region once thought to be a Martian desert. The largest of these volcanic mountains is Nix Olympica (Figure 3-13). This volcano had been observed and named by

Figure 3-12
Old impact craters on Mars with a rille that shows possibility of erosion by water. (NASA photograph.)

those making detailed maps from visual observations. It was then classified as a junction between several canals, and as all such junctions, it was called an oasis. (It was felt that the canals carried water from the polar regions to the arid deserts of the equatorial belt.)

Nix Olympica, however, is a shield volcano similar in structure to Hawaiian volcanoes such as Mauna Loa (see Figure 2-7). Shield volcanoes are formed of lava with a low viscosity, that is, lava which flows readily (molasses in January has a high viscosity). Consequently, shield volcanoes have sides that slope rather gradually. Nix Olympica is actually larger than any volcano on Earth; it is 310 miles across at the base. In comparison, the mountains that form the Hawaiian Islands are only 140 miles across. The height of Nix Olympica has been estimated to be almost 15 miles above the surrounding plain. The summit of Mauna Loa is some 30,000 feet above the ocean floor.

A different type of volcano is shown in Figure 3-14. The concentric rings of this Martian volcano indicate that it collapsed. For some reason the inside of the crater became hollow (perhaps the lava flowed back down the volcanic vent) and the mountain walls, unable to support themselves, collapsed. Such a volcano is

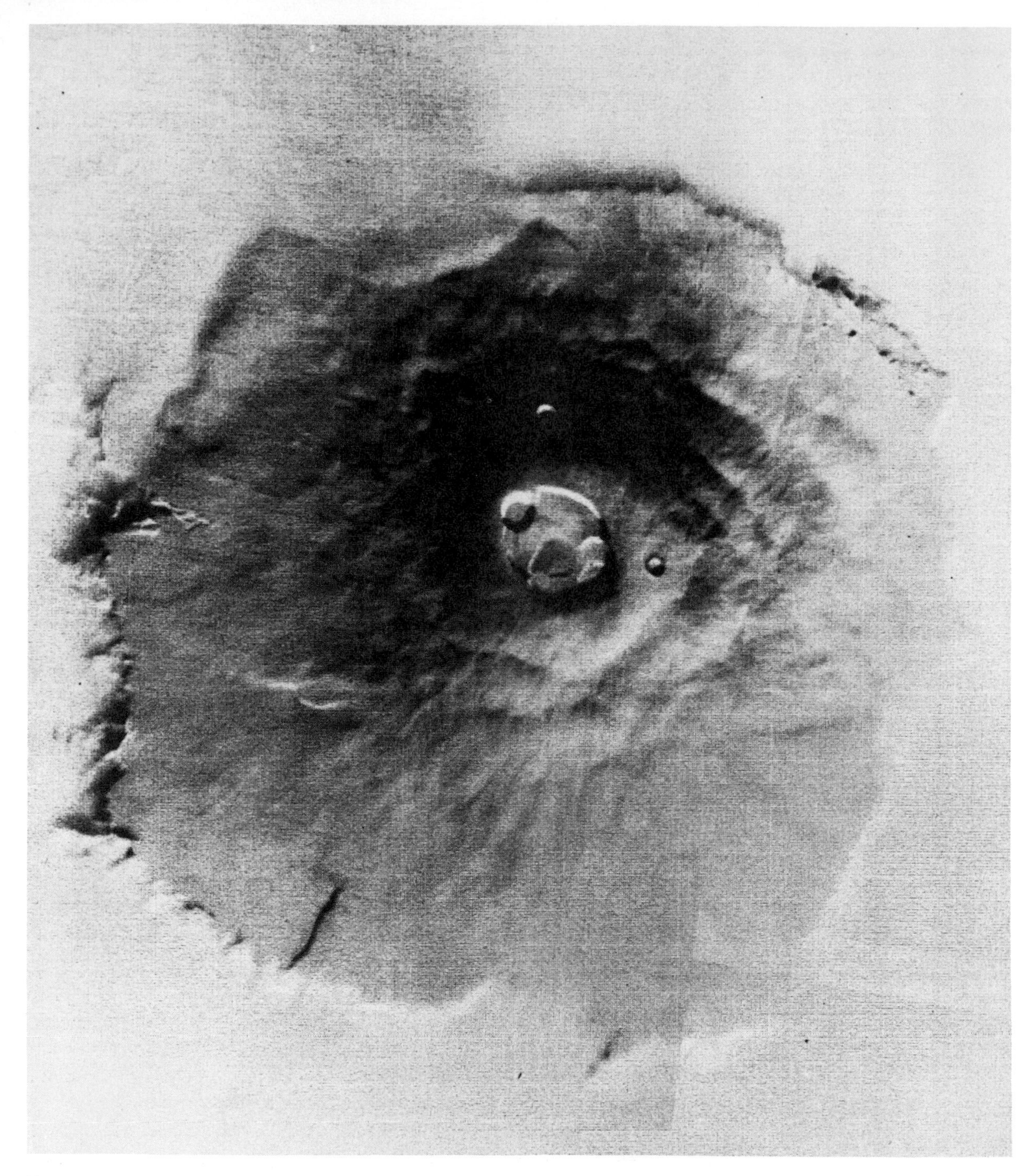

Figure 3-13
Nix Olympica, a shield volcano on Mars (NASA photograph.)

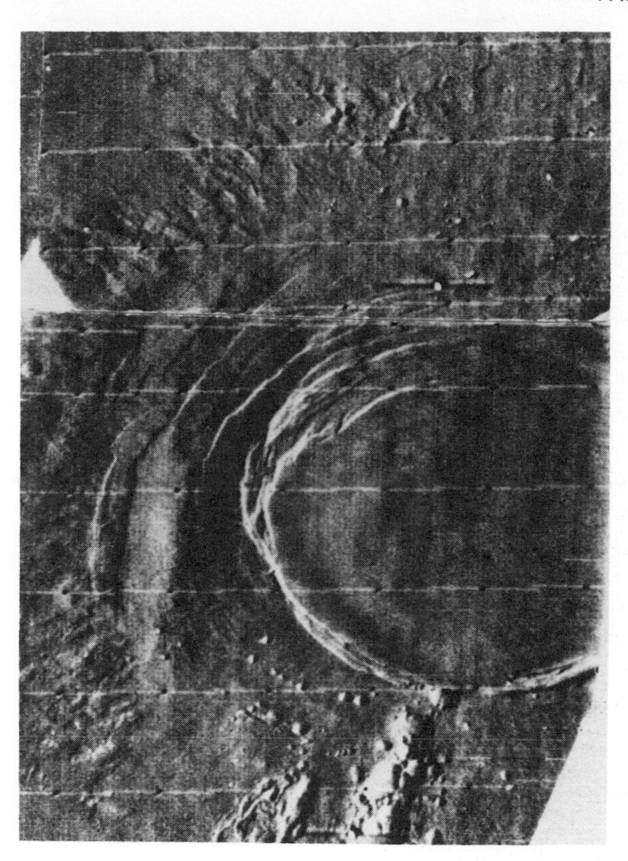

Figure 3-14
A Martian volcanic crater that may have collapsed. (NASA photograph.)

called a *caldera*. Crater Lake in Oregon is a caldera. The crater in the Martian caldera of Figure 3-14 is 75 miles in diameter; the crater of Crater Lake is only 6 miles in diameter.

3. The Martian Canyons The greatest surprise of all the Mariner 9 photographs has been the piecing together of a gigantic canyon system, part of which is shown in Figure 3-15 and all of which is shown in Figure 3-16. The entire canyon, which runs close to and roughly parallel to the equator of Mars, is 2500 miles long, 75 miles wide at the widest, and about 20,000 feet deep. The Grand Canyon in Arizona would form only a small spur of this huge Martian canyon.

The branching canyons that feed into the major canyon from the side have an appearance that can most easily be interpreted as having been formed by running water! Running water as an interpretation brings two reactions. First, there is very little water presently on the surface of Mars, so if running water formed the canyons at some previous epoch, where did the water go?

Second, if there was once so much water on Mars that giant canyons 2500 miles long could be extensively eroded along their sides, and furthermore, since the living things on Earth could have formed only in a water environment, did any life form on Mars when it had so much water?

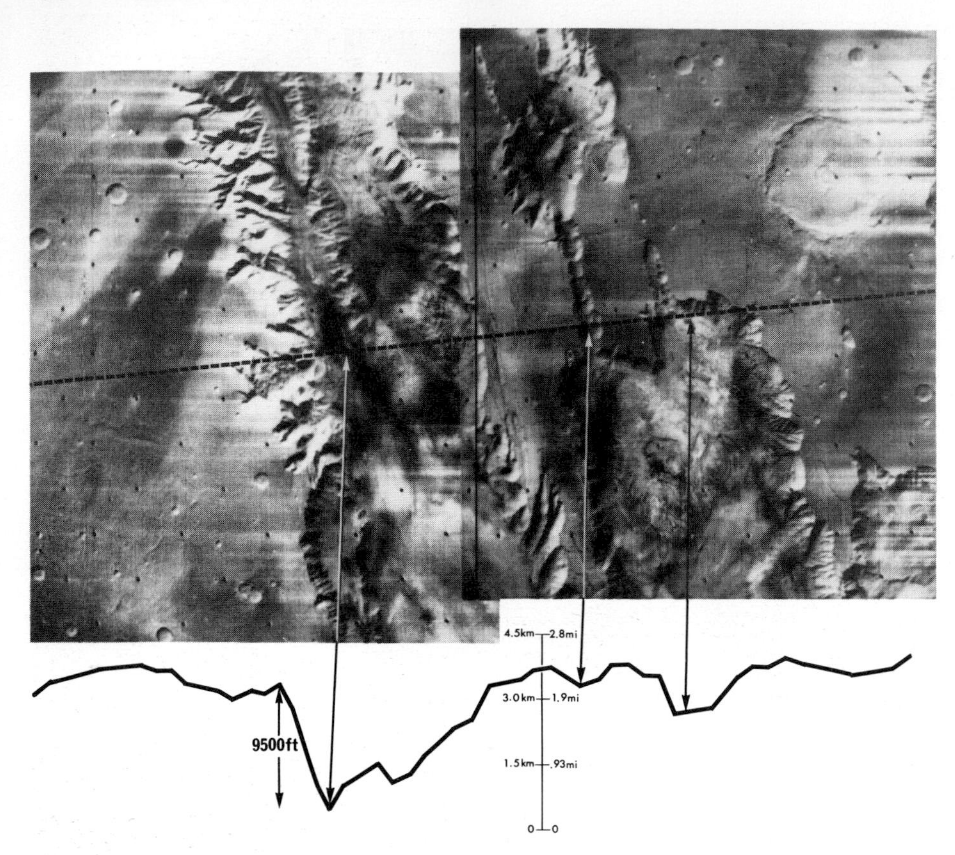

Figure 3-15
A portion of the huge Martian equatorial canyon. (NASA photograph.)

The question asked for so long: "Is there life on Mars?" has not yet been answered, but these canyons give hope to the belief that there may have been life on Mars at one time. What form could that life have taken? Would it have been similar to the life on Earth?

4. *The Polar Regions* The white polar caps seen in Figure 3-11 and in the photograph that opens this chapter have been shown to be composed of solid carbon dioxide, commonly called dry ice. (The term dry ice arises from the fact that solid carbon dioxide does not melt into a liquid as does frozen water. Instead carbon dioxide turns from solid directly into a vapor, a process called *sublimation.*) Because solid carbon dioxide does turn from a solid directly into a vapor, even at the temperatures and atmospheric pressures on the surface of Mars, astronomers know that the canyons cannot have been cut by liquid carbon dioxide.

The edge of the south polar cap of Mars was photographed by Mariner 7 and indicates a white dusting of solid carbon dioxide that seems to be heavier inside some of the craters (Figure 3-17).

Closer to the pole, however, the craters are less in evidence. Instead, there are massive layered regions such as the "table-mountain" near the south pole shown

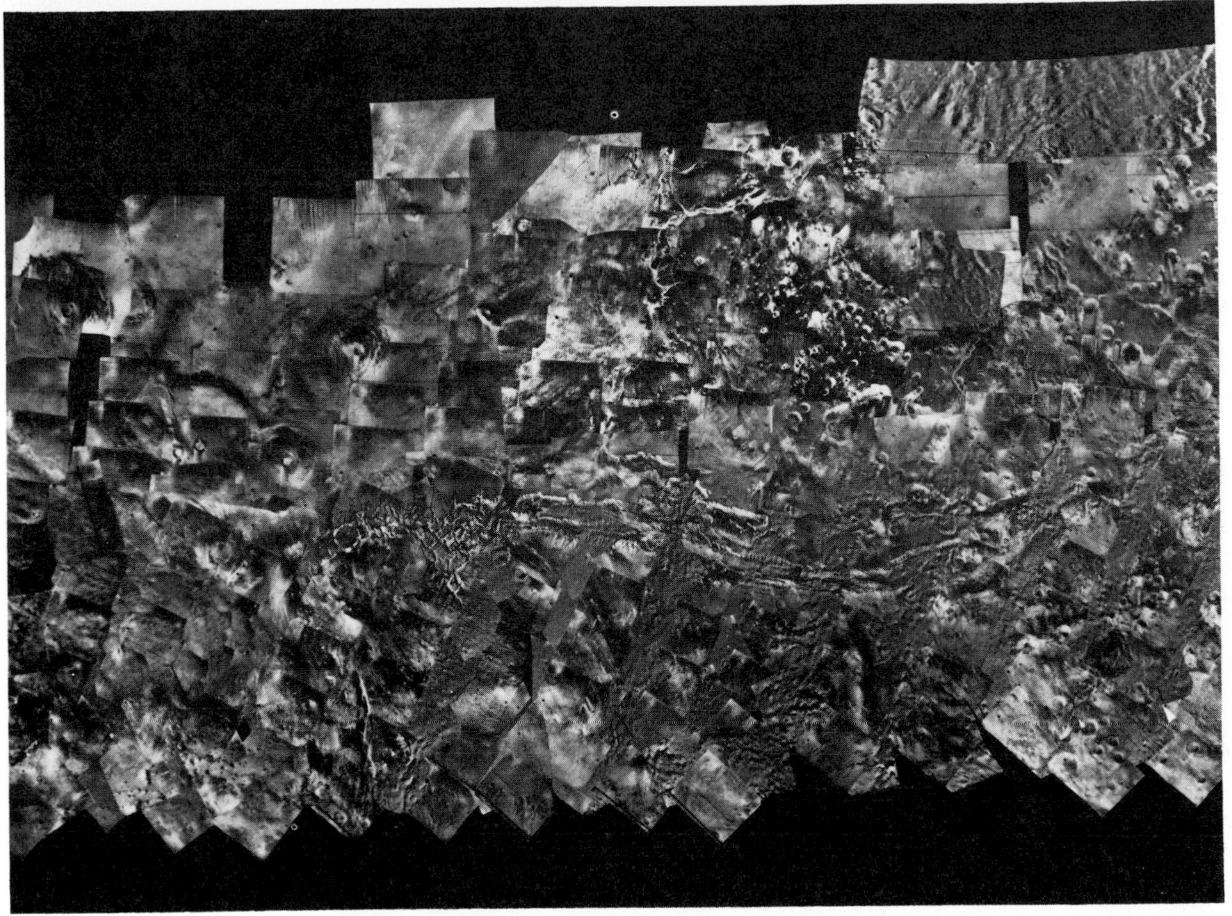

Figure 3-16
A photomosaic of the equatorial region of Mars. Nix Olympica is in the upper left. The great equatorial canyon is in the center. A field of impact craters is in the north just right of center. (NASA photograph.)

in Figure 3-18. The layering could be volcanic in origin, for instance, either lava flows or repeated collections of volcanic cinders and dust. It has also been suggested that the layering is sedimentary rock formed from repeated freezing and thawing or from some previous body of water. The south pole itself seems to be surrounded with ridges that suggest similar layering of material. Some layers appear to be 100 meters thick.

During the autumn season on either pole, clouds form and dissipate (Figure 3-19) before they settle in for the winter. Once established they remain through the winter. In the spring, the clouds disappear to reveal the whiter sharper polar cap.

As spring progresses the polar cap retreats as the solid carbon dioxide sublimes into vapor. Successive stages of one small area of the retreating polar

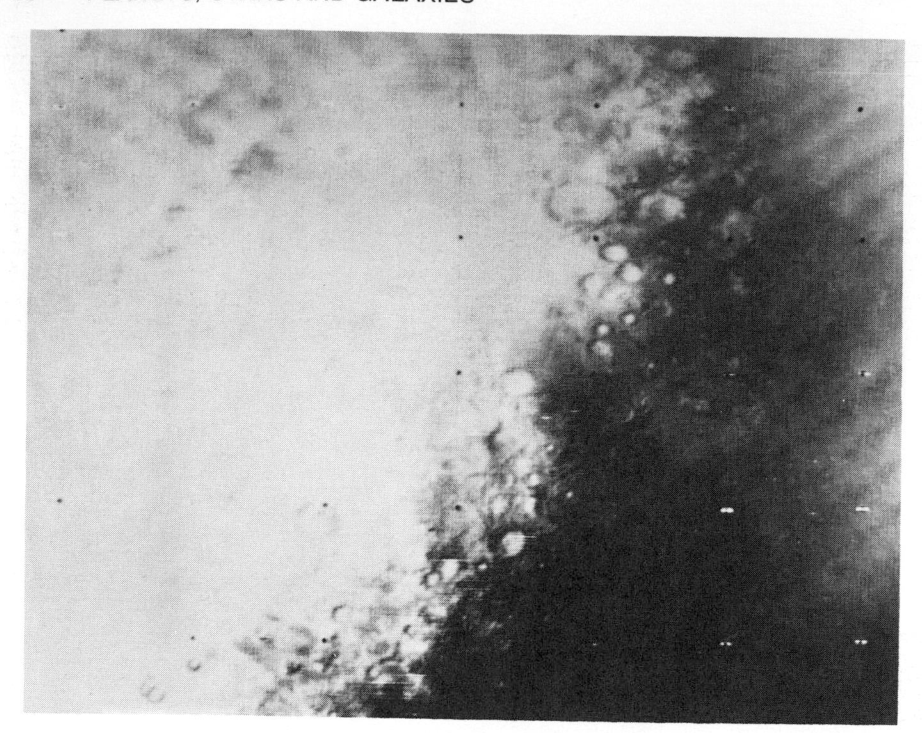

Figure 3-17
The edge of Mars' south polar cap photographed by Mariner 7. (NASA photograph.)

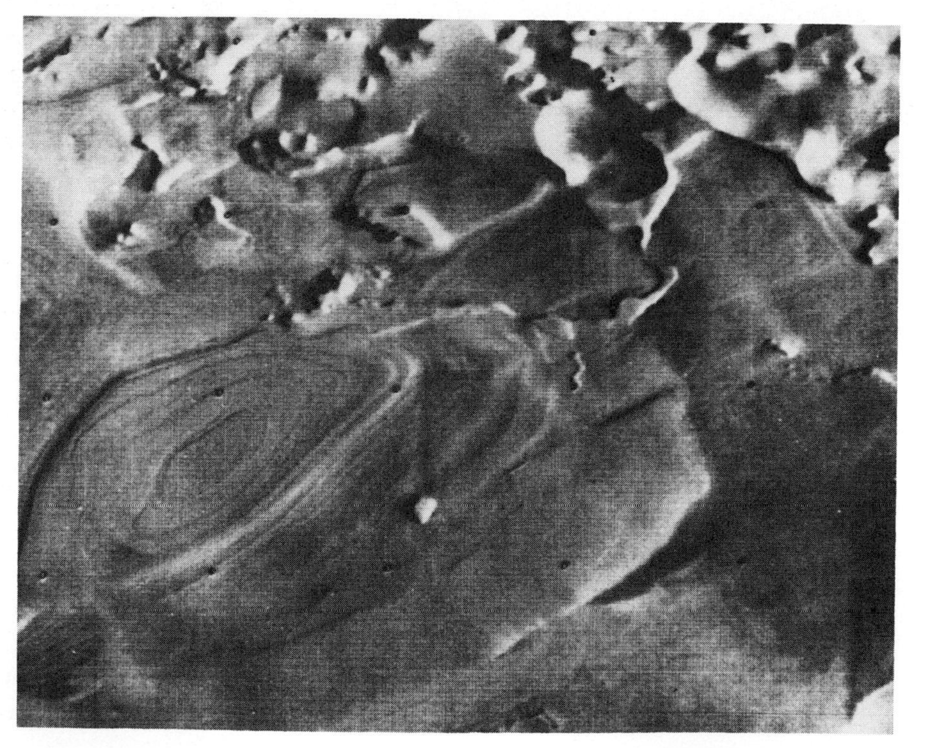

Figure 3-18
A table mountain feature near the Martian south pole may contain layered deposits of dust and volcanic ash, possibly held together by carbon dioxide and water ices. The sunlight is shining from the lower right. (NASA photograph.)

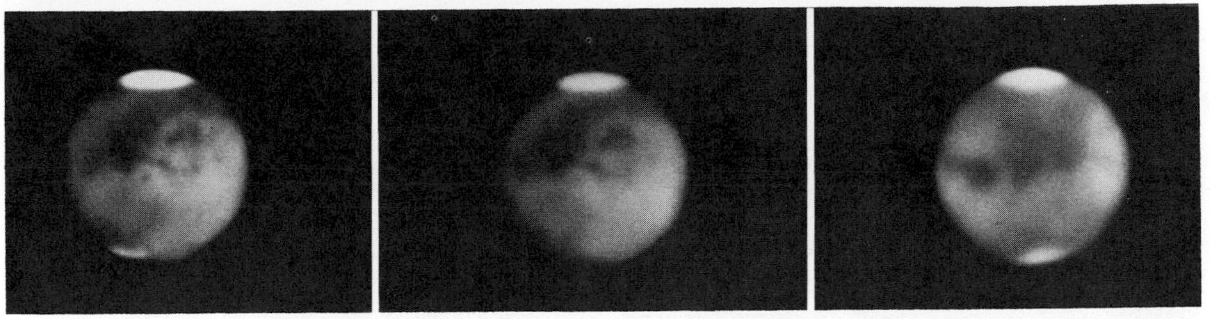

Figure 3-19
Three photographs of Mars taken on three different dates in 1939 — July 23, July 25, and July 30 — to show the formation, disappearance, and reformation of clouds over the north polar region in just a few days. (Lowell Observatory.)

cap can be seen in Figure 3-20. It is expected that no more than an inch or so of carbon dioxide snow could have sublimed during the time interval of the three closeup photographs. Those parts over which the carbon dioxide snow disappears regularly must be fairly flat lands; the dark riverlike band must be a canyon of some sort. Along what appears to be the north bank of the canyon, layered rock becomes evident as the white cap recedes.

Mariner 9 photographs have verified the existence of residual polar caps; a region of white that remains throughout the summer near each pole. Because of the temperatures and atmospheric pressures, it is felt that much of these residual polar caps are water-ice. This is especially true of the south polar ice cap, which may be composed entirely of water-ice. Water vapor has been detected in the atmosphere over the poles, and the reflection spectra of the polar caps shows evidence of the presence of water-ice. The clouds that form over each pole in autumn may well be water vapor clouds.

C. THE TEMPERATURE The temperature of the surface has been determined both by ground-based telescopes, radio telescopes, and by observations from Mariners 6, 7, and 9. The maximum temperature recorded at noon on the equator is close to 82° F; the coldest, recorded at the south polar cap by Mariner 9, is −207° F. The usual daytime temperatures at the equator, however, vary from a high of about 70° F at noon to a low of perhaps −100° F just before sunrise. This is a daily change in temperature of about 170° F. It has been observed that the dark markings are warmer than the rest of the surface of the planet.

D. THE ATMOSPHERE One of the reasons for the extreme daily changes in temperature on Mars is the rarefied atmosphere. Observations by the Mariner space probes indicate an atmospheric pressure at the surface of Mars of only 8 millibars. The atmospheric pressure on the surface of the Earth is about 1000

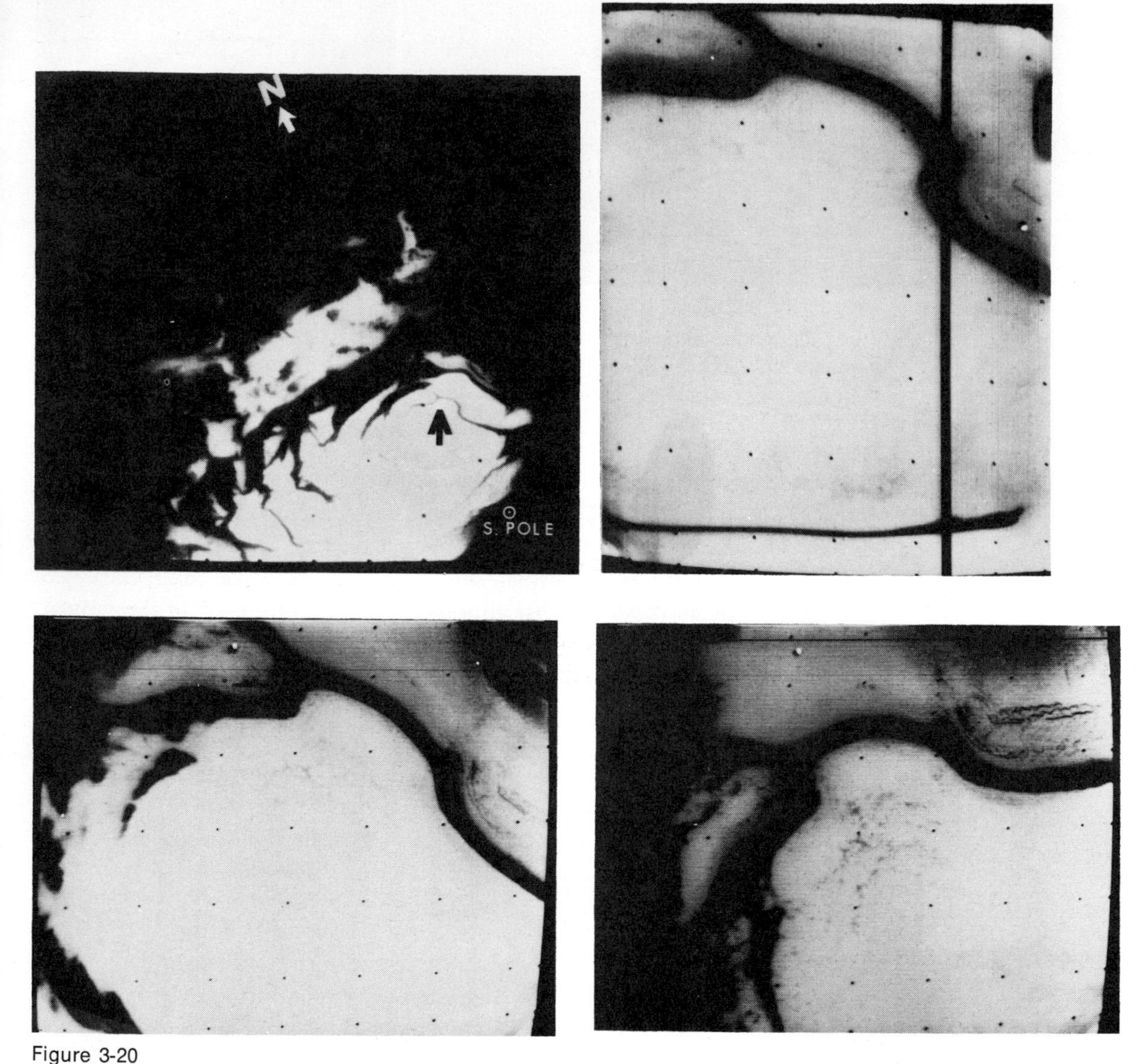

Figure 3-20
(a) The south pole of Mars with a fork in the "stream" shown by the black arrow. (b), (c), and (d) show detail about that stream as the carbon dioxide snow disappears in the spring. (NASA photograph.)

millibars. A pressure of 8 millibars is reached in the Earth's atmosphere only at a height of about 20 miles.

The main constituent in the Martian atmosphere is carbon dioxide, although oxygen, hydrogen, and water vapor have been detected in small amounts. There is, surprisingly enough, no evidence of nitrogen. It would appear, therefore, that the atmospheres of Venus and Mars have similar chemical compositions; it is the Earth's atmosphere that is unique in our solar system with its high percentage of nitrogen and oxygen. This makes it appear that the composition of the Earth's atmosphere is a result of life. It may be that the existence of life on Earth has altered the original atmosphere here — a sort of natural pollution. Detailed studies of other planetary atmospheres, therefore, will certainly help us to understand our own atmosphere better. And the better we understand our own atmosphere the better we shall be able to preserve its natural qualities.

E. THE CLOUDS Apparently every planetary atmosphere has clouds. The surface of Venus is hidden from our view by its clouds. The Earth has more clouds than we thought before artificial satellites and astronauts told us how cloudy our planet really is. And Mars has clouds of several sorts.

At times, the entire planet is shrouded in a yellowish colored cloud that is actually a dust storm. The great dust storm of 1971 greeted Mariner 9 when it first went into an orbit about the planet. The dust settled soon afterward, however, revealing the surface beneath. Mariner 9 even photographed localized dust storms.

Not all of the dust settles out as a smooth layer on the surface. The formation of dunes inside one crater are clearly revealed in the photograph of Figure 3-21. Dunes form as a result of winds, and the sizes of the dunes reveal something of the sizes of particles composing the sand as well as something of the speed and direction of the prevailing winds.

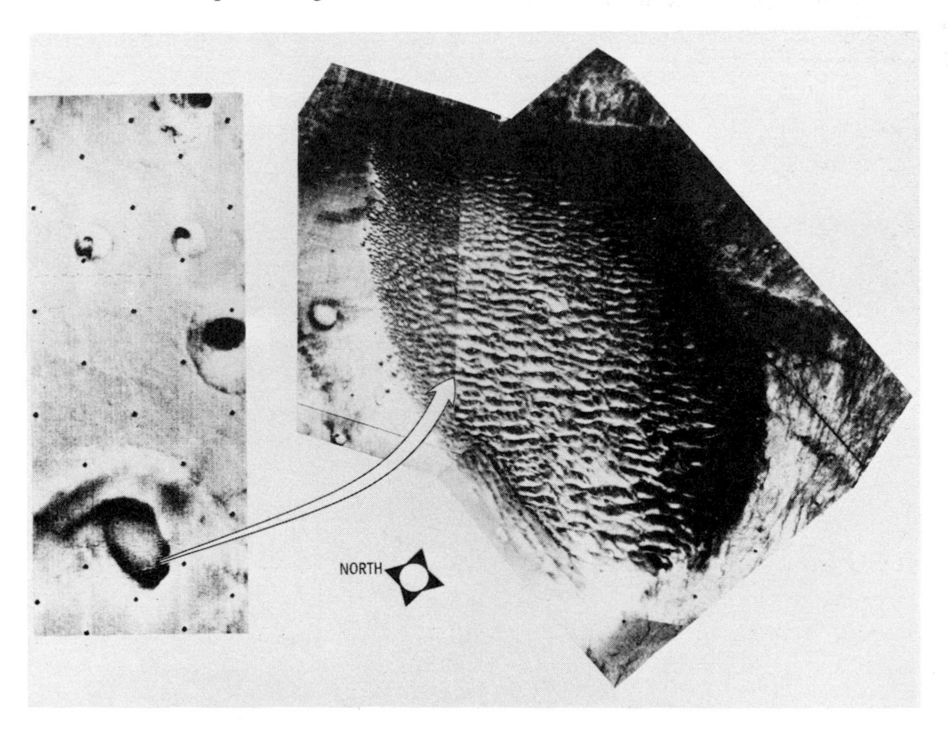

Figure 3-21
Sand dunes inside of a Martian crater. (NASA photograph.)

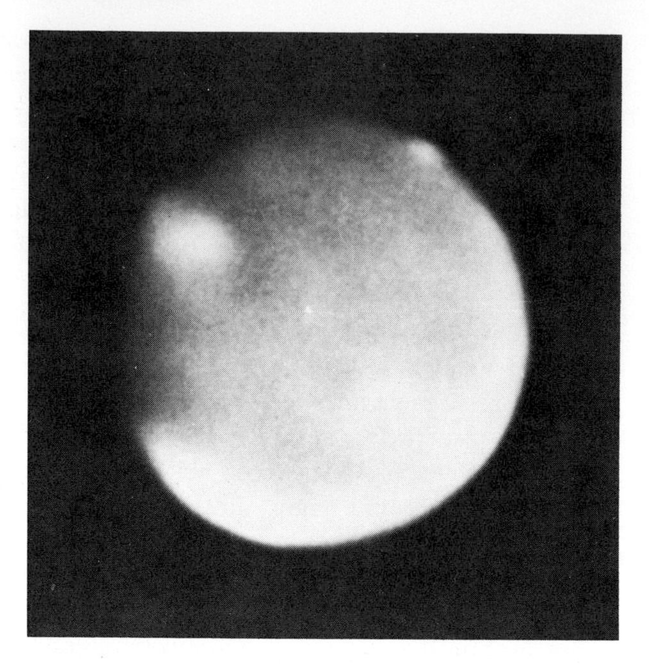

Figure 3-22
White clouds seen along the terminator of Mars. (Lick Observatory photograph.)

White clouds also appear from time to time. They are usually seen near the *terminator* (the line separating day from night) at sunrise but dissipate as morning progresses (Figure 3-22). White clouds also form in the late afternoon, so they may persist through the night. From 140 nights of earth-based observations of Mars in 1964 to 1965, when the Earth and Mars were on the same side

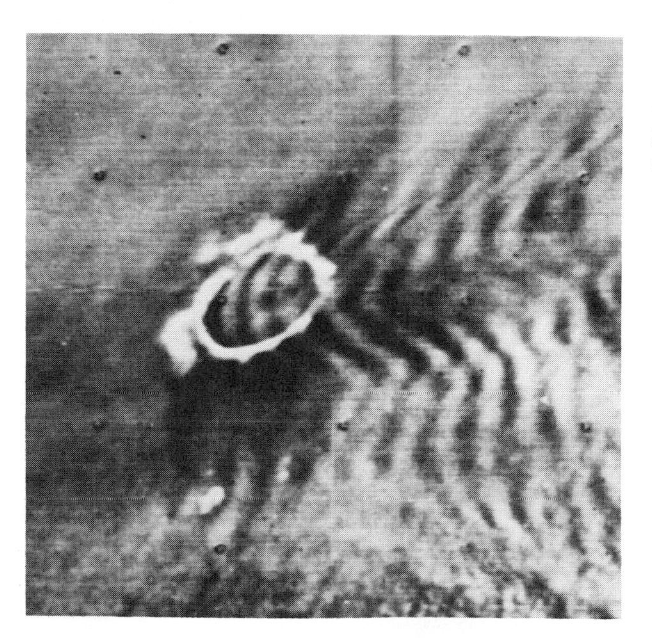

Figure 3-23
Clouds form on the crest of atmospheric waves after passing a frost-rimmed crater. (NASA photograph.)

of the sun and consequently close together, the morning clouds were observed on two thirds of the Martian days, the evening clouds on about one half.

Some clouds photographed by Mariner 9 even form on the crests of atmospheric waves in the wake of winds passing over a crater (Figure 3-23). Similar clouds form on the Earth.

The excitement raised by the Mariner 9 observations exemplifies progress in science. With a startling advance in observations, old ideas are thrust aside, old questions settled once and for all. But these observations raise new questions, sometimes more than they answer. These new questions, however, become more significant as our understanding deepens. The mysterious gives way to reason as the observations become more exacting.

BASIC VOCABULARY FOR SUBSEQUENT READING

Astronomical unit
Aurora
Caldera
Escape velocity
Greenhouse effect
Impact crater
Radar
Retrograde rotation
Revolution
Rotation
Scattering of light
Shield volcano
Solar system
Solar wind
Terminator

QUESTIONS AND PROBLEMS

1. Draw the nine planets to scale. One possible scale is to let 10,000 miles be represented by 1.0 centimeter. For the diameters of the planets see Table 4-1, page 78.

2. Compare the region about the Nix Olympica and the other volcanoes on Mars (Figure 3-16) with the area pockmarked with impact craters. (a) Why, do you suppose, is only one area so pockmarked with craters? (b) Why does the region about the volcanoes have very few impact craters?

3. Compare the surface of Mars with that of the Earth, by considering the surface features and materials.

4. Discuss the possibility of life on any of the other inner planets. For what reasons do you suspect life could or could not exist on Mercury, Venus, or Mars?

FOR FURTHER READING

Glasstone, S., *Sourcebook on the Space Sciences,* D. Van Nostrand Co., Princeton, N.J., 1965, Chapter 10.

Glasstone, S., *The Book of Mars,* National Aeronautics and Space Administration, Washington, D.C., 1968.

Moore, Patrick and Charles A. Cross, *Mars,* Crown Publishers, New York, 1973.

Slipher, E. C., *A Photographic Study of the Brighter Planets,* Lowell Observatory, Flagstaff, Ariz., and the National Geographical Society, Washington D.C., 1964.

Smith, A. G., and T. D. Carr, *Radio Exploration of the Planetary System,* D. Van Nostrand Momentum Book, Princeton, N.J., 1964.

Struve, O., and V. Zebergs, *Astronomy of the 20th Century,* Crowell, Collier, and Macmillan, New York, 1962, Chapter VIII.

Whipple, F. L., *Earth, Moon and Planets,* 3rd ed., Harvard University Press, Cambridge, Mass., 1968.

Mars as Viewed by Mariner 9, National Aeronautics and Space Administration, 1974.

Akasofu, S.-I., "The Aurora," *Scientific American,* p. 55, December 1965.

Bolt, B. A., "The Fine Structure of the Earth's Interior," *Scientific American,* p. 24, March 1973.

Cahill, L. J., "The Magnetosphere," *Scientific American,* p. 58, March 1965.

Eshleman, V. R., "The Atmosphere of Mars and Venus," *Scientific American,* p. 78, March 1969.

Leighton, R. B., "The Surface of Mars," *Scientific American,* p. 26, May 1970.

Mariner 9 Television-Experiment Team, "The New Mariner 9 Map of Mars," *Sky and Telescope,* p. 77, August 1972.

Murray, B. C., "Mars from Mariner 9," *Scientific American,* p. 48, January 1973.

Murray, B. C., "Mercury," *Scientific American,* p. 58, September 1975.

Parker, E. N., "The Solar Wind," *Scientific American,* p. 66, April 1964.

Pollack, J. B., "Mars," *Scientific American,* p. 106, September 1975.

Shapiro, I. I., "Radar Observations of the Planets," *Scientific American,* p. 28, July 1968.

Siever, R., "The Earth," *Scientific American,* p. 82, September 1975.

Strom, Robert G., "The Planet Mercury as Viewed by Mariner 10," *Sky and Telescope,* p. 360, June 1974.

Strong, J., "Infrared Astronomy by Balloon," *Scientific American,* p. 28, January 1965.

Van Allen, J. A., "Interplanetary Particles and Fields," *Scientific American,* p. 160, September 1975.

Young, Andrew and Louise, "Venus," *Scientific American,* p. 58, September 1975.

"Bright Flares on Mars," *Sky and Telescope,* p. 83, February 1970.

"Carbon Dioxide in the Martian Polar Caps," *Sky and Telescope,* p. 230, April 1972.

"Mars Photographs and Observations from Mariner 9," *Sky and Telescope,* p. 14, January 1972; p. 208, April 1972; p. 300, May 1972; p. 77, August 1972.

"Venus Observed by Mariner," *Sky and Telescope,* p. 235, April 1974.

CHAPTER CONTENTS

4.1 JUPITER

4.2 SATURN

4.3 URANUS

4.4 NEPTUNE

4.5 PLUTO

4.6 SIZES AND DISTANCES

LEARNING OBJECTIVES

BE ABLE TO:

1. DESCRIBE HOW JUPITER'S ATMOSPHERE HAS BEEN EXPLORED BY AN OCCULTATION.
2. COMPARE THE ATMOSPHERE OF JUPITER WITH THAT OF VENUS, EARTH, AND MARS.
3. DESCRIBE THE GREAT RED SPOT AND GIVE THE CURRENT EXPLANATION OF ITS EXISTENCE.
4. COMPARE THE INTERIOR OF JUPITER WITH THAT OF THE EARTH.
5. COMPARE THE ORIGIN OF MICROWAVES FROM JUPITER WITH THAT OF THE DECAMETER WAVES.
6. GIVE THE CAUSE OF ANY MAGNETIC FIELD AND IN A GENERAL WAY, APPLY THIS TO EARTH, JUPITER, AND VENUS.
7. EXPLAIN WHY JUPITER AND SATURN ARE SO MUCH MORE OBLATE THAN THE EARTH.
8. GIVE THE CURRENT THEORIES ON THE BELTS OF JUPITER.
9. DESCRIBE SATURN'S RINGS AND EXPLAIN HOW OBSERVATIONS INDICATE THE PARTICLE SIZE.
10. EXPLAIN HOW THE STUDY OF THE MOTION OF URANUS LED TO THE DISCOVERY OF NEPTUNE.
11. EXPLAIN HOW WE KNOW THE PERIOD OF ROTATION OF THE PLANET PLUTO.

Chapter Four
THE OUTER PLANETS

4.1 JUPITER

The fifth and largest planet is Jupiter. Its mass is 318 times that of the Earth and its diameter is 89,000 miles, over 11 times the Earth's diameter. It is about 5.2 A. U. from the sun. Because of its enormous distance from the source of heat its temperature is low. There are markings that enable us to determine its period of rotation near the equator at 9 hours, 50 minutes (Figure 4-1). The period of rotation increases at latitudes nearer the poles, to about 9 hours and 56 minutes. From this observation we conclude that the surface visible to us is not solid and that solid material does not occur until a considerable depth under the visible cloud layer.

Jupiter is not only the largest planet but also rotates on its axis in a shorter period than any other planet. The material at the equator whirls about the axis at about 30,000 miles per hour. (The Earth's equator whirls at about 1000 miles per hour.) As a result, Jupiter is flattened at the poles to the extent of presenting a slightly elliptical rather than a circular disk.

A. TECHNIQUES OF OBSERVATION Jupiter has been and will be investigated much as the inner planets. Telescopes, both optical and radio, have compiled a large store of information from which astronomers can form ideas about the nature of this largest planet in our solar system. Space probes are being used. In March 1972 Pioneer 10 was launched and put into an orbit that carried it only 81,000 miles above the cloud tops of Jupiter on December 3, 1973. After its flyby with Jupiter, it will continue traveling away from the sun into interstellar spaces. In 80,000 years it will have traveled away from the solar system a distance that is less that the distance to the nearest star! Pioneer 11 traveled even closer to the cloudy visible surface of Jupiter in December 1974. Both Pioneer space probes encountered heavy radiation in the vicinity of Jupiter.

Since space probes of Jupiter will not be a common occurrence, astronomers are using every optical and radio observation their ingenuity can devise. For example, stars in the path of Jupiter's motion are *occulted* by Jupiter (an *occultation* is the term applied when a planet or the moon passes in front of a star even if the term eclipse would apply). A moment before and after occultation the star's light passes through the upper atmosphere of Jupiter. The star's light is altered by Jupiter's upper atmosphere, and the way it is altered yields information on the characteristics of that atmosphere.

B. TEMPERATURES Temperature measurements have been made in the infrared region of the spectrum with the larger telescopes revealing temperature

Chapter Opening Photo
Sculptor Ralph Turner of the Lunar and Planetary Laboratory, University of Arizona. Turner holds a model of Saturn; the model of Jupiter hangs nearby. (Courtesy, Dennis Milon, *Sky and Telescope*.)

Figure 4-1
Jupiter as photographed by television cameras on board Pioneer 10 on December 2, 1973 as the spacecraft approached the giant planet. (NASA photograph.)

differences across Jupiter's surface. As might be expected for a planet warmed by the sun, the warmest regions of the disk are along the equatorial regions. Temperatures here average about −60° F, while nearer the polar regions they average about −120° F. These observations are made at a wavelength that may penetrate the outer cloud layers and, if so, these would be temperatures of regions beneath the visible surface. Observations made with longer wavelength infrared have yielded temperatures of −190° F; while observations at still longer wavelengths yield temperatures as low as −220° F. It appears that the temperature decreases with increasing heights in the atmosphere. It may be that whatever keeps the interior of the Earth hot (it may be radioactivity) also keeps Jupiter's interior hot.

The surprising observation is that the cloudy surface of Jupiter has distinct "hot spots" on it. Hot spots have been observed along the equator with temperatures as high as +70° F. Hot spots also appear in regions north and south of the equator as well. It is hoped that future observations will permit astronomers to identify these hot spots with visible features on Jupiter's changing disk.

C. THE ATMOSPHERE The atmosphere of Jupiter is completely different from that of any of the inner planets. The atmosphere is composed primarily of two gases: hydrogen 82% and helium 17%, leaving only 1% for the other gases such as ammonia, NH_3; water, H_2O; and methane, CH_4. It may be presumed that since there is so much hydrogen on Jupiter, it has combined with all the carbon, nitrogen, and oxygen available.

At the low temperatures that range over Jupiter's cloudy surface, most of the ammonia and water vapor are frozen into crystals, but the methane is gaseous. Some of the ammonia exists in the gaseous state, however, for spectral lines absorbed by that molecule have been detected. Absorption lines of methane and hydrogen have also been detected.

Figure 4-2
Jupiter as viewed by Pioneer 10. Notice the white spots in the cloud tops. The one just north of the equator is being smeared out into a minor white belt. (NASA photograph.)

It is the dark and light bands that give Jupiter its distinct appearance. The light bands are about 12 miles higher than the darker bands and are about 15° F cooler. They are believed to be gases welling up from the interior. These gases then form the dark bands as they sink back into the deeper parts of the atmosphere (Figure 4-2).

Since the period of rotation of Jupiter decreases from the equator toward each pole, these dark and light bands move east and west relative to each other. These relative velocities may be as high as 300 to 400 miles per hour (Figure 4-3).

The bands change over the years, and that change is sometimes fairly rapid. Photographs taken at New Mexico State University Observatory reveal that the changing character of the belts can be dramatic in just over one year (Figure 4-4).

D. THE GREAT RED SPOT There is one notable feature in these clouds that does remain fairly constant. It is a large spot, often a dull brick red, which is about 7000 miles wide and 30,000 miles long. It is shown in Figures 4-1 and 4-4. The Great Red Spot was first seen in the 1660s and has not changed appreciably since, although its brightness and color do vary. The Great Red Spot rotates with Jupiter, although its period is 8 seconds longer than the period of rotation of the dense clouds about it. It therefore moves relative to the clouds in the South Tropical Zone in which it resides. It makes one less rotation than the clouds in about 5 years.

Recent observational and theoretical work strongly suggest that the Great Red Spot is a cyclonic storm not unlike the tropical cyclones and high rising thunderstorms (cumulonimbus clouds) here on Earth. Since the atmosphere and heating conditions on Jupiter are vastly different from those on the Earth so are the storms. Meteorological studies adjusted to the conditions on Jupiter have led to a theory of long-lasting Jovian* storms.

The origin of Jovian storms probably lies in the hot gases underneath the immediately visible and colder surface. Studies of Jupiter using infrared radiation have revealed temperatures as high as 20° F at some depth in the cloud layer. This difference in temperature could cause the warmer gases to rise; the rapid rotation of the planet would then cause these rising gases to whirl. The Great Red Spot does, in fact, move as one would expect a tropical cyclone on Jupiter to move. Furthermore, it extends above the surrounding cloud layer as do the high rising cumulonimbus clouds here on Earth. The oval shape of the Great Red Spot would then be akin to the anvil head of the cumulonimbus clouds that pelt us with lightning and hail. It will be recalled (see Figure 4-2) that white spots, too, appear from time to time on Jupiter. These may also be cyclonic storms.

**Jovian:* Pertaining to Jupiter, the planet as well as the chief deity of the ancient Romans. Recall our expression *by Jove* and the word *jovial*. It was supposed by the Roman astrologers that those born under Jupiter were happy people.

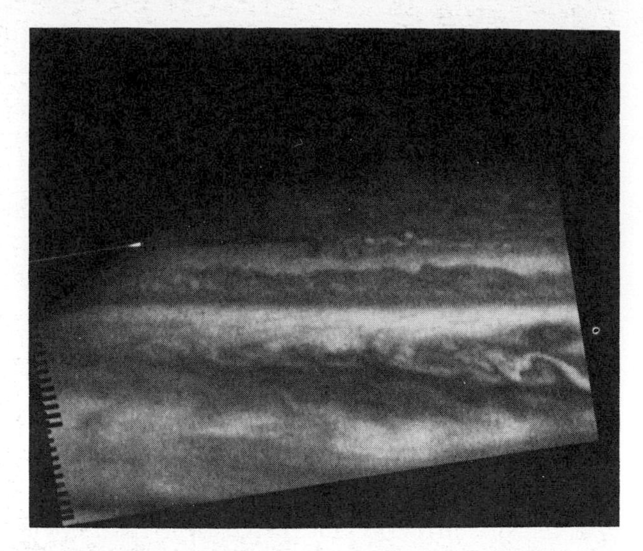

Figure 4-3
The swirling clouds of Jupiter were photographed by Pioneer 10 on December 3, 1973. It appears from this and other photographs (see Figure 4-2) that the white clouds are at a higher elevation than the dark clouds. (NASA photograph.)

Figure 4-4
(a) Photograph of Jupiter taken on October 23, 1964 shows the Equatorial Zone very dark. The Great Red Spot is obvious.

(b) By December 12, 1965 the Equatorial Zone had become very bright. Even the belt containing the Great Red Spot has changed in the intervening months between these two photographs. (New Mexico State University Observatory photographs.)

E. THE INTERIOR The motions of the clouds indicate that they are not bound by any underlying solid surface. This gives us a clue in our study of Jupiter's interior. Although Jupiter has a mass 318 times the mass of the Earth, its volume is 1312 times that of the Earth. Consequently, the material of the Earth, volume for volume, must have a mass that is about four times as great as that of Jupiter. This can be expressed in terms of *density*.

The density of a substance is its mass per unit volume. The density of water is generally taken as 1 and all other materials are then compared to water. The average density of the Earth is 5.5 and the density of Jupiter is only 1.3. Consequently, volume for volume, the Earth has a mass that is 5.5 times as great as water, whereas Jupiter is only 1.3 times as massive. This low density indicates that Jupiter is about three fourths hydrogen.

According to recent observations and theory, the atmosphere of Jupiter may be about 600-miles thick and may be composed largely of hydrogen and helium with mixtures of ammonia, methane, and water-ice. Beneath this atmosphere may be a hot interior composed largely of liquid hydrogen. Temperatures in the interior may reach as high as 54,000° F in the core. The core may well be composed of rock.

The very high temperatures of the interior are predicted for both observational and theoretical reasons. One of the strongest observations in support of a hot interior is the energy radiated by that planet. When all of the radiation, radio waves, microwaves, infrared, and the visible region of the spectrum were considered, it was discovered that Jupiter radiates about twice as much energy out into space as it receives from the sun. It can do this only if the interior is very hot. Whether the high temperature of the core is a remnant of the process by which Jupiter was formed, or whether it is being heated by present internal processes is a question now being considered.

F. RADIO SIGNALS Besides reflecting the sun's light, Jupiter emits its own radiation in the form of radio waves. This discovery, made in 1955 by K. L. Franklin and B. Burke, then with the Carnegie Institute, opened up a whole new field of planetary radio astronomy, and gave us a new tool with which to probe the secrets of our solar system. Jupiter has been studied intensively since 1955, not only with radio telescopes, but with increased optical interest. The attempt is to correlate as many observations as possible.

There are two basically different radio emissions from Jupiter: the *microwaves*, with a wavelength between 3 and 60 centimeters, and the *decameter waves* with wavelengths between 10 and 50 meters. The decameter range is more often referred to by its frequency between 5 and 25 megacycles per second. These will be considered one at a time.

The microwaves are difficult to observe, and there is some question as to their nature. For example, by using the radio energies in this range to compute the temperature of Jupiter, discordant calculations result. At a wavelength of 3 centimeters, the surface temperature is calculated to be −210° F, in close agreement with optical studies. But as longer and longer wavelengths are used, the temperatures become more appropriate for a star than for a planet. Therefore, it is necessary to assume that this radio energy is emitted by nonthermal sources, perhaps by electrical discharges, or more regular electron motion. By making simultaneous observations with two radio telescopes placed several miles apart, the source of the microwaves has been determined to be not on the surface of Jupiter, but in a radiation belt similar to the Van Allen radiation belts about the Earth.

The radiation belt about Jupiter was the first evidence that this planet also has a magnetic field, and one much stronger than the Earth's. Pioneer 10 and 11 have since verified its existence. The radiation belt about Jupiter extends out to a distance of more than 750,000 miles, far enough to include the innermost satellite, Number V. (It was the fifth satellite to be discovered and happens to be closer to the planet than the four that Galileo discovered.)

It has become clear that the decameter waves are associated more closely with the surface than the microwaves. In fact, the decameter waves exhibit a periodicity close to the period of rotation of the planet itself. They would appear to originate from a definite region of the planet.

A mysterious correlation between the radio bursts in the 10-meter range and the motion of the second satellite, Io, was detected in 1964. It appears that radio bursts in this wavelength range are much more apt to occur when Io is on one side of the planet as seen from the Earth, and not when it is on the other side. This correlation is not understood. More observations and theoretical study are needed.

G. THE MAGNETIC FIELD The heavy radiation encountered by the Pioneer space probes and the bursts of radio energy received from Jupiter result from its magnetic field. Charged particles, principally electrons and protons ejected by the sun, are trapped in the magnetic field. These charged particles traveling at high speeds constitute the radiation encountered by the Pioneer space probes. The acceleration of these charged particles, as in a bolt of lightning, causes the bursts of radio energy.

Magnetic fields exist about some planets and not about others. Mercury has a small magnetic field, about 0.03 times that of the Earth. Venus and Mars have no appreciable magnetic field. The Earth has a significant field, and Jupiter a very strong one. The questions aroused by such a comparison are many, but the key question is: Why do planets have magnetic fields at all?

A magnetic field of the kind observed about the planets (and incidentally about many stars as well), can be produced only by charged particles moving in circular paths. A coil of wire carrying an electric current will produce a magnetic field similar to those of the planets. In a planet, the circular paths that these charged particles follow can be accounted for by the planet's rotation. Venus rotates more slowly than any of the planets, and consequently has a very small magnetic field. The cause of the magnetic field, then, might be ascribed in part to the rotation of a planet.

The rotation of a planet, however, cannot cause a magnetic field unless the negative electrons and positive ions inside the planet somehow become separated. Just how they become separated may be the major question. Presumably it results from material in the planet's core that is so hot that it becomes slightly fluid despite the tremendous pressures encountered deep inside a planet. Perhaps motion of this hot, slightly fluid, material separates those charges somehow or another. Mars rotates as rapidly as the Earth, but apparently does not have any hot material moving in its core. Jupiter rotates very rapidly indeed and unquestionably has a very hot core.

4.2 SATURN

Saturn, 9.5 A. U. from the sun, is perhaps the most spectacular astronomical sight in a telescope, because of the fantastic rings that encircle it, and that have long been used by cartoon artists to set the scene out in space. Saturn revolves about the sun in 29½ years. Its oblateness is more extreme than that of Jupiter; its polar diameter is 67,900 miles, and its equatorial diameter is 75,100 miles. The period of rotation at the equator of Saturn is 10 hours, 14 minutes, but with increasing latitude its period of rotation increases even more rapidly than with Jupiter. The density of Saturn is more extreme than that of Jupiter; at 0.68 it is less than that of water — even less than that of butter! Its mass, however, is 95 times that of the Earth.

Saturn's atmosphere is similar to that of Jupiter except that since it is colder on Saturn (−290° F) some of the ammonia has crystallized out of the gaseous state. Therefore, Saturn's atmosphere has in the gaseous state relatively more methane and less ammonia. (The solid particles, of course, do not add any dark lines to its spectrum.)

Saturn too has belts or bands somewhat like Jupiter's, though they are much less pronounced. These also change from time to time. Occasionally, white spots appear that merge into a white belt and then disperse into the atmosphere.

The multiple rings are what make Saturn unique in our solar system. They have a maximum diameter of 170,000 miles, are concentric in a single plane, cannot be much thicker than 10 miles and may be thinner. Looking through a large telescope one can see gaps between them. The outer gap, called Cassini's division, is seen in Figure 4-5. There is also an inner division.

The composition of the rings was a major puzzle for astronomers until the advent of the spectrograph. Around the turn of the century an astronomer named Keeler placed the image of the planet formed by his telescope across the slit of a spectrograph in order to find the rate of rotation of the rings. If the rings are solid then the outer part should rotate faster since it has farther to go in one rotation. If they are not solid, however, then the inner portion should have a higher velocity than the outer. This is derived from the same principle (discussed in Chapter 5) that explains why Mercury revolves about the sun faster than the other planets, and why Pluto, the planet most distant from the sun, revolves the slowest in its orbit.*

*The orbital velocity of Mercury is 29.7 miles per second whereas Pluto's orbital velocity is only 3.0 miles per second.

Figure 4-5
A photograph of Saturn with its ring system showing Cassini's division. (Lick Observatory photograph.)

Keeler determined the velocities of approach and recession of the edges of both Saturn and its rings by the Doppler principle. This critical test revealed that the inner portion of the rings rotates more rapidly than the outer portions and therefore the rings are not solid.

But if the rings are not solid, what are they? So far as we know they yield no spectral lines of their own and therefore cannot be gaseous. Since they do reflect the sun's light, however, the spectrum of the light reflected can be examined. As two paints can be matched by comparing their reflection spectra, so the material on the surface of the particles composing Saturn's rings can be identified by comparing the reflection spectrum of Saturn's rings with materials in terrestrial laboratories. Studies of this nature reveal that the reflection spectrum of the rings closely matches the reflection spectrum of ice at $-190°$ F. Recent radar observations lead to the conclusion that at least some of the particles composing the rings are jagged and fairly large with some flat surfaces. The amount of radio energy reflected from the rings was more than would be expected from little round ice-coated rocks. Some of the jagged rocks may be more than a meter across.

As viewed from the Earth, the rings present different aspects because of the orientation of the three planes: the plane of the Earth's orbit, the plane of Saturn's orbit, and the plane of the rings. At most times, these planes are such that the rings of Saturn appear tilted as seen from the Earth. Sometimes, however, the rings appear edge on, and since they are so thin, they are then scarcely visible.

4.3 URANUS

The planets discussed thus far are bright enough to be seen with the naked eye and consequently have been known since prehistoric times. Uranus, the first planet discovered in recorded history, is the next to be seen as we go out from the sun. One of the greatest astronomical observers of all time, William Herschel, discovered it in 1781. The story of the discovery indicates the value of better instruments and experience in observing.

Herschel built his own telescopes as an amateur and devoted his life to observing. Those were days when but few of the spectacles of the sky were known, and it was Herschel who discovered many of the objects that today are being investigated so that we may better understand how the universe is put together and why it stays together — if it is staying together.

Herschel spent much of his time scanning the skies looking for objects that were not stellar in appearance, objects such as the large nebulae we discuss in a later chapter. One of these objects presented a disk-like structure which made it appear different not only from the stars but from the nebulae as well. On magnifying it further Herschel thought it was a comet, but later it was recognized as the first discovered planet. Moreover, it had already been seen by one other astronomer who at that very time was preparing a star map of the sky and had, in fact, recorded the planet's position 12 times. But he had thought of it only as a star, partly because his telescope was not as large as Herschel's and partly because it was stars with which he was concerned.

Uranus is about 19.2 A. U. from the sun, with a mass 14 times as large as that of the Earth. It revolves about the sun in 84 years and rotates on its axis in 10 hours, 49 minutes. Its diameter is about 32,000 miles. Because of its great distance from the sun its surface temperature, at −300° F, is even less than that of Saturn. Its atmosphere is similar to that of Jupiter and Saturn, except that even more of the ammonia has crystallized, leaving an atmosphere with an even higher percentage of methane gas. Recently molecular hydrogen has been discovered on Uranus as well as on Jupiter.

Uranus is so far from the Earth that any surface markings it may have are very difficult to see, so its period of rotation had to be determined by the spectrograph.

The most unusual thing about Uranus is the inclination of its axis of rotation. The Earth's axis of rotation is not perpendicular to the plane of its orbit but is tipped 23½° from this perpendicular. The axis of rotation of Uranus, however, is tipped 98° from the perpendicular to its orbit. Or, to put it another way, its axis is tipped by only 82°, if we consider it as rotating in retrograde motion (Figure 4-6). Just how Uranus got this way is a puzzle which must be answered by any satisfactory theory of the origin of the solar system.

After Uranus was discovered the astronomers observed its position carefully

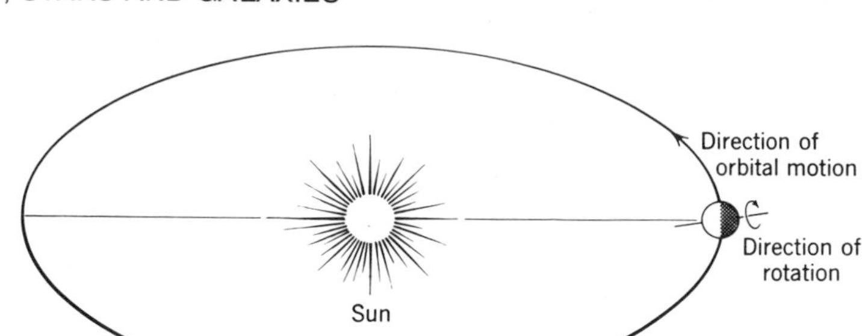

Figure 4-6
The axis of rotation of Uranus is tipped by 98° from the perpendicular to the plane of its orbit. To compare with the Earth, see Figure 5-6.

over a period of years in order to determine its orbit, that is, where it goes and how fast. Newton's laws of motion and his law of gravity (see Chapter Five) describe the motion of the planets about the sun very well and can be used to predict the position of a planet or comet once its orbit has been determined, but Uranus did not move as expected. Two astronomers, Adams and Leverrier, noted this discrepancy and independently both concluded that Uranus was not following its prescribed path because an assumed planet, still more distant from the sun, was disturbing its motion. By Newton's law of gravity they both calculated where such a planet should be.

Leverrier made contact with an observatory that had been preparing a star map. By comparing the position of the predicted planet on the star map with the actual sky, the new planet was seen within a matter of minutes. This was a tremendous triumph for Newton's law of gravity; not only could it account for the motions of the planets but it enabled the astronomer to detect heretofore unknown planets by their gravitational effect on already known planets, *and* to determine their positions!

4.4 NEPTUNE

The new planet was named Neptune. It is about 30 A. U. from the sun, has an equatorial diameter of 31,000 miles (it is not as oblate as Jupiter), and revolves about the sun in 165 years. The mass of Neptune is 17 times that of the Earth. It has a period of rotation a little longer than 15 hours and a surface temperature of about −330° F. Aside from knowing that Neptune has stronger spectral lines of methane than any other planet, astronomers know little about the planet.

Uranus was discovered in 1781 and Neptune was discovered in 1846. These planets travel so slowly (their average orbital velocities are respectively 4.2 and 3.4 miles per second), that it takes a long time to determine an accurate orbit for them. By the early 1900s Lowell, the founder of the Lowell Observatory in Flagstaff, Arizona, had reobserved Uranus and concluded that the discovery of

Neptune could not account for all the discrepancies in the orbit of Uranus. He therefore predicted that there must be another planet even farther away than Neptune. The new planet was discovered by Clyde Tombaugh in 1930 at the observatory in Flagstaff and named Pluto.

4.5 PLUTO

Pluto travels around the sun in 248 years and with an average velocity of 3.0 miles per second. It is about 40 A. U. from the sun and its surface temperature seems to be about −348° F. Because it is so very far away, it is extremely difficult to observe. Nevertheless, its period of rotation has been determined by brightness measurements by Robert Hardie of the Dominion Observatory, Ottawa. Apparently its surface is not uniform, for the light it reflects varies with a consistent period of 6 days, 9 hours, 17 minutes. This is assumed to be its period of rotation.

The diameter of Pluto remains a bit of a mystery, although its maximum size has been estimated to be 3600 miles. This estimation is based on the knowledge of which stars Pluto's tiny apparent disk does *not* eclipse as it moves across the sky.

Since Pluto's spectrum does not contain any molecular absorption bands, we assume that it is quite different from the giant planets between Mars and Pluto. In fact, it has been speculated that Pluto was once a satellite of Neptune that escaped and began circling the sun in an orbit of its own. (Since Pluto now revolves about the sun and not about Neptune, it must be classified as a planet and *not* as a satellite.) The main justification for this hypothesis is Pluto's eccentric orbit. The planet actually travels inside of Neptune's orbit for a portion of its journey around the sun.

4.6 SIZES AND DISTANCES

We have spoken of the sizes and distances of the planets, but it is difficult to grasp the dimensions involved because they are so large. Figure 4-7 shows the planets with respect to their sizes. The sun is added in the figure to show how much larger it is than any of the planets (the sun's diameter is nearly 10 times that of Jupiter). This is fairly easy to imagine, but conceiving of their distances is more difficult. Mercury is 36 million miles from the sun and Pluto is about 100 times as far away or 3662 million miles. This can be drawn to scale, but to fit such a model of the solar system on a page the size of this book it has been necessary to break the solar system into two parts and to draw each to a

Figure 4-7
The largest members of the solar system drawn to scale. The sun's diameter is about 10 times that of Jupiter, which in turn is about 10 times that of the Earth.

Table 4-1
Data on the planets

	Mean distance from sun							
	Astronomical units	Million miles	Period of revolution	Period of rotation	Diameter in miles	Mass in terms of Earth's mass	Density in terms of water	Escape velocity, miles per second
Mercury	0.39	36	87.96 d	58.6 d	3,100	0.054	5.4	2.6
Venus	0.72	67	224.69	−243 d	7,600	0.82	5.1	6.5
Earth	1.00	93	365.24	23 hr 56 min	7,917	1.00	5.5	7.0
Mars	1.52	142	686.95	24 hr 37 min	4,200	0.11	4.0	3.2
Jupiter	5.20	484	11.86 yr	9 hr 50 min	88,700	317.8	1.3	37.1
Saturn	9.56	889	29.46	10 hr 14 min	75,100	94.2	0.7	22.3
Uranus	19.2	1,782	84.01	10 hr 49 min	32,000	14.5	1.3	13.9
Neptune	30.1	2,784	164.8	15 hr 40 min	31,000	17.2	1.7	15.4
Pluto	39.4	3,662	247.7	6 d 9 hr 17 min	$<$3,600	0.2?	?	?

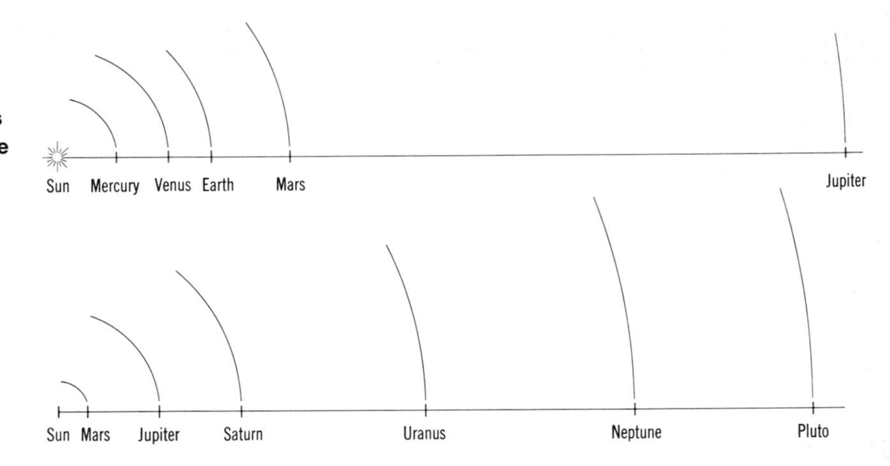

Figure 4-8
The Jovian planets are so much farther from the sun than the terrestrial planets that two scale drawings are needed to show both. The planets themselves would be microscopic on either scale.

different scale. This is shown in Figure 4-8. On both scales each planet is too small to be seen except by a good microscope (Table 4-1).

The Earth seems large to us; and so, upon reflection, do the other planets. All are complex in their structure and fascinating in their details. But what we know of them, even of those we have studied the most, is little compared to the complete picture we would like to have. It is space and the distance between the planets that keeps us from this knowledge — distances so vast that even Jupiter, 87,000 miles across, is microscopic in comparison to the 480 million miles that separate it from the sun, or the 3700 million miles that separate the sun from Pluto.

BASIC VOCABULARY FOR SUBSEQUENT READING

Astronomical unit
Decameter waves
Density
Great Red Spot
Mass
Microwaves
Occultation
Terminator

QUESTIONS AND PROBLEMS

1. Explain how an occultation of a star by Jupiter can be used to learn something of Jupiter's atmosphere.

2. Describe the radio signals received from Jupiter. What aspect of those signals is correlated with one of Jupiter's satellites?

3. Which of the planets are known to have a magnetic field at least as strong as the Earth? Theoretical considerations lead astronomers to conclude that planetary magnetic fields exist only about planets that have molten cores and that rotate rapidly. Do the observations of the planets out as far as Jupiter lend support to this conclusion?

4. Discuss the possibility that Jupiter has an aurora.

5. Explain how the period of rotation of Pluto was determined.

6. What are the significant differences between the Jovian planets (Jupiter, Saturn, Uranus, and Neptune) and the terrestrial planets (Mercury, Venus, Earth, and Mars)? Consider their size, mass, density, and chemical composition.

FOR FURTHER READING

Glasstone, S., *Sourcebook on the Space Sciences,* D. Van Nostrand Co., Princeton, N. J., 1965, Chapters 10 and 11.

Jackson, J. H., *Pictorial Guide to the Planets,* Thomas Y. Crowell Co., New York, 1965.

Page, T., and L. W. Page, ed., *Neighbors of the Earth,* The Macmillan Co., New York, 1965.

Slipher, E. C., *A Photographic Study of the Brighter Planets,* Lowell Observatory, Flagstaff, Arizona, and The National Geographic Society, Washington D. C., 1964.

Smith, A. G., and T. D. Carr, *Radio Exploration of the Planetary System,* D. Van Nostrand Momentum Book, Princeton, N. J., 1964.

Struve, O., and V. Zebergs, *Astronomy of the 20th Century,* Crowell, Collier and Macmillan, New York, 1962, Chapter VIII.

Whipple, F. L., *Earth, Moon and Planets,* 3rd ed., Harvard University Press, Cambridge, Mass., 1968.

Chapman, C. R., "The Discovery of Jupiter's Red Spot," *Sky and Telescope,* p. 276, May 1968.

Goodman, J. W., "The Edgewise Presentation of Saturn's Rings," *Sky and Telescope,* p. 128, September 1965.

Hartman, William K., "A 1974 Tour of the Planets," *Sky and Telescope,* p. 78, August 1974.

Hunten, D. M., "The Outer Planets," *Scientific American,* p. 130, September 1975.

Keay, C. S. L. et al., "Infrared Maps of Jupiter," *Sky and Telescope,* p. 296, November 1972.

Kuiper, G. P., "Lunar Planetary Laboratory Studies of Jupiter," *Sky and Telescope,* part I, p. 4, January 1972; part II, p. 75, February 1972.

Sagan, C., "The Solar System," *Scientific American,* p. 22, September 1975.

Wolfe, J. H., "Jupiter," *Scientific American,* p. 118, September 1975.

"Pioneer Observes Jupiter," *Sky and Telescope,* p. 79, February 1974.

"Pluto's Diameter," *Sky and Telescope,* p. 213, October 1965.

"Radar Echoes from Saturn," *Sky and Telescope,* p. 214, April 1973.

CHAPTER CONTENTS

LEARNING OBJECTIVES

BE ABLE TO:

1. APPLY KEPLER'S THREE LAWS TO ANY ORBIT OR PAIR OF ORBITS TO DESCRIBE THE RELATIVE SPEEDS OF PLANETS.
2. DESCRIBE HOW THE FORCE OF GRAVITY DEPENDS ON THE MASSES OF TWO OBJECTS AND THEIR SEPARATION.
3. DESCRIBE THE DIFFERENCES BETWEEN A PROOF AND A CONSEQUENCE OF THE EARTH'S MOTION.
4. DESCRIBE HELIOCENTRIC PARALLAX AND EXPLAIN WHY IT IS A PROOF OF THE EARTH'S MOTION.
5. DESCRIBE THE REASONS FOR THE SEASONS AND EXPLAIN WHY THEY ARE ONLY A CONSEQUENCE OF THE EARTH'S MOTION.
6. DESCRIBE ONE PROOF, OTHER THAN HELIOCENTRIC PARALLAX, OF THE EARTH'S MOTION.
7. DESCRIBE HOW THE SUN APPEARS TO MOVE THROUGH THE SKY DURING THE DAY AND DURING THE YEAR.

Chapter Five
PLANETS IN MOTION

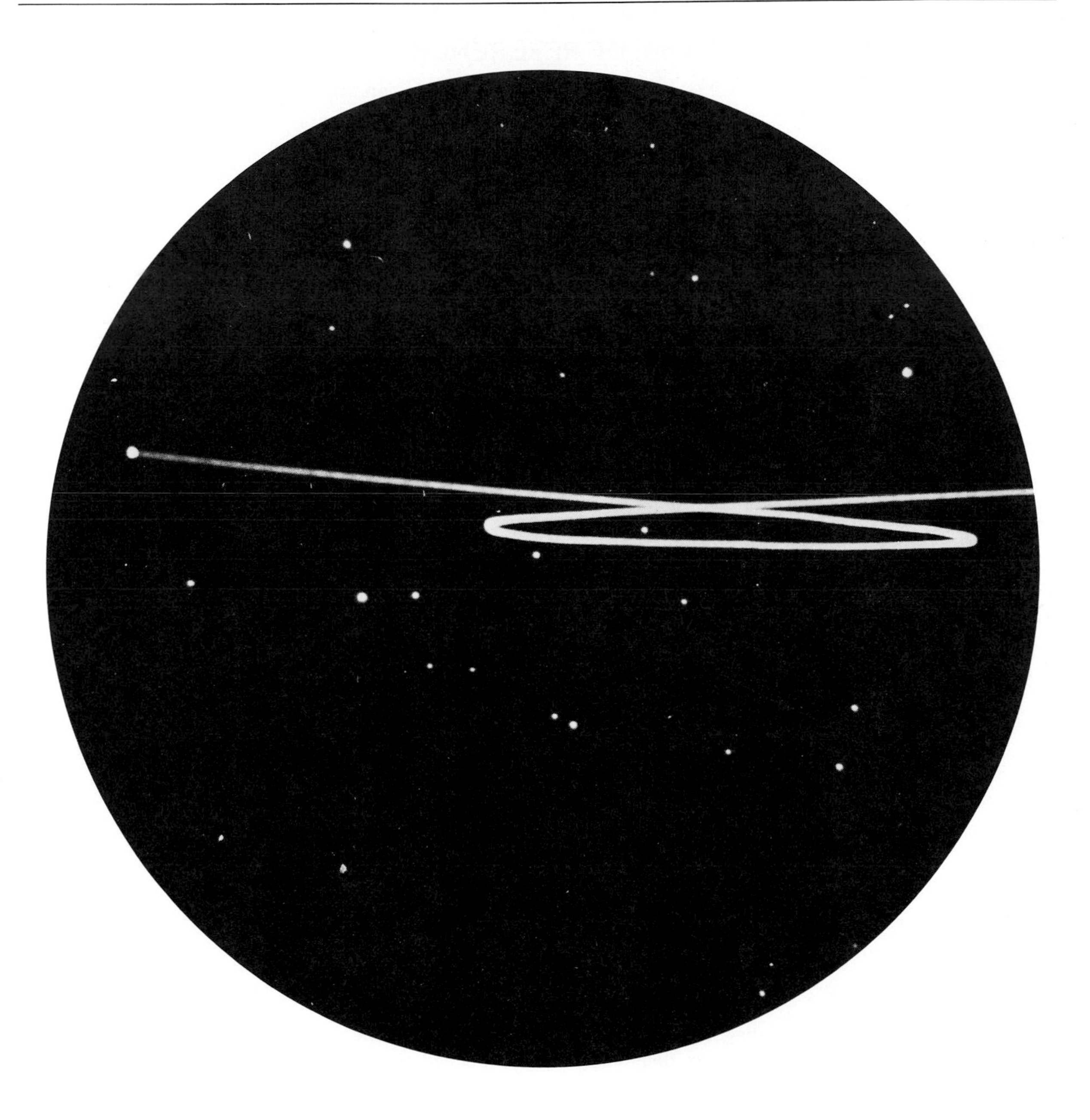

Chapters 3 and 4 gave us some idea of the physical characteristics of the planets. But this is not a complete description of the solar system, or even of the planets. To complete their story it is necessary to discuss their motions.

5.1 OBSERVATIONS AND A FRAME OF REFERENCE

"Planet" — derived from a Greek word meaning "wandering" — is an appropriate name for these objects in the sky that appear to be continually moving stars. Their motions were observed many thousands of years ago and can be observed today by looking at the sky with some care.

Few people, however, pay much attention to the sky. If they did they would realize that the planets move against the background of stars over the weeks and months. They would also be able to see that the moon moves against the same backdrop in just an hour's time, and that even the sun moves differently in the sky than do the distant stars.

It should be noticed that each of these motions has been related to the background of stars, for these apparent motions become more obvious when compared with the "fixed" system of stars. But that to which we must ultimately refer all these motions is the Earth itself, for the sun, the moon, and the planets would appear to move as they do if the stars were not there. In other words, the Earth (the local horizon) is the *frame of reference* against which we measure all these apparent motions.

Every motion must be referred to some frame of reference. The motion of a car is tacitly referred to the Earth when we say that it is traveling 60 miles per hour, for it is traveling much faster than this when referred to celestial frames of reference.

The apparent *diurnal* (daily) motion of the stars is from east to west. The moon moves differently than the stars in that it moves westward more slowly than they do. Consequently it appears to move eastward relative to the stars (that is, it lags behind in the "diurnal race") approximately one of its diameters (½°) every hour. Thus the moon sets on the average 49 minutes (1° = 4 minutes of time) later each day.

The sun, too, moves eastward relative to the stars, but this motion is only about 1° each day. To put it another way (since we reckon civil time by the sun), the stars move west relative to the sun about 1° each day. Thus each star sets about 4 minutes earlier on succeeding nights.

Chapter Opening Photo
The retrograde motion of Mars during August 1971 as photographed with a time exposure in the Medford (Oregon) Schools Planetarium. Since the speed of the planet changes, its brightness in a time exposure also changes. The path is brighter where the planet moves more slowly. (Courtesy Jack Fink, Director, Medford Schools Planetarium.)

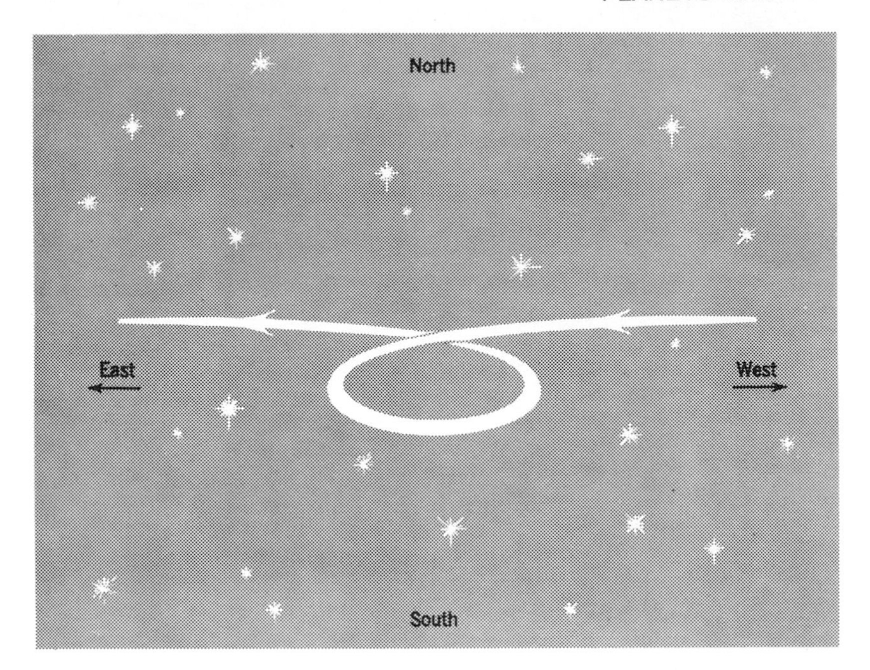

Figure 5-1
The westward motion of Mars (or any planet) is called retrograde motion.

The planets, generally speaking, also move eastward through the field of stars, but there are times when each planet moves westward. Since this motion is contrary to the usual motion it is called *retrograde motion.* Figure 5-1 and the photograph that opens this chapter show how Mars may move during retrograde motion. It is brighter during retrograde motion than at any other time.

5.2 KEPLER'S LAWS

Since the time of Plato, in ancient Greece, there have been a number of theories to explain the motion of the planets. Plato, in effect, challenged the mathematicians to explain the apparently erratic motion of the planets, and he went so far as to specify that the motion must be circular, at a steady speed, and *geocentric* (from the Greek, *ge,* Earth). The circular orbits and the steady speed were selected essentially because of their basic symmetry and "perfection." The geocentrism, however, was undoubtedly favored because it is more comforting to believe that the Earth is *not* moving, and more pleasing to man's ego to think that the Earth is at the very center of the universe. In A.D. 150, Claudius Ptolemy created a detailed geocentric theory accounting for the motions of the planets. The Ptolemaic theory survived until the 17th century. The *heliocentric* theory (from the Greek, *helio,* sun) originated by Aristarchus of Samos in the third century B.C. and developed in detail by Copernicus in the 16th century, finally became established in the 17th century, but only after great arguments. Galileo led the arguments favoring the Copernican development. Antagonists of the heliocentric theory, however, shouted, "Prove it to me!" But theory must precede proof.

In the first two decades of the seventeenth century, Johannes Kepler published what have become known as Kepler's three laws of planetary motion. The first two laws describe the shape of a planetary orbit and the motion of the planet in that orbit. The third law compares the motion of any two of the planets revolving about the sun. Although the laws were derived specifically for the planets revolving about the sun, they apply to satellites revolving about the Earth and planets revolving about any star.

A. THE FIRST LAW The shape of each planetary orbit is described by the first law as an ellipse. An *ellipse* is a figure that can be drawn by taking a piece of string, 8 in. long, for example, tacking the ends down by tacks that are placed 6 in. apart, and running a pencil inside the string, always keeping the string taut. The position that each tack occupies is called a *focus*. *The sun occupies one of the foci of each planetary orbit.*

Since the sun is at one of the foci, the distance of each planet from the sun varies (Figure 5-2). When the planet is in position *A*, *aphelion*, it is at its farthest from the sun. When it is in position *P*, *perihelion*, the planet makes its closest approach to the sun. The corresponding terms in the orbit of a satellite of the Earth are *apogee* and *perigee*.

B. THE SECOND LAW The second law is equally important for an understanding of the motions of objects about a central body when under gravitational attraction. The line joining the planet with the sun is called the *radius vector* and, according to Kepler's second law, *the radius vector sweeps out equal areas in equal lengths of time*. This means that while the planet goes from point *a* to point *b* (both near perihelion) in a certain length of time, say one month, its radius vector will sweep out the area that is shaded (Figure 5-2). Starting from point *c*, its radius vector will sweep out the same area in the same length of time, but the planet will move more slowly since the radius vector is longer near aphelion.

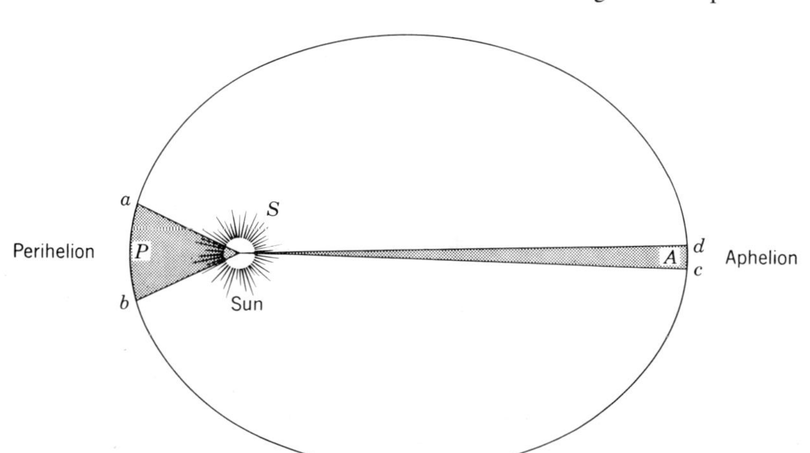

Figure 5-2
According to Kepler's second law, the line joining the sun with any planet sweeps out equal areas in equal intervals of time. If the planet takes the same length of time to move from *a* to *b* as it does from c to *d* then the areas *Sab* and *Scd* must be equal.

Thus, when a planet is near perihelion, it travels faster than when it is near aphelion. The average velocity of the Earth in its orbit is 18.6 miles per second; at perihelion it travels 18.9 miles per second, and at aphelion it travels 18.3 miles per second.

C. THE THIRD LAW Kepler's third law describes how the periods of revolution of the planets about the sun depend on the average distance of each planet from the sun. In effect, the 3rd law tells us that the closer a planet is to the sun the faster it travels. Mercury's average orbital velocity is 30 miles per second, that of Pluto is 3 miles per second.

5.3 NEWTON'S LAW OF GRAVITATION

Newton observed the motion of many objects, but apparently pondered most over the motion of a pendulum, of an apple, and of the moon. He concluded that every object will travel in a straight line unless a force makes it change its direction of travel. That force may be a sudden thing making the direction of travel change abruptly such as a billiard ball bouncing off the side of the table, or it may be a steady force like the force the string exerts on a rock whirled in a circular path above your head.

Since the moon revolves in a nearly circular orbit with the Earth at the center, Newton concluded that it is the Earth which supplies the required force on the moon to prevent it from flying off in a straight line path. Correspondingly, an artificial satellite travels in an orbit about the Earth because the Earth's gravitational pull prevents it from flying off into space (Figure 5-3).

Furthermore, the moon with its mountains and valleys which Galileo had observed, ought to be made of the same stuff the Earth is made of. So, if the Earth exerts a force on the moon, then the moon should also exert a force on the Earth.

If the moon and the Earth exert an attractive force on each other, then so

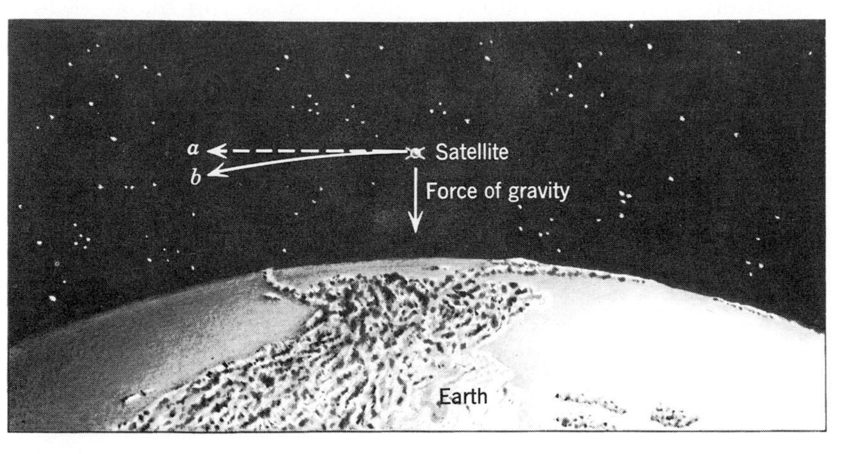

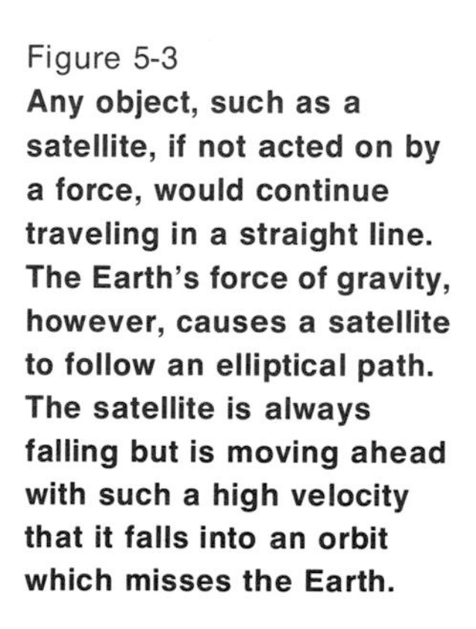
Figure 5-3
Any object, such as a satellite, if not acted on by a force, would continue traveling in a straight line. The Earth's force of gravity, however, causes a satellite to follow an elliptical path. The satellite is always falling but is moving ahead with such a high velocity that it falls into an orbit which misses the Earth.

must the sun and the Earth, the sun and Saturn, the Earth and Saturn, the Earth and you, and so forth. Newton then derived the expression relating the force of gravitational attraction between two bodies to the mass of each body and to the distances separating them. This relationship is *Newton's law of gravity: every object in the universe attracts every other object in the universe with a force that is proportional to the product of the masses of the two objects, m_1 and m_2, and inversely proportional to the square of the distance d between them.*

$$F = G\frac{m_1 m_2}{d^2}$$

where G is the universal gravitational constant that changes the proportionality expressed above into an equality. In effect, G tells us something about the universe, something about the strength of the gravitational force when compared with other forces such as electrical and magnetic forces.

But how does this gravitational force operate? Newton was not able to say, but the fact that it does operate as described by his law of gravity and that it alone accounted for the motion of the planets (until the introduction of the theory of general relativity nearly 230 years after Newton) makes it a bold and decisive step in man's description of the universe.

But even if Newton's law of gravity was generally accepted within a few decades after he published it in 1687, men still looked for observational *proof* that the Earth rotates on its axis and revolves about the sun.

5.4 A PROOF AND A CONSEQUENCE

To prove formally that the Earth moves and revolves about the sun was not easy. For millennia man had nourished beliefs of his superiority with the geocentric theory. The geocentric theory places mankind in the auspicious position of being in command of the Earth at the very center of the universe. This made man the most important creature in the universe; the universe was made for man. It was not so much a geocentric system as it was an anthropocentric (man-centered) system. Nevertheless studies and observations of both Galileo and Newton had tremendous weight in encouraging man's acceptance of the heliocentric theory. Newton, it might be added, called attention to Roemer's observations of Jupiter's satellite. The explanation of the changing period of that satellite was based on the heliocentric theory.

The problem of a formal proof of the heliocentric theory, however, revolves about the question: What constitutes *a proof* of the Earth's motion and what constitutes *a consequence* of that motion. A proof is established if observations are made that can be accounted for *best* by the theory in question. The word "best" used here not only means that the theory is complete, that is, the theory accounts for *all* motions in the solar system; but that the theory is the simplest and the most straightforward. If a theory requires that a number of *ad hoc* theories be added to the main theory to account for the observations, then that theory is not straightforward; it is not the best theory.

If, on the other hand, an observation can be easily explained by two theories, then it cannot be used to prove either. Those observations may then be said to be a consequence of the actions.

A. A PROOF OF THE EARTH'S MOTION It was realized rather early that if the Earth revolves around the sun and if the stars are at different distances, then as a result of changes in the Earth's position there should be a shifting back and forth of the nearest stars with respect to those more distant. This apparent annual shift of position for the nearest stars is called *heliocentric parallax* (Figure 5-4).

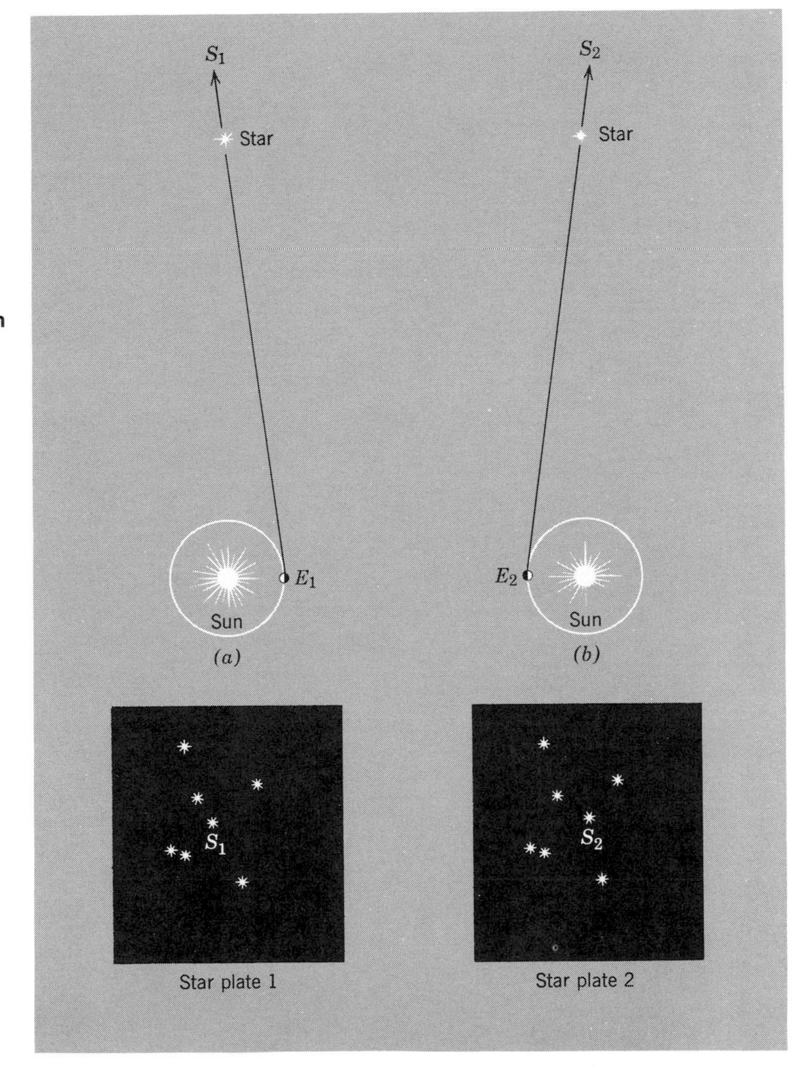

Figure 5-4
At one time of the year a nearby star (a) will be seen in one direction, S_1; 6 months later (b) the same star will be seen in a slightly different direction, S_2. In this figure the difference in direction is highly exaggerated, for in reality it is always less than 1 second of arc. Star plate 1 shows the position S_1 against the more distant stars. Star plate 2 shows the new position of the nearby star.

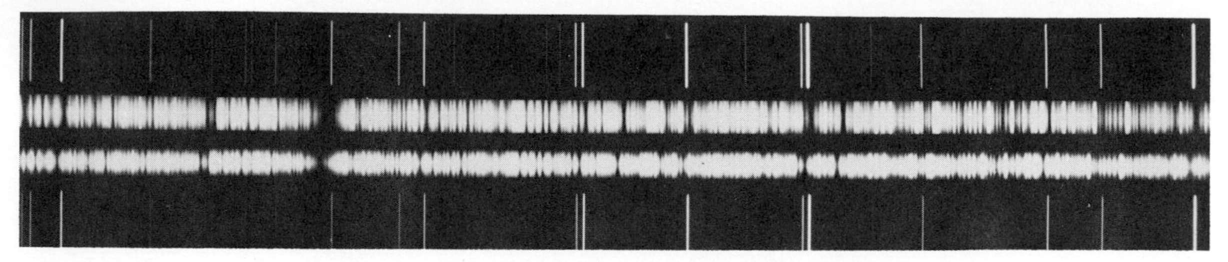

Figure 5-5
Doppler shift in the spectrum of the star Arcturus due to the Earth's revolution about the sun. In the top spectrum the Earth is receding from Arcturus, in the bottom it is approaching. (Photograph from the Hale Observatories.)

When the Earth is in position E_1, shown in *(a)*, the nearer star appears in the direction indicated, and a photograph of the region might appear something like star plate 1. In *(b)*, 6 months later, the Earth has changed its position to E_2 and the nearer star is seen from a different direction. The corresponding photograph might appear as star plate 2.

The result is that the nearest stars will appear to oscillate back and forth against the background stars in a period that is equal to the period of the Earth's revolution. This periodic shifting of stellar positions cannot be accounted for by the geocentric theory unless we assume some peculiar motions for the stars themselves. But this would amount to an *ad hoc* theory concocted for the sole purpose of salvaging the geocentric theory and, hence, is not acceptable. Since the stars are so far away in comparison with the radius of the Earth's orbit, the resulting change in their positions is extremely small and thus was not detected until the nineteenth century.

A second proof of the Earth's revolution can readily be made from the study of spectroscopy. The Earth has an average speed in its orbit of 18.6 miles per second, which is enough to cause a measurable Doppler shift in the spectrum of a star that is in or nearly in the plane of the Earth's orbit (Figure 5-5). This has not only been observed but becomes a nuisance when an astronomer wants to measure the radial velocity of a star. To obtain consistent radial velocities for any star, it is convenient to use the sun as a frame of reference rather than the revolving Earth; thus, the Earth's motion must be subtracted out of each measurement of radial velocity.

As with the heliocentric parallax, this annual change in the Doppler shift of stars that lie near the plane of the Earth's orbit cannot be accounted for by the geocentric theory unless we concoct an *ad hoc* theory about some peculiar motions of the stars themselves, motions that would differ from star to star depending on their position relative to the Earth's orbit.

B. A CONSEQUENCE OF THE EARTH'S MOTION It is common knowledge today that the Earth revolves about the sun and as *a consequence* we have our seasons. The seasons are considered a consequence of the Earth's motion, since

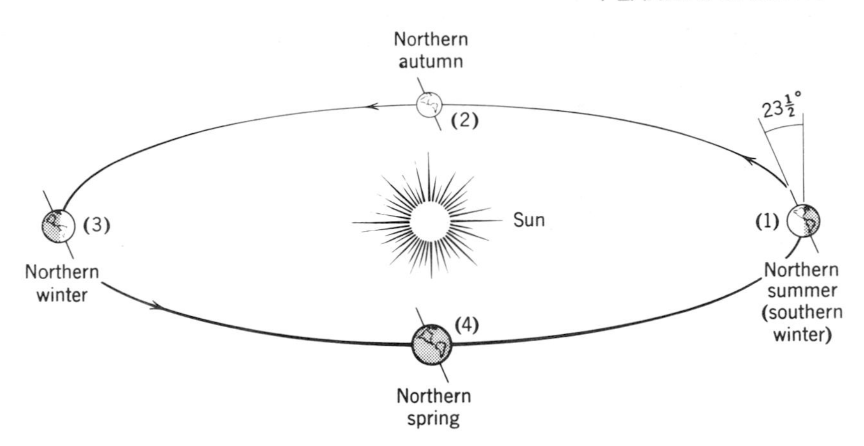

Figure 5-6
The seasons result from the fact that the Earth's axis of rotation is not perpendicular to the plane of the Earth's orbit, and because this axis of rotation points to essentially the same point in the sky during the year.

they can be easily accounted for by either the heliocentric or the geocentric theory. They cannot, therefore, be used as a proof of either.

It was pointed out that the axis of rotation of the Earth is not perpendicular to the plane of the Earth's orbit (see page 75). For purposes of this discussion only, the axis of rotation may be said to maintain the same direction with respect to the stars (Figure 5-6). The darkened portion of the Earth is in the shadow and, therefore, in nighttime. In position 1 the north pole is tipped toward the sun. Hence, not only do the sun's rays strike the northern hemisphere more directly than the southern hemisphere but, as the Earth rotates, any point on the northern hemisphere remains in the sunlight for more than 12 hours, while any point in the southern hemisphere remains in the sunlight for less than 12 hours. Therefore, in position 1, the northern hemisphere has its summer while the southern hemisphere has winter.

Three months later, in position 2, the sun shines equally on the northern and southern hemispheres; the day is 12 hours long (except very near the poles); it is autumn up north and spring in the southern hemisphere.

Three months later again, in position 3, the seasons are reversed from position 1; the northern hemisphere has its winter, the southern its summer. Finally, the northern hemisphere has spring in position 4 while those down under have autumn.

Referring to position 1, we can see that the sun is north of the Earth's equator, and in position 2 it is in the plane of the Earth's equator. In position 3 the sun appears to be south of the equator, and in position 4 it is again directly overhead at noon for those on the equator.

5.5 THE SUN'S APPARENT MOTION

As viewed from the Earth, therefore, the sun appears to move north and south in the sky as well as eastward. The celestial sphere is bisected by the *celestial equator* just as the Earth is bisected by the terrestrial equator; in fact, the

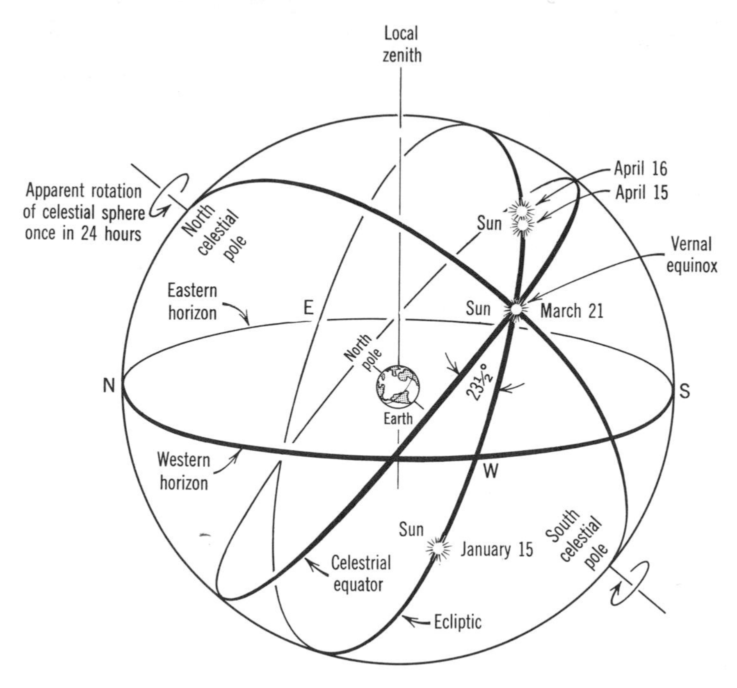

Figure 5-7
The Earth is drawn at the center of a greatly decreased celestial sphere. The Earth's axis of rotation intersects the celestial sphere at the north and south celestial poles. A projection of the plane of the Earth's equator on the celestial sphere is the celestial equator. The apparent path of the sun against the background stars is called the ecliptic. The sun moves eastward along the ecliptic about 1° each day.

projection of the Earth's equator onto the celestial sphere defines the celestial equator.

As the Earth revolves about the sun, we see the sun appear to revolve about us. This is the apparent motion of the sun eastward against the background of stars referred to in the first part of this chapter. The sun's apparent path through the field of stars is called the *ecliptic*, which is the imaginary circle formed by the plane of the Earth's orbit cutting the celestial sphere. The sun, like all celestial objects, appears projected on this sphere (Figure 5-7).

The plane of the Earth's equator (and thus the celestial equator) is tipped from the plane of the Earth's orbit (and thus the ecliptic) by that famous 23½°. As a consequence, as the sun appears to travel along the ecliptic, it moves first north and then south while it slowly progresses eastward (Figure 5-7). When the sun crosses the celestial equator on its apparent journey north, about March 21, the northern hemisphere begins spring and the southern hemisphere begins autumn. This intersection of the ecliptic and the celestial equator is called the *vernal equinox*. The other intersection, that is, where the sun crosses the celestial equator on its way south, is called the *autumnal equinox*. The sun makes this crossing about September 21.

The point in the sky where the sun reaches its most northerly position, about June 22, is called the *summer solstice*, the most southerly point is called the *winter solstice*. It should be pointed out that these points were named several thousand years ago, and so they apply for the seasons in the northern hemisphere.

Another consequence of revolution is that puzzling retrograde motion of the planets. Since Mercury travels faster than Venus, Mercury overtakes Venus when they are on the same side of the sun. Similarly Venus overtakes the Earth and the Earth overtakes Mars, etc. When we overtake Mars it appears to us as if Mars is going backwards, just as passing a car on the highway makes the slower car appear to go backwards.

By definition, the sun's motion is precisely along the ecliptic, the plane of the Earth's orbit extended to the celestial sphere. Furthermore, the plane of the Earth's orbit is nearly coincident with the plane of the solar system. Therefore the apparent motions of the planets are confined to a band around the celestial sphere centered on the ecliptic. This band is called the *zodiac.* Consequently each planet, as well as the moon, appears to move north and south of the celestial equator as each traces out its motion on the celestial sphere.

The sun's apparent motion in the sky along with the apparent motion of the planets makes a complex system for study and analysis. Further study reveals that the stars also move, as do even the distant galaxies. In fact, it has been concluded that in the universe there is no one principal frame of reference that can be considered at rest, and against which all motions can be referred. Each observer must choose his own frame of reference, which may be the Earth, the sun, or the distant stars which seem "fixed" in place on the celestial sphere above our heads.

BASIC VOCABULARY FOR SUBSEQUENT READING

Aphelion
Autumnal equinox
Celestial equator
Celestial sphere
Diurnal
Ecliptic
Ellipse
Frame of reference
Geocentric
Heliocentric
Heliocentric parallax
Perihelion
Retrograde motion
Summer solstice
Vernal equinox
Winter solstice

QUESTIONS AND PROBLEMS

1. Explain why, when Mercury and Venus are observed in the sky, they are never far from the sun.

2. From where on the Earth can Mercury and Venus (a) never be seen at midnight, (b) be seen at midnight?

3. If the sun moves east relative to the stars, why does it set in the west?

4. Draw the planetary distances from the sun on a scale such that 1 A. U. equals 0.5 centimeter. Can you draw the planets on the same scale?

5. On June 22, the sun is 23½° north of the celestial equator. Does it then rise north of due east?

6. Describe one proof of the Earth's motion.

7. Explain why the Earth has seasons.

8. Describe the appearance of the sky both day and night through each of the four seasons as seen by (a) explorers on the North Pole, (b) people on the equator.

9. Plot the location (right ascension and declination) of Mars throughout the year from its positions as listed in the current *American Ephemeris and Nautical Almanac.* Its motion can readily be seen by plotting positions for the tenth, twentieth, and the last day of every month.

FOR FURTHER READING

Cohen, I. B., *The Birth of a New Physics,* Doubleday Anchor Books, New York, 1960.

Geymonat, L., *Galileo Galilei,* McGraw-Hill Book Co., New York, 1965.

Glasstone, S., *Sourcebook on the Space Sciences,* Van Nostrand Co., Princeton, N. J., 1965, Chapter 2.

Hawkins, G. S., *Stonehenge Decoded,* Doubleday and Co., New York, 1965.

Inglis, S. J., *Physics: an Ebb and Flow of Ideas,* John Wiley & Sons, Inc., New York, 1970, Chapters 3 and 4.

Keston, H., *Copernicus and His World,* Roy Publishers, New York, 1945.

Koestler, A., *The Watershed,* Doubleday Anchor Books, New York, 1960.

Kuhn, T. S., *The Copernican Revolution,* Modern Library Paperback, New York, 1959.

Munitz, M. K., *Theories of the Universe,* The Free Press, New York, 1957.

Page, T., and L. W. Page, ed., *Wanderers of the Sky,* The Macmillan Co., New York, 1965.

Santillana, G. de, *The Crime of Galileo,* The University of Chicago Phoenix Books, Chicago, Ill., 1959.

Christianson, J., "The Celestial Palace of Tycho Brahe," *Scientific American,* p. 118, February 1961.

Franklin, K. L., "The Astronomer's Odd Figure 8," *Natural History,* p. 8, October 1962.

Franklin, K. L., "The Gravitational Forces and Effects," *Natural History,* part I, p. 12, October 1963; part II, p. 44, November 1963.

Gingerich, Owen, "Copernicus and Tycho," *Scientific American,* December p. 87, 1973.

Muul, Illar, "Day Length and Food Caches," *Natural History,* p. 22, March 1965.

Rawlins, D., "The Mysterious Case of the Planet Pluto," *Sky and Telescope,* p. 160, March 1968.

Ronan, C. A., "Phoenix of Astronomers," *Natural History,* p. 52, January 1965.

Rosen, E., "Copernicus' Place in the History of Astronomy," *Sky and Telescope,* p. 72, February 1973.

Rothrock, G. A., "Steps to New Astronomy," *Natural History,* p. 64, May 1965.

"Galileo Galilei," series of articles, *Sky and Telescope,* February 1964.

CHAPTER CONTENTS

LEARNING OBJECTIVES

BE ABLE TO:

1. GIVE THE RELATIVE POSITIONS OF THE EARTH, MOON, AND SUN FOR EACH PHASE OF THE MOON.
2. DESCRIBE OBSERVATIONS THAT MIGHT HELP DISTINGUISH IMPACT CRATERS FROM VOLCANIC CRATERS.
3. GIVE OBSERVATIONAL EVIDENCE OF EROSION ON THE MOON.
4. GIVE THREE CAUSES OF EROSION ON THE MOON.
5. DESCRIBE HOW TINY METEORITES CAUSE LUNAR EROSION.
6. DESCRIBE DIFFERENCES BETWEEN THE MARE AND THE LUNAR HIGHLANDS ABOUT CRATER TYCHO.
7. EXPLAIN HOW THE AGE OF A ROCK CAN BE DETERMINED.
8. EXPLAIN HOW MASCONS WERE DISCOVERED.
9. EXPLAIN HOW SOME RILLES ON THE MOON RESEMBLE LAVA TUBES HERE ON EARTH.
10. EXPLAIN THE DIFFERENCE BETWEEN THE MOON'S SYNODIC AND SIDEREAL PERIOD.
11. DESCRIBE THE RELATIVE POSITION OF THE EARTH, MOON, AND SUN DURING A LUNAR AND DURING A SOLAR ECLIPSE.
12. EXPLAIN WHAT IS MEANT BY THE ECLIPSE SEASON.
13. DESCRIBE CONDITIONS FOR A TOTAL AND A PARTIAL ECLIPSE.
14. EXPLAIN WHY THERE ARE TWO HIGH AND TWO LOW TIDES EVERY DAY.

Chapter Six
SATELLITES

Our discussion of the planets has covered one phase of the description of the solar system. But of the nine planets, six are known to have satellites circling them as each pursues its complicated motions through space.

Every satellite, since it revolves about a planet, in turn has one more motion than those characteristic of the planets. For us here on the Earth the most obvious satellite in the universe is our own natural one, the moon, which has an average distance from us of 239,000 miles, and which we see going through its phases during the month.

It is not completely correct to say that the moon revolves about the Earth, for in reality they both revolve about a common point called their *center of mass*. The Earth is about 81 times as massive as the moon, and the center of mass is therefore about 3000 miles ($^1/_{81}$ of 239,000 miles) from the center of the Earth on a line joining that center with the center of the moon. Since the center of mass is inside the Earth, the Earth's motion about this center is barely perceptible. The corresponding motion of the other planets is even less noticeable, for all other satellites are much smaller when compared with their mother planets.

6.1 PHASES OF THE MOON

Since we see the moon only by reflected sunlight, it is understandable that its phases result from its revolution about the Earth (Figure 6-1). Starting with the new moon and following it through the lunar month, the phases progress in order: new moon, first quarter, full moon, and third quarter (Figure 6-2). Starting with each new month, the length of time within the lunar month is reckoned by the "age" of the moon. That is, the full moon is said to be 14 days old, etc. (Figure 6-3).

The moon moves eastward in its orbit about the Earth. The new moon sets in the west right along with the sun. Due to the moon's motion eastward, however, the 1-day-old moon sets about 50 minutes after the sun sets; the 2-day-old moon sets about twice as late or 1 hour and 40 minutes later than the sun. Each succeeding day the moon sets about 50 minutes later. The first quarter moon (7 days old) is high in the sky at sunset; it sets about midnight. The full moon is just rising in the east at sunset; it sets at sunrise. The third quarter moon is not in the sky at sunset, for it rises at midnight and sets at noon.

From new moon to first quarter, the phase (and shape) of the moon is crescent; from first quarter to full moon its phase (and shape) is gibbous; from full moon to third quarter its phase is again gibbous; from third quarter to new moon, its phase is again crescent.

Chapter Opening Photo
Mount Hadley showing stratification. The footprints are those of the Apollo 15 crew. (NASA photograph.)

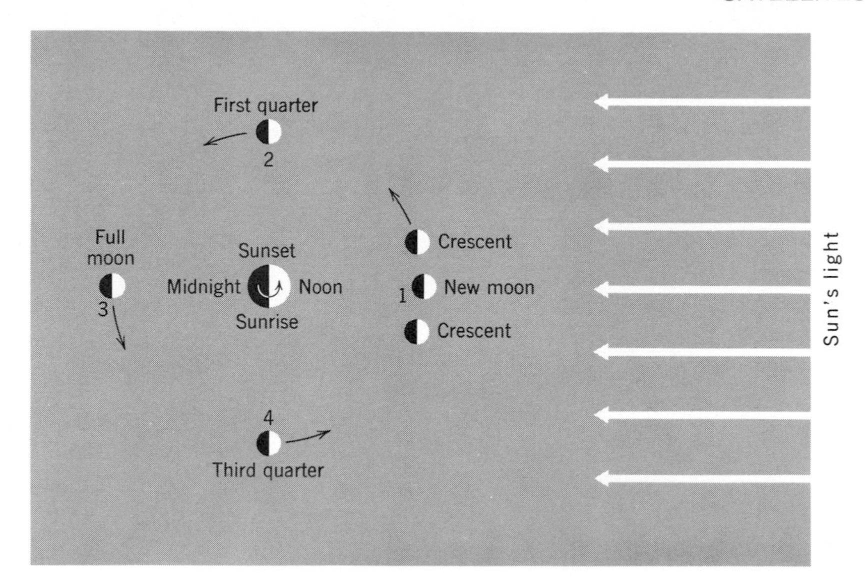

Figure 6-1
The phases of the moon.

The crescent moon has one unique feature. Most of the side that faces the Earth is dark, for the moon is nearly between the Earth and the sun. The side of the Earth that faces the crescent moon, however, is fully lighted. Consequently the moon is faintly illuminated by the Earth's reflected light, or *earthlight*. When the moon is two or three days old, the earthlit portion of the moon nestled in the bright crescent is sometimes called the old moon in the new moon's arms.

The location of some of the major craters and maria (darker areas) is shown in

Figure 6-2
(a) The moon, 4 days after new, is said to be 4 days old,

(b) the moon 7 days old (first quarter), and

(c) the 10-day old moon (gibbous phase). (Lick Observatory photographs.)

Figure 6-3
Full moon showing impact craters Tycho, Copernicus, Kepler, and others with rays about them. (Lick Observatory photograph.)

Figure 6-4. Figure 6-4*a* is a composite of the first and third quarter moon to produce shadows so that the features are more easily visible. Figure 6-4*b* is a drawing with some of the features labeled.

6.2 TECHNIQUES OF OBSERVATION

As Galileo's telescope opened up a new era for astronomical investigation, so have electronics, computers, and rocket propulsion opened up a new and exciting era. The successful space voyages that have put both men and equipment on the moon have — as did Galileo's telescope — answered old questions but asked new ones as well. The identification and study of lunar features is now following lines very different from the studies made as recently as the middle of the 20th century.

Figure 6-4

(a) A mosaic of the first and third quarter moon. (Lick Observatory photograph.)

(b) A map of the moon. 1. Clavius; 2. Tycho; 3. Arzachel; 4. Alphonsus; 5. Ptolemaeus; 6. Albategnius; 7. Abulfeda; 8. Copernicus; 9. Eratosthenes; 10. Archimedes; 11. Autolycus; 12. Aristillus; 13. Cassini; 14. Plato; 15. Julius Caesar; 16. Flamsteed; 17. Kepler.

Figure 6-5
One of the Apollo 12 astronauts took this photograph of their lunar module (far left) close to the Surveyor 3 craft. Notice the "fresh" crater in the upper right. (NASA photograph.)

A series of Ranger spacecraft made hard (destructive) landings on the moon and took photographs as they descended (see Figure 6-8). Surveyor spacecraft made soft (nondestructive) landings, took close-up photographs of the lunar surface, and dug into the soil a bit (Figure 6-5). Lunar orbiters carried cameras into orbits around the moon, taking wide angle photographs (Figure 6-11) and

Figure 6-6
Photograph taken by a member of the Apollo 15 crew from inside a "relatively recent" crater. (NASA photograph.)

narrow angle photographs with a telescopic lens to record detail (Figure 6-12). Finally, the Apollo rockets have sent men to the moon to photograph surface features, to set up equipment for special studies, and to bring surface material back to the Earth for detailed studies in our laboratories.

From Earth-based telescopes, studies are still being made, however, in the infrared, microwave, and radio region of the spectrum. Radar and laser observations are also being made. From all of these studies a great deal of information is being compiled that will increase our understanding of the moon, its origin, and history. These studies will, we hope, add to our understanding of the solar system and even of the universe.

6.3 THE LUNAR SURFACE

A. CRATERS That craters scar the moon's surface has been obvious since the 18th century, when telescopes large enough to resolve detail on the moon were constructed. The use of still larger telescopes in the 19th and 20th centuries has made clear that small craters far outnumber large ones. With the advent of travel to the moon, even smaller craters have been observed, and their number is greater than even the smallest craters seen with the biggest telescopes. Many small craters are seen to line the larger crater photographed by the Apollo 12 crew (Figure 6-5).

1. Impact Craters The crater just above and to the right of the Surveyor 3 craft in Figure 6-5 has more rubble around it than the others. There is also more rubble in and around the crater photographed by the Apollo 15 crew (Figure 6-6). Why should some craters have more rubble around them than others?

Even some of the big craters have more rubble around them than others. For example, the crater Tycho (see Figures 6-3, 6-4, and 6-7) has rubble strewn out as great *rays* that extend for a quarter of the way around the moon in many directions. Other craters also have these rays, for example, Copernicus and Kepler (see Figure 6-3 and 6-4). A careful consideration of craters with rubble and of those with rays has led astronomers to conclude that these craters were formed by the impact of chunks of rock on the lunar surface. Since we have considered impact craters on both the Earth and Mars, it is not surprising to learn of their existence on the moon, too.

2. Volcanic Craters Some craters on the moon, however, appear to be volcanic

Figure 6-7
The Crater Tycho in the lunar highlands. (Lick Observatory photograph.)

Figure 6-8
(a) The crater Alphonsus and the edge of Mare Imbrium to the left. Ranger 9 struck at the head of the rille marked by the circle. Notice the "dark-haloed" craters about the periphery of the floor of Alphonsus. Taken 265 miles above the surface. (NASA photograph.)

(b) A closer view of one of the "dark-haloed" craters taken 107 miles above the floor of the crater. (NASA photograph.)

in origin. Consider the "dark haloed" craters photographed by Ranger 9 as it fell to a crash landing on the moon. Three or four of these dark haloed craters line the larger crater Alphonsus (Figure 6-8*a*). In Figure 6-8*b*, one of them was photographed from only 107 miles above the surface, revealing the dark material more closely. It seems to have been ejected from the crater and to have filled in the smaller craters in the immediate vicinity as well as part of the rills on which it resides.

Before Ranger 9 took these photographs, it was suspected that there was some volcanic activity about the crater Alphonsus. Spectrograms taken by the Russian astronomer Kosyrev indicate the occasional presence of luminous gases, for *bright* spectral lines appeared superimposed on the solar spectrum reflected from the lunar surface. It is supposed that gas from one of these dark-haloed craters emitted light of its own when struck by solar radiation.

B. EROSION Some craters have rubble about them, others do not, even if they did when they were created by the explosion that was caused by the impact of a rock. What happened to that rubble? The dark halo about only some volcanic craters indicates that it, like the rubble, disappears with time. This disappearance can only mean that the moon suffers the effects of erosion. The sharpness of the crater Tycho when compared with the smoothed-over craters around it is

Figure 6-9
This photograph taken from Lunar Orbiter 5 gives direct evidence of lunar erosion; two rocks have rolled down the central peak of a crater. The larger rock is about 75 feet in diameter. (NASA photograph.)

clear evidence that erosion alters the lunar surface. Since the rays of Tycho overlie the craters about it, it is obviously younger.

Direct evidence of erosion is seen in Figure 6-9. Those two rocks must have become dislodged from some place higher on the slope down which they rolled. These two rocks, plus many more, mean that central peaks, crater walls, etc., all become lower and more rounded with time.

The causes of erosion must be several. The surface of the moon does suffer a large change in temperature between its day and night. The highest temperature measurement is +283° F, the lowest temperature is probably about −300° F. This large change in temperature causes the surface of rocks to expand and contract more than their interiors, which in turn causes a cracking and flaking off of material.

The electrons and protons that make up the solar wind must cause changes on the surface of the rocks on the moon. These changes can only result in a weakening of the crystalline structure of the rocks. Ultraviolet radiation, X rays, and gamma rays from the sun must also cause a breakdown of the crystalline structure of the surface of the rocks.

Another agent of erosion is cosmic rays. *Cosmic rays* are the nuclei of atoms that are accelerated somewhere in the universe to velocities very close to that of light. Protons, the nuclei of hydrogen atoms, make up most of the cosmic rays, but nuclei of other atoms, such as helium and iron, are also found in cosmic rays. When these very high energy particles strike the surface of rocks on the moon, they leave little streaks that alter slightly the structure of the rock. These cosmic ray tracks have been found in rocks brought back by the Apollo astronauts. It may well be that the record of cosmic ray tracks in rocks on the moon will tell us a great deal about the history of the sun (it ejects some low energy cosmic rays) and of the universe.

The observation that the number of impact craters on the moon increases for ever smaller craters tells us something about the material that is bombarding the surfaces of the moon, the Earth, and Mars. The number of bombarding particles increases as chunks of ever smaller diameters are considered. When one of those small particles strike the Earth's upper atmosphere, we see it flash as a *meteor* (or shooting star). The flash is the grain-of-sand sized particle burning up in a fraction of a second. But the moon has no atmosphere, so those particles must strike the actual surface of the moon. They are small and, hence, thousands and millions must strike the lunar surface each day. Each must chip away a little bit of that surface. Over the millions and billions of years, this must amount to a reasonable sandblasting job. No wonder the surfaces are so well rounded. Those craters that still have rubble or rays about them must, indeed, be younger than the other craters. Those rays and the sharpness of that rubble will fade with time, as the edges are bombarded with cosmic rays, electrons and protons, and with the bits of sand particles falling onto the surface from interplanetary spaces.

C. MARE Each of the large dark areas of the moon is called a *mare* (plural *maria*). Observations of the color of the maria have been made by photographing them with film sensitive to the ultraviolet and again with film sensitive to the infrared. From the difference in color it can be concluded that the maria are lava flows. In fact, some of the maria are composed of many lava flows.

To substantiate the lava nature of the maria, some of the material brought back from the moon by the Apollo astronauts has been examined. Much of it is lava-type rocks (Figure 6-10). These rocks have been dated by using radioactive techniques, and their ages reveal that changes took place on the moon but apparently not continuously. Rocks from the Apollo 11, 12, 14, 15 and 16 flights reveal ages that cluster around 4.4 billion, 4.0 billion, 3.9 billion, 3.6 billion, and 3.3 billion years old.

Figure 6-10
A piece of lava rock brought back from the moon by the Apollo 15 crew. (NASA photograph.)

These ages are obtained by the radioactive conversion of one element (or isotope), called the *parent* element, into another called the *daughter* element. This conversion takes place at a constant and known rate. Once the rock solidifies by forming crystals, the daughter element begins to accumulate. The age of the rock is then determined by measuring the percentage of parent and daughter element that exist side by side. The older the rock, the higher the percentage of daughter element.

It would appear that if the moon was formed 4.6 billion years ago (the consistent age of meteorites), then there was some melting and recrystallization at later intervals, yielding the rocks of apparently these younger ages. One of the rocks had a crystallization age of 3.93 billion years with some individual crystals as old as 4.48 billion years. It would appear that the melting was only partial. The cause of the melting, of course, must also be the cause of the maria.

D. MASCONS Further evidence of past changes on the moon in the region of the maria has been obtained from an unsuspected source: the variations in the orbits

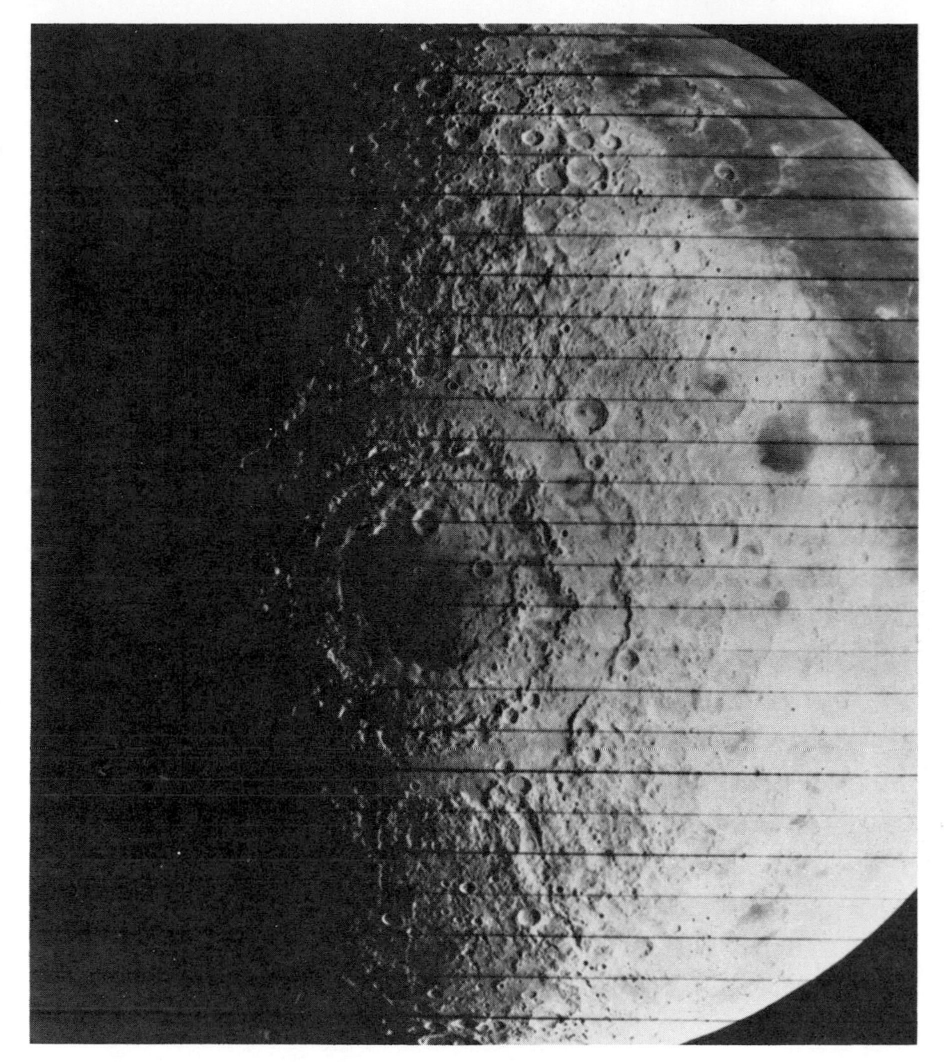

Figure 6-11
The Orientale Basin, photographed by Lunar Orbiter 4. The largest ring in this huge impact basin is 600 miles in diameter. (NASA photograph.)

of the Lunar Orbiter and Apollo spacecraft. These orbits depend on the mass of the moon as well as the height of the spacecraft. Very small changes in the Doppler shift of the radio signal sent back to Earth indicate clearly that the spacecraft occasionally make slight dips in their orbits. Each dip recorded was about 3 to 4 feet and each occurred over a mare. Astronomers could only conclude that the gravitational fields above, at least, some of the maria are stronger than the gravitational field above the rest of the moon.

The best way to account for this increased gravitational field is to assume that there is material under the maria whose density is greater than the material just under the surface of the rest of the moon. This greater density means that there is more mass per unit volume and, consequently, a greater gravitational pull on the spacecraft. These concentrations of mass have, quaintly enough, become known as *mascons*. It has been tentatively concluded that they are large deposits of high-density lava. Some of the maria are certainly circular and, in fact, the Apennines (see Figure 6-4) form part of a circular wall about Mare Imbrium.

The Orientale Basin (Figure 6-11) seems to be the youngest of the large circular maria on the moon. The outer circular ring of mountains is some 600 miles in diameter, with peaks rising to more than 20,000 feet above the surrounding plains. There seems little doubt but that this is an impact crater, and it lends support to the idea that all the maria are impact craters with their ancient rims greatly eroded.

E. THE ALPINE VALLEY A striking feature of the moon's surface has been named the *Alpine Valley* and has long puzzled astronomers (Figure 6-12). Lunar Orbiter 5 photographed this feature with a telescopic lens and revealed that the long straight valley, which cuts through the lunar Alps just south and east of the crater Plato, has what looks like a meandering stream running down the middle. The cause of this valley with its crooked rille running down the middle is the

Figure 6-12
The Alpine Valley photographed by Lunar Orbiter 5. (NASA photograph.)

Figure 6-13
The rille on the floor of Alphonsus is seen to be a string of collapse craters; compare with Figure 6-8. Taken from 12.2 miles above the surface. (NASA photograph.)

Figure 6-14
The Hadley-Apennine area as photographed from the command module of Apollo 15. Compare features with Figure 6-4. Crater Aristillus (34 miles in diameter) is near the horizon, Autolycus (25 miles in diameter) is south of Aristillus. The graben are the fairly wide flat-bottomed "valleys" near the center of the photograph. Hadley rille meanders in the lower center. The Apollo 15 landing site was to the right of the "chicken beak" in Hadley rille. The Apennine Mountains are in the right foreground, Caucasus Mountains in the right background. (NASA phtograph.)

Figure 6-15
Hadley Delta showing inclined stratification. (NASA photograph.)

subject of speculation right now, but we do not expect to find stream-worn pebbles along its banks.

F. RILLES The *rilles* are another common feature of the moon. They are frequently found on the floor of craters, such as those in Alphonsus (Figure 6-8 and 6-13). These were photographed by Ranger 9 as it plunged to the surface. It can be seen from this illustration that these particular rilles are actually a series of craters resulting from a collapse of the surface material. These *collapse craters*, as they are called, could have resulted from lava flowing out from under the surface, reducing the support of the surface. This lack of support permitted the surface to collapse. Similar features are found in lava formations on the Earth: when not collapsed, they are called lava tubes.

G. GRABEN The *graben* are features of the moon that result from a sinking in of a portion of the surface (Figure 6-14). These features have been named after similar features on the Earth that have been shown to be the result of earthquakes. If the crust of the Earth cracks along two parallel seams, and if the portion in the middle sinks, a graben is formed.

H. LUNAR MOUNTAINS Some of the largest mountains are those that rim some of the maria, such as the Apennines which form the southeast rim of Mare

Imbrium (see Figure 6-4). Photographs of some of the Apennines were made by the Apollo 15 crew that surveyed an area at the base of Hadley Mountain, which rises up 14,000 feet from the surrounding plain (see the photograph that opens this chapter). Their lunar module landed near Hadley Rille (Figure 6-14),

Figure 6-16
The Hadley rille with astronaut David Scott standing beside the lunar rover in the foreground. (NASA photograph.)

not far from the craters Autolycus and Aristillus (see Figure 6-4 and 6-14), and their mode of transportation on the surface was the lunar buggy. Both Hadley Mountain and Silver Spur reveal a layered structure indicating that each was formed by successive lava flows (Chapter opener and Figure 6-15). That the strata of each mountain is tilted, indicates they suffered a catastrophic upheaval that might have been caused by the large meteorite that formed Mare Imbrium, since the Apennine Mountains form part of the rim of that mare.

I. HADLEY RILLE Hadley Rille has an average width of about 1 mile, and near the Apollo 15 landing site it is 1200 feet deep. From Figure 6-16 it is clear that some rock has tumbled down the sides to the bottom of the sinuous canyon. Figure 6-17 reveals that the land at the base of the Apennines and into which Hadley Rille cuts is formed of layers of material, again indicating successive lava flows.

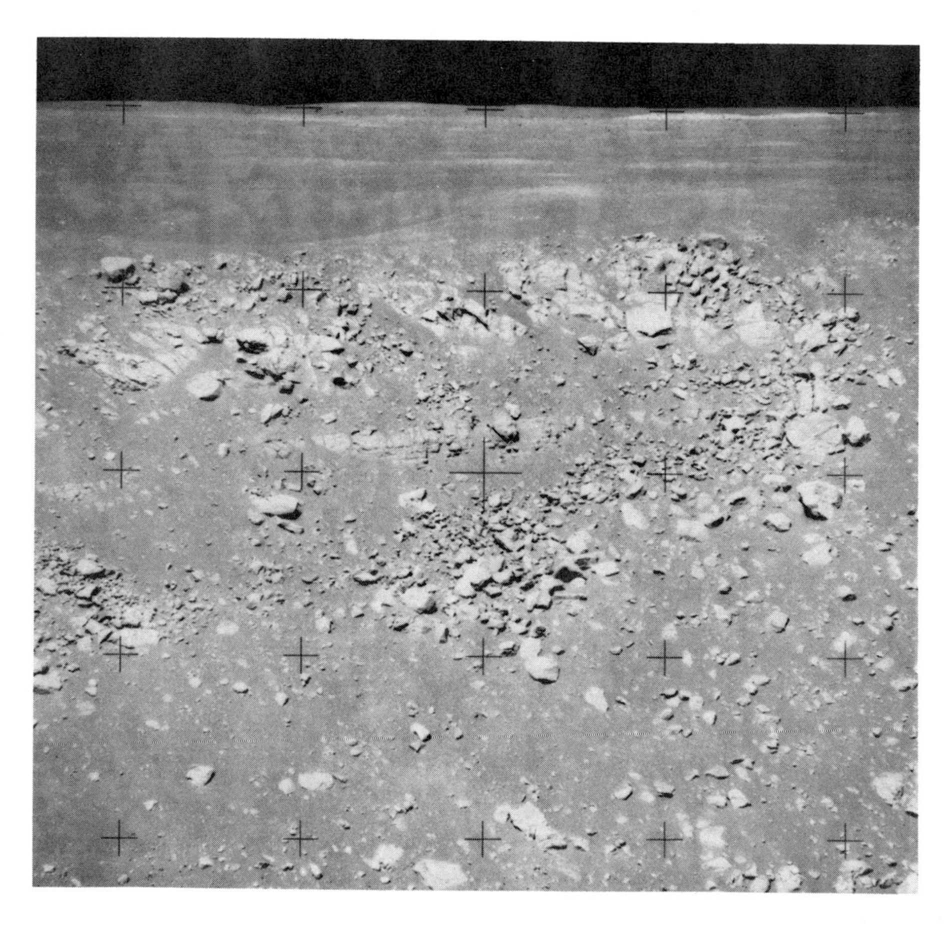

Figure 6-17
View across Hadley rille reveals that layered rock makes up part of the lunar surface. (NASA photograph.)

6.4 THE MOON'S MOTION

We have mentioned that the moon revolves about the Earth, but its motion needs to be described in greater detail. Although the moon's diurnal motion is westward, it actually revolves in its orbit from west to east (see page 98). Its orbit is elliptical, as are the orbits of all celestial bodies. The point of its closest approach to the Earth is called *perigee*, and the point farthest from the Earth is called *apogee*.

Our evidence for the moon's eastward motion lies in the fact that it sets about 50 minutes later every time it sinks below the western horizon. If the stars are used as reference, the moon has a period of 27⅓ days, during which time it completes a circuit of 360° in the sky. But at the same time the Earth has continued its motion about the sun; thus the moon needs more than 2 additional days, 29½ days in all, to complete its period of phases from new moon through

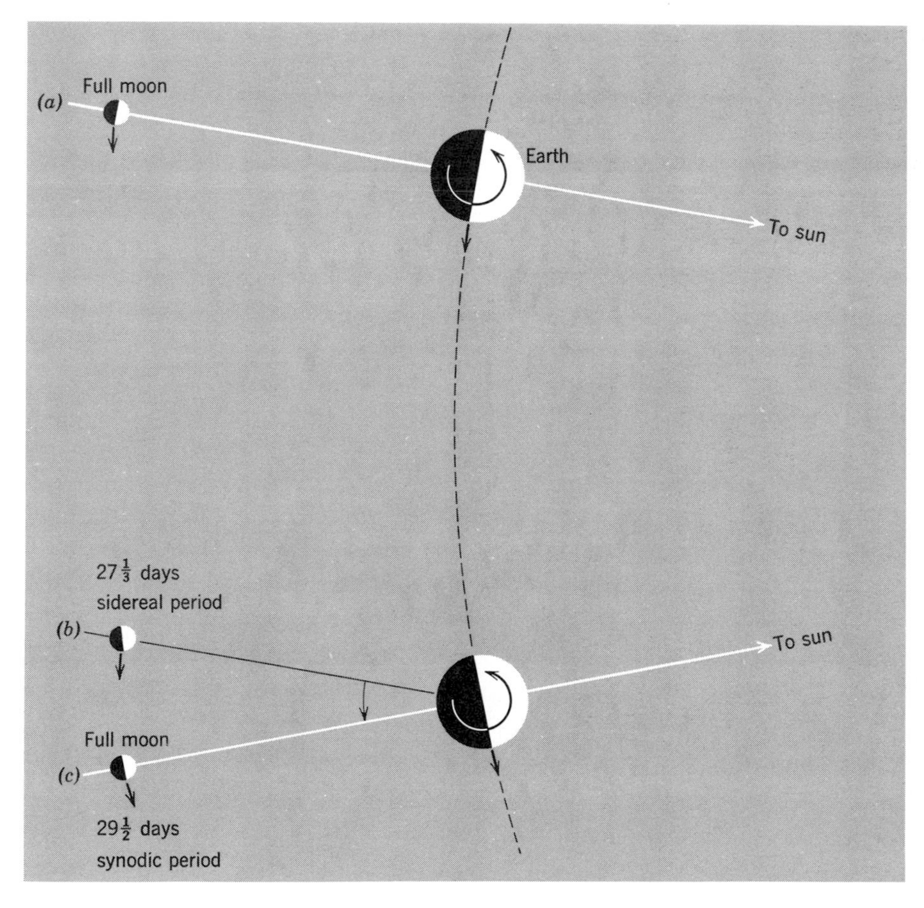

Figure 6-18
The sidereal period of the moon, 27⅓ days, is the time required for the moon to make one complete revolution of 360° with respect to the stars. The synodic period, 29½ days, is the time required for the moon to revolve from one phase to that same phase again, such as from full moon to full moon.

full and back to new moon again. The difference is seen in Figure 6-18. One complete revolution of the moon with respect to the stars (the 27⅓ days) is called a *sidereal period*, while one complete revolution with respect to the line joining the sun and the Earth (29½ days) is called a *synodic period*. This latter is the basis for the month.

To calculate how far the moon moves in its orbit in one day, consider that it revolves through 90° from full moon to first quarter when the moon is 7 days old. In 7 days it revolves through 90°. Therefore it must revolve through an angle of 90°/7 in one day, or close to 13°. Revolving 13° in one day means that it revolves 13° in 24 hours or close to ½° in one hour. The moon's apparent diameter, however, is just that: ½°. The moon moves eastward in its orbit one of its diameters every hour. During the next first-quarter moon, go outside at one-hour intervals and by lining the moon up with several stars see if you can detect this motion.

The plane of the moon's orbit is coincident neither with the plane of the Earth's orbit (the ecliptic) nor with the plane of the Earth's equator. It is closer to the ecliptic, however, from which it is tipped by some 5°.

6.5 ECLIPSES

It is because the moon's orbit is not coincident with the ecliptic that the moon does not go into the Earth's shadow every full moon to give us a *lunar eclipse;* nor does the new moon pass in front of the sun every month to give us a *solar eclipse*. Eclipses can occur only during two seasons of the year.

In order for the moon to pass in front of the sun they both must be in the same part of the sky. This means that the moon must be on or very near the ecliptic because, by definition, the sun is always on the ecliptic. The two points in the sky where the moon's path crosses the ecliptic are called the *lunar nodes*. The moon passes through each node once a month, but the sun passes through each of the nodes only once a year. Thus, a *solar eclipse* can occur only during those two seasons of the year when the sun is at or near one of the nodes during the new moon.

In order for the Earth to cast its shadow on the moon, a *lunar eclipse*, the moon must be at one of its nodes and the sun at the other. Again this can happen only during two seasons of the year.

The number of eclipses (lunar and solar) each year varies. It cannot be less than two nor more than seven, although the usual number is four. Solar eclipses are more common than lunar eclipses, but since a lunar eclipse can be seen from nearly half of the Earth's surface, more people have seen a lunar eclipse than have seen a solar eclipse.

A total solar eclipse is a beautiful sight. It occurs when the disk of the moon

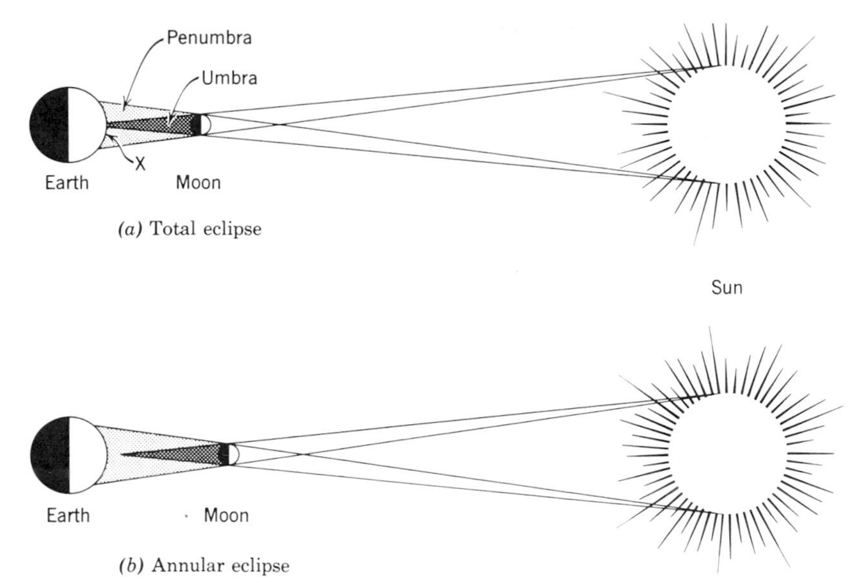

Figure 6-19
(a) The umbra of the moon just reaches the Earth when the moon is near perigee. (b) When the moon is near apogee the umbra does not reach the Earth, and thus a total eclipse cannot occur, only an annular eclipse.

completely covers the disk of the sun (Figure 8-15). But a total solar eclipse is seen only over a small area of the Earth; over a larger area it appears as a partial eclipse. The reason for this is apparent from Figure 6-19*a*. (It has been necessary to draw the sizes and distances of the sun, moon, and the Earth out of proportion to each other.)

In the black region, the sun is completely hidden from view by the moon. This part of the shadow is called the *umbra*, meaning shade or cover (see our word "umbrella"). In the part of the shadow that is gray the sun is not completely hidden. For example, a person standing on the Earth at *X* would be able to see part of the sun. Where this portion of the shadow, called the *penumbra*, strikes the Earth, a partial eclipse may be seen. If an observer is near the umbra, a large portion of the sun's disk is covered by the moon, but if he is near the outer edge of the penumbra, only a small section of the sun will be obscured from view. This small section is a silhouette of part of the moon.

The umbra is actually a cone, and it is sheer coincidence that the length of this cone is very nearly the same as the distance from the Earth to the moon. When the moon is near perigee and the Earth near aphelion, the cone of the umbra projects a dark circular shadow about 167 miles in diameter on the face of the Earth. When the moon is near apogee, however, the tip of the cone does not even reach the Earth (Figure 6-19*b*). The resulting visible effect makes the disk of the moon appear slightly smaller than the disk of the sun. Consequently the moon does not cover the sun completely. The ring of the sun that overlaps the moon's disk gives rise to the term *annular eclipse*.

Since the umbra follows the moon in its motion and is projected on a nearly

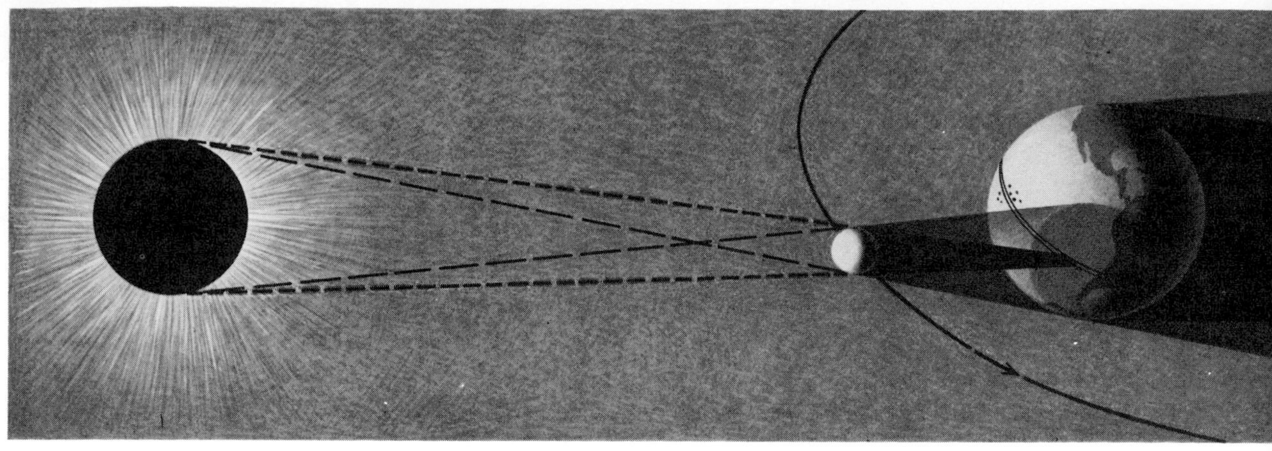

Figure 6-20
As the moon revolves in its orbit and the Earth rotates on its axis, the umbra traces the path of totality across the surface of the Earth. (Courtesy of *Sky and Telescope*)

spherical rotating Earth, it traces a rather complicated path, called the *path of totality*, on the Earth's surface. Anyone located in this path may see a total eclipse that can at best last 7 minutes, 40 seconds. The penumbra may stretch for a distance of several thousand miles on either side of the path of totality. The path of totality for the eclipse of October 12, 1958, is seen in Figure 6-20.

At the time of totality the outer regions of the sun become visible. (Because they are fainter than the blue sky they cannot normally be seen.) The pearly white area that extends farthest from the sun is called the *corona*. Very close to the surface of the sun is a region called the *chromosphere* because it is reddish. Prominences that appear to be gases erupting from the sun may be seen standing out from its surface. The nature of these is discussed in Chapter 8.

A total lunar eclipse, although not as spectacular as a total solar eclipse, is a sight long remembered. Even though the moon's disk may be entirely covered by the umbra of the Earth's shadow it is never completely black. The Earth's atmosphere acts like a lens and focuses some of the sunlight on to the moon. Since the red light can travel through a long path in the atmosphere better than the blue, the moon takes on a deep reddish or copper color. The blue light has been scattered out into the surrounding atmosphere.

Eclipses, both lunar and solar, have been beneficial to the historian and archeologist as well as to the astronomer. In the records of ancient peoples there are many references that not only enable the historian to date an event, but also give the astronomer observations extending over many centuries of time. With the help of historic eclipses he has been able to describe the motion of the moon with greater accuracy.

6.6 TIDES

Not only through eclipses does the moon affect us here on the Earth; the moon is also the main cause of the ocean tides. These tides result from the fact that the gravitational pull of any body, including the moon, decreases with increasing distance from that body. If there were no moon (or sun) the water layer over the Earth would be uniform as shown in Figure 6-21*a*. When the moon is put back in place (shown by the arrow in Figure 6-21*b*) the Earth responds to the additional gravitational force. Those portions of the Earth nearest the moon are subjected to a stronger gravitational pull, and thus the top of the ocean near the moon is pulled closer to the moon than is the bottom of the ocean. On the far side of the Earth, the bottom of the ocean, being nearer the moon than the top, is pulled closer to the moon. Thus the ocean bulges on two sides: the side facing the moon and the side directly opposite the moon. It is in these regions that we experience high tides. Halfway between the high tides are the low tides, the regions which supply the water for the high tides.

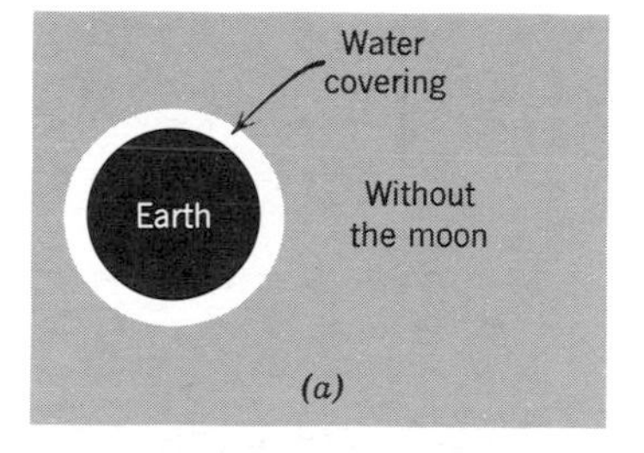

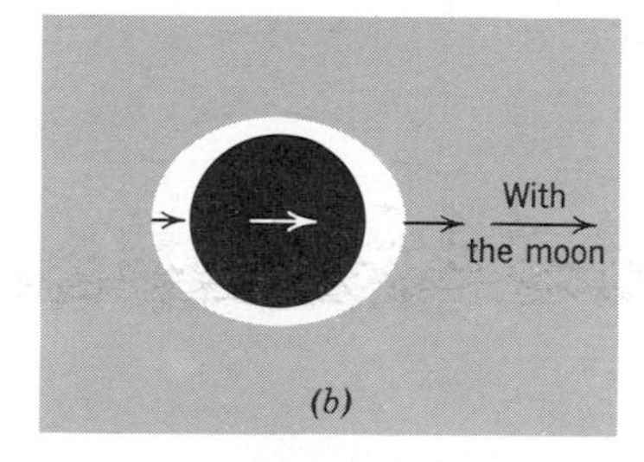

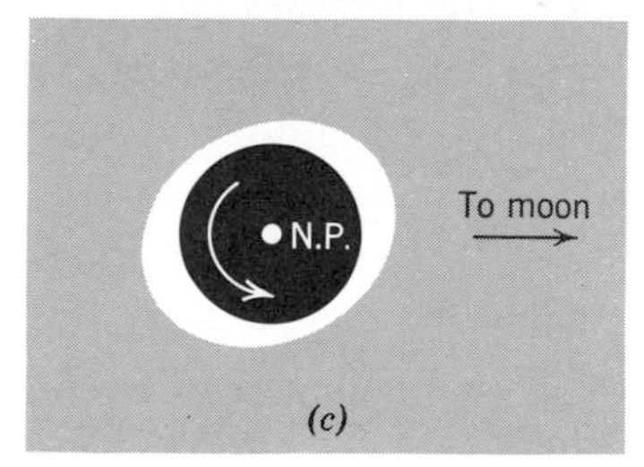

Figure 6-21
(a) The layer of water about the Earth in the absence of the moon (and the sun) would be uniform about the equator. (b) The moon is there, however, and pulls the Earth toward it. Those parts of Earth closer to the moon are pulled with a stronger force. (c) Because the Earth rotates faster than the moon revolves about the Earth, the continents of the Earth drag the tides with them.

The net effect of this, and all tidal forces, is that the moon tries to pull the Earth apart but fortunately never succeeds. The Earth tries the same with the moon, and were the moon closer the Earth might succeed.

The Earth rotates underneath the tidal bulges to give us two high and two low tides every day. During one rotation of the Earth, however, the moon has moved eastward in its orbit; consequently the Earth must rotate farther so that the cycle of tides may be completed. Thus there are two high and two low tides every 24 hours, 49 minutes.

The variation in the level of the water differs with locality. Tides of 3 to 6 feet are fairly common on the shores of the oceans, but in some places, because of funneling, the water level may differ by 50 feet between high and low tide. In such places (for example, the Bay of Fundy, which is the gulf between Nova Scotia and New Brunswick) special harbors must be built.

It is bays such as these and shallow waters such as the Bering Strait that cause friction between the tides and the Earth. As a consequence the tidal bulge is dragged by the Earth, which rotates faster than the moon revolves. Hence the tidal bulge does not line up with the moon, but precedes it (Figure 6-21*c*).

Since the Earth is not a rigid body it, too, gives slightly with the tides. Although small, this movement has been measured — evidence that having your feet on "solid ground" means something quite different, for that ground is moving up and down every 12 hours, 25 minutes.

6.7
SATELLITES OF OTHER PLANETS

Many of the characteristics of our natural satellite can be generalized to apply to satellites of the other planets. Of the nine planets, only Mercury, Venus, and Pluto do not, to our knowledge, have satellites. Those planets that do will be considered next, with emphasis on the unusual satellites.

After the Earth the first planet from the sun to have satellites is Mars with two: Phobos, the inner one, and Deimos. Photographs of these two satellites taken by Mariner 9 reveal that each is irregular in shape and each has craters over it.

Phobos (Figure 6-22), the larger of the two, is only 13 by 16 miles and has a rougher surface than Deimos. The number of craters on it leads astronomers to believe that it is old and has suffered many collisions with chunks of rock smaller than it. That it is not broken into pieces by any of these collisions indicates that it has some internal strength. Its irregular shape, however, could indicate that it was once part of a larger spherical body that was broken into smaller chunks by a collision.

Jupiter, because of its gigantic mass, has more satellites than any other planet. The first ones to be discovered were the four discovered by Galileo in 1609. The fifth satellite was discovered in 1892 and is closer to the planet than the

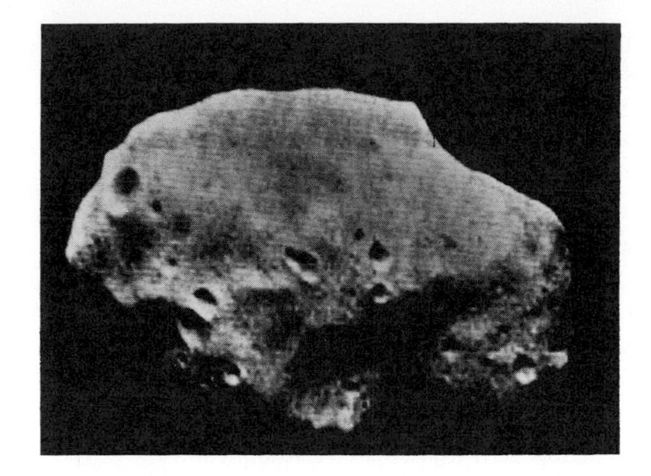

Figure 6-22
Phobos, a satellite of Mars as photographed by Mariner 9. (NASA photograph.)

innermost Galilean satellite. The thirteenth satellite was discovered in 1974 by Charles Kowal of the Hale Observatories.

The four Galilean satellites are visible with binoculars. By watching them from night to night an interested observer can follow the motion of these largest Jovian satellites. The largest two, with diameters of 3280 miles and 3100 miles are slightly larger than the planet Mercury with a diameter of 3030 miles. Observations have been made definitely establishing the existence of water frost on two of the Galilean satellites.

Since the four Galilean satellites revolve in planes that are very nearly parallel to the plane of Jupiter's equator, they not only go behind Jupiter during part of their cycle but also pass in front of it. During a transit of one of these satellites its shadow is visible on the surface of Jupiter (Figure 6-23).

Figure 6-23
Jupiter with one of its satellites casting its shadow on the cloud cover. Photograph by the 200-in. telescope on Mount Palomar. (Photograph from the Hale Observatories.)

Of the ten satellites of Saturn (we do not count the tiny particles that compose its rings), Titan is the most interesting because of all the satellites in the solar system, it alone has a spectrum with absorption lines indicating a substantial atmosphere. Its atmosphere is composed of at least methane, though there may be other gases present.

Uranus, whose axis of rotation is nearly parallel to the plane of its orbit, has five satellites revolving about it. Their planes of revolution are nearly parallel to the plane of Uranus' equator. Since these planes are more or less fixed in space, we on the Earth can at certain times (as in 1944) see the satellites moving in circles about the planet; that is, at certain times our line of sight is nearly perpendicular to the planes of their orbits. By the time Uranus moved in its orbit one-quarter the way around the sun (by 1967), the satellite orbits were seen edge on and the satellites appeared to oscillate in straight lines.

Neptune has two satellites, one of which when compared with our moon illustrates the dependency of the period of revolution on the mass of the central body. Neptune is 14.5 times as massive as the Earth. One of its satellites, Triton, which has an orbital radius of 220,000 miles (nearly the same as our moon's orbital radius of 239,000 miles), has a period of revolution of only 6 days as compared with our moon's 27⅓ days.

The sun rotates, the planets all revolve and rotate, and most of the satellites revolve and rotate all in the same direction. Triton, however, like four satellites of Jupiter and one of Saturn, moves in retrograde motion, that is, each of these satellites revolves about its mother planet in a direction contrary to the principal motions of the solar system. It may be presumed that these satellites were captured after their planets were formed.

BASIC VOCABULARY FOR SUBSEQUENT READING

Annular eclipse
Apogee
Cosmic rays
Crescent moon
Dark-haloed craters
Earthlight
First quarter moon
Full moon
Gibbous moon
Graben
Lunar eclipse
Lunar nodes
Mare
Meteor
New moon
Partial eclipse
Penumbra
Perigee
Rilles
Sidereal period
Solar eclipse
Synodic period
Tides
Total eclipse
Umbra

QUESTIONS AND PROBLEMS

1. Draw the Earth-moon system to scale and indicate the location of the center of mass of the system.

2. After the next full moon, follow the moon from night to night through at least one complete cycle of phases (weather permitting).

3. Describe the location of the Earth, moon, and sun for the following phases of the moon: (a) new moon, (b) first quarter, (c) gibbous moon, (d) full moon.

4. What time of the day is the moon high in the sky for each of these phases of the moon: (a) new moon, (b) first quarter, (c) full moon, (d) third quarter.

5. At what time of the day does the moon set during each of the following phases: (a) new moon, (b) first quarter, (c) full moon, (d) third quarter.

6. Describe the appearance and possible origin of three features of the lunar landscape.

7. What are four causes of erosion on the lunar surface?

8. What evidence indicates that the maria are extensive lava flows?

9. What is a mascon?

10. What is the difference between the sidereal period and the synodic period of the moon?

11. Why are there two high and two low tides every day?

FOR FURTHER READING

Alter, D., *Pictorial Guide to the Moon,* Thomas Y. Crowell Co., New York, 1963.

Cherrington, E. H. Jr., *Exploring the Moon Through Binoculars,* McGraw-Hill Book Co., and Peter Davies Ltd., 1969.

Glasstone, S., *Sourcebook on the Space Sciences,* Van Nostrand Co., Princeton, N.J., 1965, Chapter 9.

Whipple, F. L., *Earth, Moon and Planets,* 3rd ed., Harvard University Press, Cambridge, Mass., 1968.

Cannon, P.J., "Lunar Landslides," *Sky and Telescope,* p. 215, October 1970.

Cruikshank, D. P., and D. Morrison, "Titan and its Atmosphere," *Sky and Telescope,* p. 83, August 1972.

Dyal, P., and C. W. Parkin, "The Magnetism of the Moon," *Scientific American,* p. 62, August 1971.

Eglinton, G. et al., "The Carbon Chemistry of the Moon," *Scientific American,* p. 80, October 1972.

Faller, J. E., and E. J. Wampler, "The Lunar Laser Reflector," *Scientific American,* p. 38, March 1970.

Goldreich, P., "Tides and the Earth-Moon System," *Scientific American,* p. 42, April 1972.

Greenacre, J. A., "A Recent Observation of Lunar Color Phenomenon," *Sky and Telescope,* p. 316, December 1963.

Hess, W. et al., "The Exploration of the Moon," *Scientific American,* p. 54, October 1969.

Hillenbrand, R., "Apollo 17 Exploration and Experiments," *Sky and Telescope,* p. 372, December 1972.

Hillenbrand, R., "The First Men on the Moon," *Sky and Telescope,* p. 144, September 1969.

Hillenbrand, R., "The Apollo 12 Explorers on the Moon," *Sky and Telescope,* p. 95, February 1970.

Kuiper, G. P., "The Lunar and Planetary Laboratory," *Sky and Telescope,* part I, p. 4, January 1964; part II, p. 88, February 1964.

Page, Thornton L., "Notes on the Fourth Lunar Science Conference," *Sky and Telescope,* part I, p. 355, June 1973; part II, p. 14, July 1973; part III, p. 88, August 1973.

Page, Thornton L., "Notes on Lunar Research," *Sky and Telescope,* p. 88, August 1974.

Pike, Richard J., "The Lunar Crater Linné," *Sky and Telescope,* p. 364, December 1973.

"Findings from a Sample of Lunar Material," *Sky and Telescope,* p. 144, March 1970.

"First Report on Apollo 16," *Sky and Telescope,* p. 353, June 1972.

"Orange Soil and Other Apollo 17 Results," *Sky and Telescope,* p. 146, March 1973.

"Orbits of Phobos and Deimos," *Sky and Telescope,* p. 87, August 1972.

"Results of Apollo 11 Research," *Sky and Telescope,* p. 226, April 1970.

"Standing Waves on the Moon," *Sky and Telescope,* p. 166, September 1969.

"Visit to Taurus-Littrow," *Sky and Telescope,* p. 79, February 1973.

CHAPTER CONTENTS

LEARNING OBJECTIVES

BE ABLE TO:

1. DESCRIBE HOW THE "NUMBERS GAME" CAN LEAD TO THE PREDICTION OF A "MISSING PLANET."
2. GIVE AN EXPLANATION FOR THE VARIATION IN BRIGHTNESS OF MINOR PLANETS.
3. EXPLAIN HOW WE KNOW METEORS ARE VERY SMALL.
4. EXPLAIN WHY METEORS ARE MORE NUMEROUS, FASTER, AND BRIGHTER JUST BEFORE DAWN THAN AFTER DUSK.
6. GIVE THE THREE TYPES OF METEORITES AND DESCRIBE THEIR PROBABLE ORIGIN.
7. EXPLAIN WHY CHONDRITES ARE OF PARTICULAR INTEREST.
8. DESCRIBE THE RELATIONSHIP BETWEEN METEOR SHOWERS AND COMETS.
9. DESCRIBE, IN GENERAL, BOTH THE ORIBT OF A COMET AND THE SPEED OF THE COMET IN THAT ORBIT.
10. DESCRIBE THE THREE MAIN PARTS OF A COMET AND HOW EACH VARIES AS THE COMET APPROACHES AND LEAVES THE SUN.
11. GIVE OBSERVATIONS SUPPORTING THE "DIRTY SNOWBALL" THEORY OF THE NUCLEUS OF A COMET.
12. EXPLAIN WHY VERY BRIGHT NAKED EYE COMETS CANNOT HAVE MADE MANY TRIPS AROUND THE SUN.
13. DESCRIBE THE SIGNIFICANCE OF THE RADIANT OF A METEOR SHOWER.

Chapter Seven

MINOR PLANETS, METEORITES, AND COMETS

Man leaves no stone unturned in his eternal search for system and law. This search has led from the Pythagorean numbers games to the bases of modern mathematics; and numbers games are still being played, with results that are sometimes as fruitful.

7.1 NUMBERS AND PLANETARY DISTANCES

Consider, for instance, the mean distances in astronomical units of the planets from the sun. Beginning with Mercury, we find that these give, naturally enough, a sequence of increasing numbers:

0.4, 0.7, 1.0, 1.5, 5.2, 9.6, 19.2, 30.1, 39.4

Such a sequence of numbers invites many to search for an arithmetical scheme that ties these numbers together. Indeed, such schemes have been deduced, but none thus far has had any theoretical significance. Nevertheless it is possible to perform a simple operation on these numbers and then make some deductions from the results to learn something of consequence about our solar system. This example also demonstrates one of the methods of science: the method of hypothesis formation and testing.

A. HYPOTHESIS FORMATION Divide the distance of each planet from the sun by the distance of the planet just inside its orbit. For example, the distance of Venus (0.7 A. U.) divided by that of Mercury (0.4 A. U.) gives

$$\frac{0.7}{0.4} = 1.9$$

Repeating this simple division for each of the successive pairs of adjacent planets yields a second sequence of numbers:

1.9, 1.4, 1.5, 3.4, 1.8, 2.0, 1.6, 1.3

A casual observation of this sequence reveals that all of these numbers lie between 1.3 and 2.0 except for that single 3.4. It would appear that with this single exception, each planet is either twice as far from the sun as the next inner planet, or less than twice as far away.

The two planets that yielded the 3.4 were Jupiter and Mars. This argument, then, would lead us to deduce that Jupiter and Mars are too far apart when compared with the other planets.

Thus, we are left with a puzzle. Are Mars and Jupiter really too far apart? Is there a gap in the solar system that separates the four inner planets from the four giant planets (neglecting poor Pluto)?

Chapter Opening Photo
Comet Bennett photographed on March 28, 1970 at 5 A.M. (Courtesy Dr. J. U. Gunter.)

B. HYPOTHESIS EXTENSION We can extend our hypothesis by considering a bit this second sequence of numbers. What is the average value of the numbers in this series, excepting, of course, that 3.4? The sum of the remaining 7 numbers yields 11.5 and this divided by 7 yields, easily enough, the average value 1.6.

How does this 1.6 compare with our 3.4? Well, 1.6 is nearly one-half 3.4. So it would seem that the 3.4 can be split into two numbers, 1.7 and 1.7, to give us a third sequence of numbers that looks like this:

1.9, 1.4, 1.5, 1.7, 1.7, 1.8, 2.0, 1.6, 1.3

and admittedly this does look more uniform than its counterpart above.

But what does it mean? It means simply that our deduction that Mars and Jupiter are too far apart has gained some weight. It could mean that there may be something between those two planets at about a distance of 2.5 A. U. from the sun, that is, 1.7 times 1.5 A. U., the distance of Mars.

C. HYPOTHESIS TESTING To test our hypothesis, we must make a search of the solar system with a telescope, or more quickly, of the literature. Either search would reveal the observation that the region between Mars and Jupiter is littered with thousands on thousands of tiny objects. Tiny, that is, when compared with Jupiter or Mars.

7.2 MINOR PLANETS (ASTEROIDS)

These objects orbit the sun and are planetlike; hence, they are called *minor planets*. Some of the literature refers to them as *asteroids*, but that term has historical significance only. In all but the biggest telescopes, the minor planets appear as points of light, that is, as stars. Planets appear as disks. So the early astronomers, working with relatively small telescopes, called the moving points of light asteroids which means "starlike." Many astronomical objects must be named before their true nature is deduced, and these names do not always match later on. The minor planets are anything but starlike.

Observations with many telescopes and studies of their orbits reveal that the minor planets are small, and their average distance from the sun is about what our hypothesis suggested.

Minor Planet	Diameter	Average Distance from Sun
Ceres	640 miles	2.8 A. U.
Pallas	340	2.8
Vesta	320	2.4
Juno	140	2.7

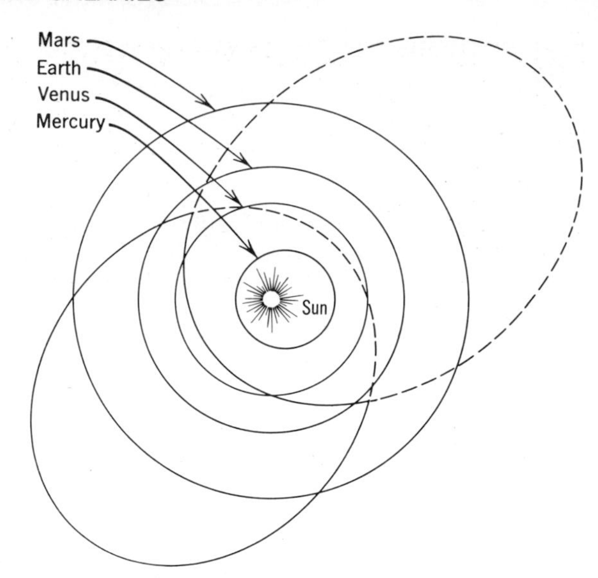

Figure 7-1
The orbits of two minor planets that come inside the orbit of Mars. The dotted portions of the orbits are south of the plane of the ecliptic.

A number of the smaller minor planets have distances from the sun that range down to less than 1.5 A. U. (Figure 7-1). Their diameters, which can be estimated by studying the amount of sunlight they reflect, decrease to less than a mile. But we find that as ever smaller minor planets are considered, their numbers increase. Indeed, it is estimated that there must be more than 50,000 minor planets in range of our present telescopes! The total mass of all these minor planets, however, is probably not much more than about one tenth the mass of the moon.

Prolonged observations of the larger minor planets reveals that the brightness of many of them varies with time. Their observed period of brightness variation ranges from about 4 hours to 9 hours and 5 minutes for Ceres. A variation in brightness of a minor planet can be accounted for by one of two theories: either the planet has sides of different reflectivity or it is oblong in shape. Judging from Phobos, the inner satellite of Mars (see Fig. 6-22), it may be assumed that these minor planets are irregular in shape. When they present their broadside to us and the sun, they appear brighter than when they present their smaller cross section.

In fact, it would be reasonable to assume that the larger minor planets look a good deal like Phobos. With so many of them in that gap in the solar system, there certainly must be occasional collisions resulting in small craters.

Minor planets are still being discovered. Since they move relative to the background stars, they leave a streak on a photograph of the stars (Figure 7-2), all of which are time exposures, some for as long as one hour. During a photograph of the stars the telescope is driven to follow the stars, that is, to counteract the Earth's rotation.

Although each minor planet is small, their total number is significantly large.

Figure 7-2
The motion of an unidentified minor planet causes it to appear as a streak on a time exposure of the stars. On the extreme left is a distant galaxy. (Courtesy of C. Wirtanen, Lick Observatory.)

This coupled with their location in the gap between Mars and Jupiter makes them an interesting group of the solar system. Their presence and location must certainly be considered by any reasonable theory of the origin of the solar system.

7.3 METEORITES

Compared to the nine major planets the minor planets are very small, but they are far from being the smallest objects that revolve about the sun. *Meteorites* are not only smaller, but since they collide with or fall on the Earth they have become the first extraterrestrial objects to be touched by man and examined in his laboratory.

A. SEEN AS METEORS Meteorites travel in orbits about the sun at velocities of 10 to 30 miles per second. When one collides with the Earth it must first penetrate our protective atmosphere. During this high-speed penetration, the air in front of the meteorite does not have time to move to the side, so the meteorite acts like a piston, compressing the air in front of it. The heat resulting from this compression causes the meteorite to melt and burn. Small meteorites never survive this burning; they filter down to the surface of the Earth as small particles of ash or dust. From the length of time it takes an average meteorite to burn — just a fraction of a second — one can estimate their size to be not much larger than an "o" on this page.

As the meteorite streaks and burns through the upper atmosphere, it may be visible from the earth as a "shooting star" or *meteor*. The term meteorite is reserved for the object itself; the term meteor, for historical reasons, has come to designate the flash of light in the nighttime sky. The word meteor derives from the fact that before the very early part of the 19th century, meteoric streaks were thought to be strictly an atmospheric phenomenon. Hence, the confusion: meteorology is the study of the atmosphere not the study of meteors.

A patient observer away from city lights may observe 5 to 10 meteors every hour. If he continues watching all night long he will find that the frequency of meteors increases as the night progresses. Not only will the number increase each hour but after midnight they will become, on the average, more brilliant.

We can understand the increase in meteor activity in the early morning hours if we recall how the Earth revolves and rotates. It rotates in the same direction in which it revolves, so the leading side of the Earth is the morning side (Figure 7-3). That portion of the Earth in dawn, is, therefore "out in front" of the Earth as it travels in its orbit about the sun. As a consequence it is able to pick up more meteors; it runs into them. Thus the early morning meteors have a higher velocity when they strike our atmosphere. The brightness of a meteor depends to a large extent on its relative velocity, so these meteors are on the average more brilliant than those seen in the evening. An evening meteor overtakes the Earth and therefore has a low relative velocity.

Most meteors are rather faint, but sometimes a very bright one rivals the brightness of the full moon and even casts shadows. Such a bright meteor is generally called a *fireball*. Since the difference between a meteor and a fireball is dependent on brightness, it is difficult to make a definite distinction between the two. In his book *Between the Planets*,* Watson calls attention to a very quaint

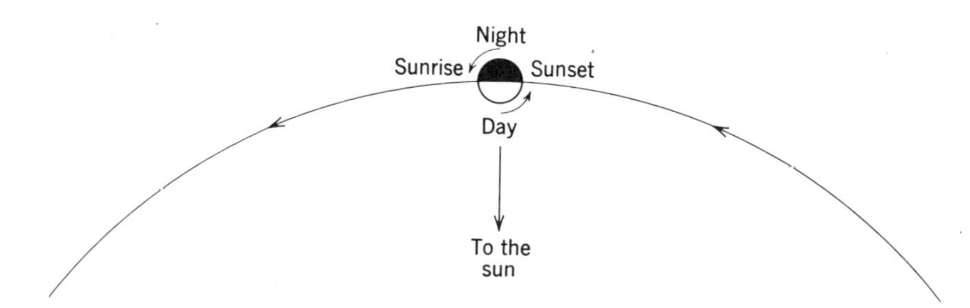

Figure 7-3
The sunrise portion of the Earth leads the way as the Earth revolves about the sun.

*Fletcher G. Watson, *Between the Planets*, Cambridge, Harvard University Press, 1956.

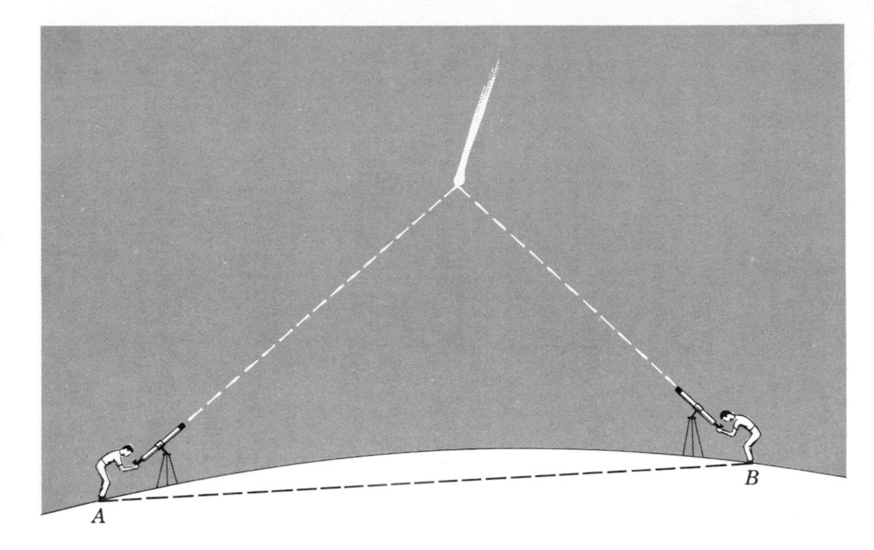

Figure 7-4
The height of the meteor can be determined if it is observed either visually (without a telescope) or photographically (with a telescope) from two well separated positions on the Earth.

way of distinguishing a meteor from a fireball. He says that, "One astronomer adopted a very practical definition: 'A fireball is a meteor sufficiently bright to make people report it.'" Some fireballs may be so bright as to be seen over the areas of several states.

At best meteors are difficult to study, for they give no warning when or where they will be seen, and they last only a fraction of a second once they strike our atmosphere. Modern observational techniques, however, have permitted a great deal of information to be gathered about these illusive objects.

B. TECHNIQUES OF OBSERVATION The first study of meteors was made in 1798 by two German students who realized that each had seen the same meteor even though they had been separated by several miles on the Earth. In following up this observation with others, they learned how to determine the height of the meteors by the method of parallax (Figure 7-4). They knew the distance *AB* and the angles formed at the points *A* and *B*, so they needed only to draw a similar triangle to scale in order to estimate the height of the meteor. Or they could make a more accurate determination by using trigonometry. Their results indicated that meteors appear at heights of about 50 miles. Figure 7-5 shows two photographs of a single meteor taken from two different locations.

Current studies are being made to determine the motion of meteorites with the hope of being able to determine their orbit before colliding with the Earth and, if possible, of recovering them after their fall. The first success in this program, carried out by the Smithsonian Astrophysical Observatory, was the recovery of a 22-pound meteorite near Lost City, Oklahoma, on January 9, 1970.

The photographs in Figure 7-6 help tell the story. Two of many well separated cameras used in the study photographed the meteor streak. Each stationary camera made a time exposure of the skies (note that the stars leave streaks), and each camera has a special shutter that interrupts the light entering the camera 20 times each second. By counting the interruptions in the meteor trail, it is possible to determine how long the meteorite was burning. By comparing the photographs of the two well separated telescopes, its height at both the beginning and end of its observed fall could be determined. From its distance and time of fall, its velocity of fall was determined.

From its speed and direction of fall, it was possible to determine not only the orbit of this meteorite before it hit the Earth, but where it was located after it

Figure 7-5

Two photographs of the same meteor in the Perseid meteor shower during the night of August 11-12, 1972. The two cameras were located 10 miles apart, and by chance were pointing to the same spot in the atmosphere. That each camera photographed the same meteor can be seen from the flaring pattern of the meteor. That the cameras were pointing in different directions when they photographed the meteor can be seen by the difference in the background stars. Photograph (a) was taken by Ronald Oriti of the Griffith Observatory, Los Angeles, California. Photograph (b) was taken by Dr. James Matteson, University of California, San Diego.

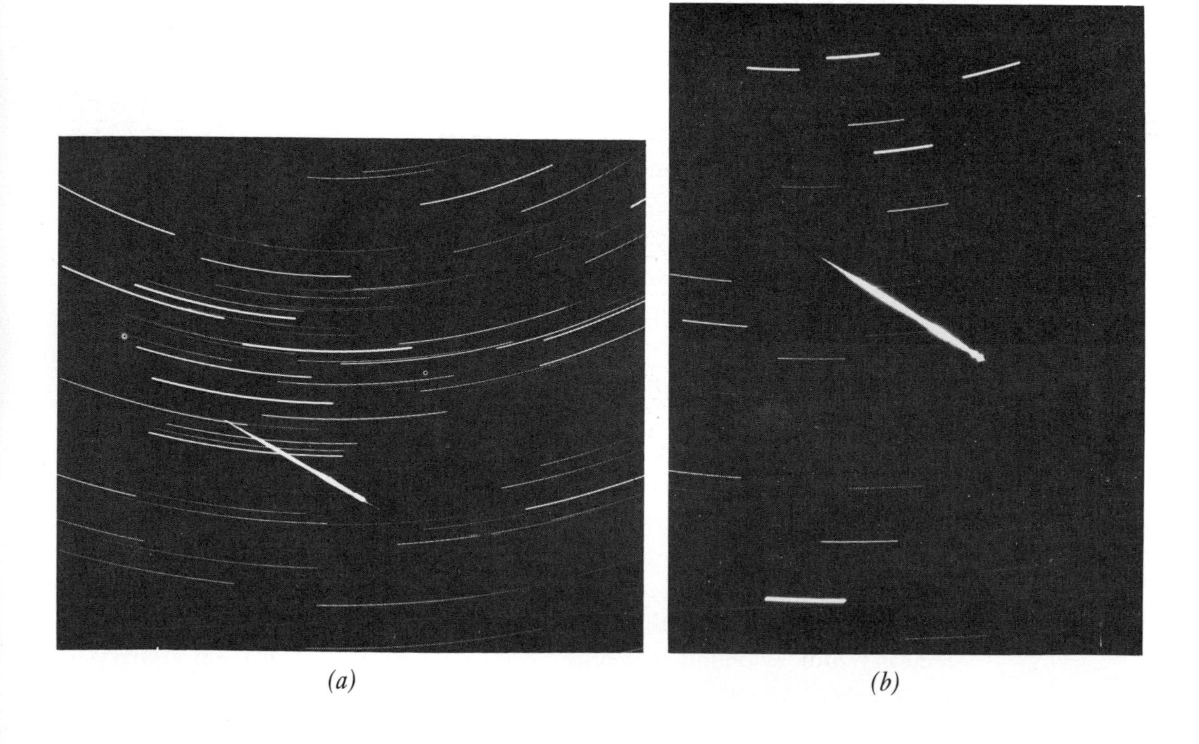

(a) *(b)*

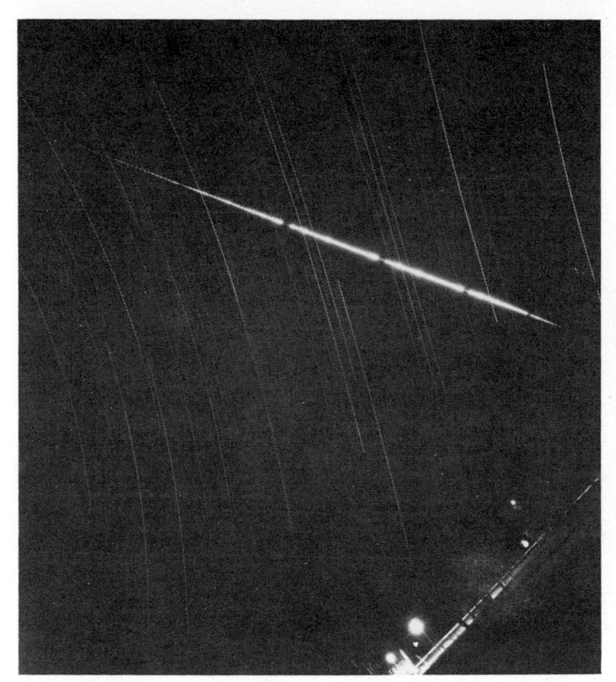

Figure 7-6

(a) The Lost City meteorite photographed during its fall by a meteor patrol camera. (Smithsonian Astrophysical Observatory.)

(b) The Lost City meteor as it was found and recovered by Gunther Schwartz in March 1970. (Courtesy Sky and Telescope.)

fell. Luckily it fell right alongside a road, and thus was easily recovered only 6 days after it fell even though a snowstorm caused some delays.

When the Lost City meteorite struck the Earth's atmosphere its speed was 8 miles per second; at a height of 15 miles, its speed had decreased to 6 miles per second. Its trail ended at a height of only 12 miles when its speed was 2 miles per second. This was not a fast meteorite, so it did not create much of a crater where it fell. During its fall, however, its surface was altered by the intense heat to form a crust (Figure 7-7). Oftentimes the crust will show flow lines indicating the orientation of the meteorite during its passage through the atmosphere.

Meteors may help us understand the upper atmosphere but meteorites are of greater interest to astronomers as extraterrestrial objects. Not only are they tangible evidence that the chemical elements on the Earth are no different from those in the solar system in general (which is certainly to be expected), but also they have been helpful in determining the solar system's age. They are, incidentally, a bundle of clues that need to be untangled to help solve the mystery of the origin of the solar system.

Figure 7-7
The Lost City Meteorite, a chondrite. Note the crust caused by the intense heating during its fall through the atmosphere. (Smithsonian Astrophysical Observatory.)

C. THREE GROUPS OF METEORITES Chemical analyses indicate that these extraterrestrial objects can be classified into three general groups: *iron-nickel*, *stony*, and *stony-iron*. The iron-nickel meteorites average about 90% iron and about 8% nickel (although these percentages vary considerably) with the balance consisting of other elements. The stony meteorites have quite a different composition; their main chemicals are 36% oxygen, 24% iron, 18% silicon, 14% magnesium, (again, these are averages and there is considerable variation), and a smattering of other elements. The stony-irons form a group that bridges the gap between the other two main groups.

More iron-nickel meteorites have been found than all others put together. But iron meteorites are easily distinguished from terrestrial rocks, even by a farmer tilling the soil, or a hiker or hunter. Iron meteorites have been found which, according to radioactive studies, have been on the Earth for as long as 800,000 years. Many of the stony meteorites would have weathered away in that time interval. Of those observed to fall and then recovered, only a little more than 40 have been iron meteorites; more than 600 stony meteorites have been seen to fall and then collected! Nevertheless, all of the big meteorites are significantly iron-nickel and they far outweigh all of the stony meteorites.

The reason for the existence of three main types of meteorites is not really known, but they have fed speculation on a possible relationship between the minor planets, the meteorites, and the gap between Mars and Jupiter. It is reasoned that perhaps there were small planets in this region while the solar system was forming. If, after those planets had formed, their interiors became hot enough, then the heavier elements such as iron and nickel would

Figure 7-8
Large crystals called the Widmanstatten structure in an iron meteorite. (Courtesy Brian Mason, U.S. National Museum.)

have settled to the center. It is then supposed that several pairs of these minor planets collided each breaking into many smaller pieces and giving rise to the thousands of small minor planets now observed. Because of the resulting explosion some of the debris would have been spread out over a fairly large region of the solar system; and some of this debris would certainly have landed on Mars, the Earth, and the moon.

The cores of those minor planets would have broken into what we now find as the iron-nickel meteorites, the outer parts would have given rise to the present stony meteorites. There is no other reasonable explanation for the separation of the heavier iron and nickel from the rest of the rocky material except that it occurred inside of a much larger body, such as a planet, with a hot interior.

However, in order for the interiors of those minor planets to become hot enough for the settling of the heavier elements, they would have to have been fairly large planets. But perhaps the heating processes in those minor planets was different from the heating processes in the Earth.

There is other evidence that leads us to suspect that the meteorites were formed inside a much larger body. When meteorites are examined in the laboratory it is found that they contain crystals. This in itself is nothing startling, but the crystals are large (Figure 7-8). If a rock contains large crystals the geologist knows that the rock cooled and the crystals formed very slowly; if they are small he knows that the rock cooled and the crystals formed more rapidly. Since the crystals in meteorites are large they may once have been a part of a much larger body the core of which took some time to cool from a molten state.

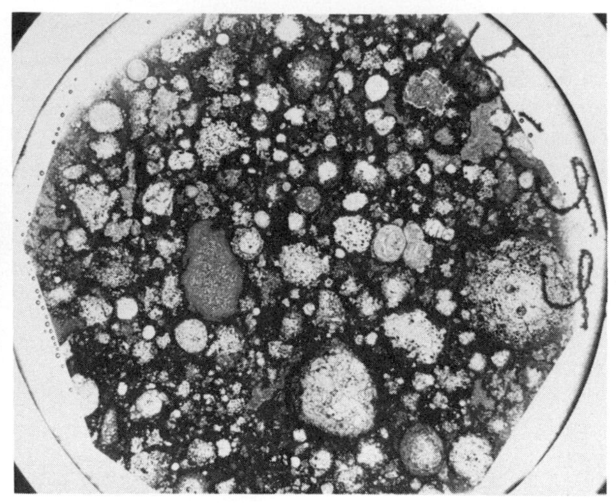

Figure 7-9

(a) A chondritic meteorite. The chondrules average about 1 millimeter in diameter. (Courtesy Brian Mason, U. S. National Museum.)

(b) A photomicrograph of the chondrules in the carbonaceous chondrite that fell in Pueblito de Allende, Mexico on February 8, 1969. (Courtesy Brian Mason, U. S. National Museum.)

Detailed studies of the crystals by J. A. Wood of the Enrico Fermi Institute for Nuclear Studies have permitted good estimates of the rate of cooling of the iron meteorites. One group of meteorites studied apparently cooled at the rate of 1° C per 100,000 years; the other group cooled 10 times more slowly. This evidence indicates that two parent bodies are involved: the faster-cooling meteorites came from a body about 60 to 120 miles in diameter, the slower-cooling meteorites from a body about 160 to 200 miles in diameter. These are not very large bodies, and it is wondered whether they are large enough for the settling of heavier elements to the center. It can only be concluded that the origin of the three forms of meteorites remains a mystery, but one that is yielding to improved techniques of observation.

D. CHONDRITES The stony meteorites have proved every bit as interesting as the iron-nickels. They have been divided into two subclasses: the *chondrites* and the *achondrites* (stony meteorites not chondrites). The chondrites contain little spherical bodies only about 0.1 cm diameter, called *chondrules*. Study of these chondrites has opened up new avenues of approach to the age-old problem of how the solar system came into being.

The study of chondrites has proceeded along lines similar to studies of the iron-nickel meteorites. Both objects, after all, are basically crystalline in structure. As such they are the subject of the geochemist. The geochemist studies the growth of crystals, the chemical reactions which lead to the particular formations, and conditions of temperature and pressure which must prevail before a particular crystal or chemical can form. With this background, he can learn a great deal about the history of the different kinds of meteorites. Studies of chondrites have, it appears, led to some enigmas.

The chondrules are held in a matrix composed to a large extent of two minerals found in terrestrial rocks: olivine and pyroxene (Figure 7-9*a*). The matrix also contains other minerals, including glass; the matrix is often rather friable and as such crumbles easily. But the principal distinction between the matrix and the chondrules is their iron content.

This distinction is particularly evident in a class known as *carbonaceous chondrites*, because of the black, loosely held matrix (Figure 7-9*b*). The matrix contains almost no free iron; nearly all of the iron has been combined chemically to form oxides and other compounds. The iron in the chondrules, however, is almost entirely free; less than 1% occurs as iron oxide.

Furthermore, it is found that the chemical composition of the chondrules is almost identical with that of the sun. This cannot be said of any rock on the Earth. During the Earth's history, rocks have been heated and melted so that chemicals have become separated. This has not happened to the chondrules in the carbonaceous chondrites.

Radioactive studies lead to the conclusion that these bodies were formed in the very earliest stages of the solar system — in the original gases from which the planets were made. Their content of free iron further indicates that they were formed during the passage of a shock wave set up by the forming and partially unstable sun, a shock wave which was transmitted through the primordial gases. It is presumed that during the shock wave some of the material would condense into liquid droplets and harden into the chondrules after the shock wave passed. Material not condensing into liquid droplets would later form the dust from which the matrix of the chondrite is formed. The chemical action of the primordial gas following the passage of the shock wave could have oxidized the iron in the fine dust particles, but not so in the larger — and thus protected — chondrules.

Some of the chondrules have suffered recrystallization. In these, as expected, some of the iron has become oxidized. It is supposed that chondrules which have been recrystallized became part of a larger body which was heated. This heating would account for the recrystallization.

It seems to be generally accepted that the round little chondrules embedded in the matrix of the chondrites were once molten. Although much study remains to be done on these very fascinating objects, their once molten condition, their chemical similarity to the sun, their free iron, make them unique in meteoritic studies. It is anticipated that they hold still more clues to the history of the solar system.

Whether chondrules are solidified droplets from the primordial gases of our solar system, and whether there is any connection between the minor planets and the meteorites is not known with certainty. But the entire story of the solar system cannot be learned in one day or one step. Speculation gives direction to observational research. If a speculation proves incorrect, progress has been made even though a new direction must be taken. Man has never seemed to be at a loss for new ideas and new directions, at least not in the past 400 years.

7.4 METEORITE CRATERS

When a large meteorite falls, it leaves a crater on the surface of the Earth. Some of these craters are large, although none is as large as many of those on the moon. The largest that is definitely known to be of meteoritic origin is the Barringer crater near Winslow, Arizona (Figure 7-10), which is about 4150 ft in diameter. The Chubb crater in Quebec, however, is 2½ miles in diameter and unlike any of the lakes in the surrounding area. These lakes were formed by

Figure 7-10
The Barringer Meteor Crater, Winslow, Arizona. (Great Meteor Crater photograph.)

glaciers and consequently are long and narrow. The lake in the crater, however, is quite circular. The crater is assumed to be meteoritic in origin, although, unlike the Barringer crater, no meteoritic fragments have been found around it. This may be because the area around the Chubb crater is covered with boulders which make it quite difficult to find fragments, whereas the area around the Barringer crater is flat and sandy.

The biggest fall in recorded history occurred in Siberia on June 30, 1908. The explosion resulting from this fall was big enough to make any man-made explosion seem small in comparison. At least ten craters were formed; presumably the meteor broke into smaller fragments when it hit the atmosphere. The trees outside the region of craters, to a distance of nearly 20 miles, were all blown down, with their tops pointing away from the explosion. Windows were broken for a distance of 50 miles, and an engineer about 450 miles away stopped his train in fright. The pressure wave traveled through the atmosphere as far as England, where it was recorded on barographs (instruments that measure atmospheric pressure). The shock waves traveled through the surface of the Earth and were recorded on seismographs (instruments that measure earthquakes) in Europe. Had the meteorite struck only 4 hours, 47 minutes later, the Earth would have turned enough so that it would have struck the city now called Leningrad!

Recent studies have shown that this explosion was too powerful to be accounted for by craters of the size found in the region of impact. It has therefore been speculated that it was caused by a comet and not by a meteorite. The gases in the comet, when heated in the Earth's atmosphere, could account for the magnitude of the explosion, for a gas such as methane would react chemically with the atmosphere.

That there are not more meteoritic craters on the Earth has caused some concern. Even with the geological changes, it seems that if the moon is so scarred and if Mars has been pelted by meteorites, then the Earth ought to have more such craters. Increased searching has revealed a number of *astroblemes*, fossil remnants of meteoritic craters. These scars of ancient collisions between the Earth and meteorites have been found on each of the great continents of the world. A number of them have been found in the Hudson Bay area of Canada in some of the oldest rock formations in the world.

7.5 COMETS AND METEOR SHOWERS

Although only seldom seen, comets have in the past made a tremendous impression on mankind. Not knowing what comets were, and being plagued with superstitions anyway, people generally blamed all sorts of disasters on comets, whether it be war, massacres, the black plague, assassinations, floods, or earthquakes. Comets have been looked upon as omens of terror, blood, evil.

Powerful men (who happened to live when a comet was seen in the sky) have been associated with them. The comet of 43 B.C. was supposed to have been the soul of Julius Caesar transported to heaven. This belief was encouraged by Augustus for political reasons. A comet was associated with the coming of William the Conqueror when it appeared in April of 1066. The comet of 1811-1812, of course, was called "Napoleon's comet."

It is true that comets are a spectacular sight when seen in the sky (Figure 7-11); little wonder that people who did not know what they were associated

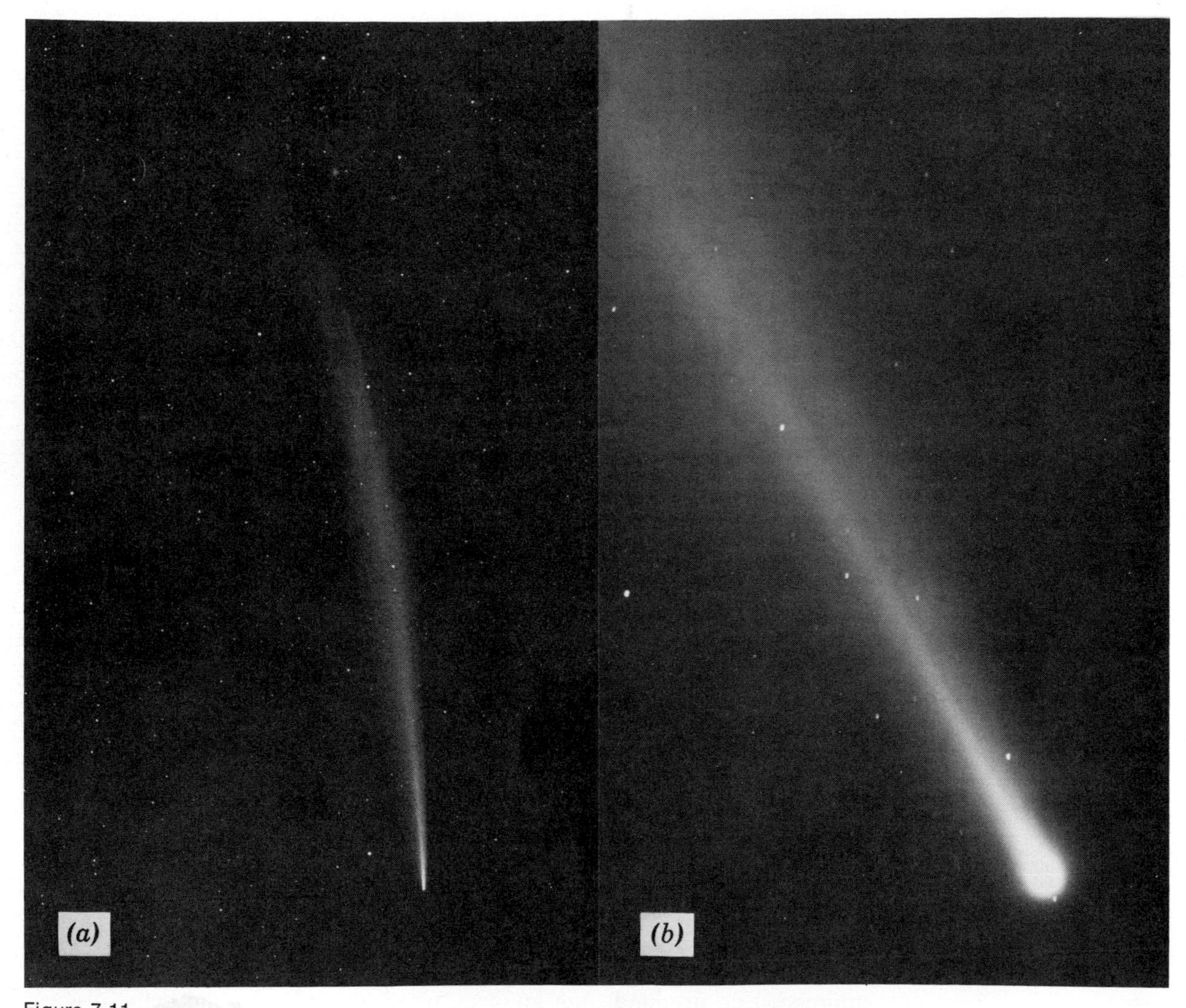

Figure 7-11

(a) Comet Ikeya-Seki as photographed with 35mm camera. (Courtesy Elizabeth Roemer.) (b) The head of comet Ikeya-Seki taken with the 40-in. reflector at the U. S. Naval Observatory, Flagstaff, Arizona at nearly the same time that (a) was taken. (U.S. Navy photograph.)

them with the evils that plagued man! When we learn more about comets, however, we find that they are of little danger to us.

A. THE ORBITS OF COMETS Edmund Halley, using the law of gravitation and the laws of motion devised by his good friend, Isaac Newton, was able to predict the return in 1758 of a comet he had observed in 1682. Halley based his prediction on the orbit he had derived and on the fact that comets had appeared in the past at times that agreed with that orbit. He cited the comets of 1607, 1531, 1456, and 1305 as evidence that these were but passages of a single comet with a period of 75 or 76 years. On Christmas Day of 1758 Halley's comet did indeed appear in the skies, although he himself did not live to see his prediction confirmed.

A *periodic comet* revolves about the sun in an elliptical orbit which is usually quite eccentric. Halley's comet, for instance, comes within about 0.6 A. U. of the sun at perihelion and goes out as far as 35 A. U. at aphelion. Since all comets move in accordance with Kepler's three laws, they must travel faster at perihelion than at aphelion.* We are able to see comets with very eccentric orbits for only a small part of their journey because when they travel far from the sun they become too faint to be seen with our telescopes. Since a comet in a very eccentric orbit is observable only during a small part of its period, such an orbit is difficult to compute.

Although the planes of the orbits of all the planets are inclined by only small angles from one another, and their motions are all direct, the same is not true of comets, Halley's comet, for example, revolves in retrograde motion, and a number of comets have orbits that are nearly perpendicular to the plane of the solar system. In fact, the planes of very elongated elliptical orbits are, more often than not, tilted by a large angle from the plane of the solar system. Since all other members of the solar system do have orbits that lie in nearly the same plane, it has been suggested that comets may be interstellar in origin and may be captured by the sun as it moves through space. This is, however, a difficult hypothesis to either prove or disprove.

B. THE ANATOMY OF A COMET The aspects of comets are nearly as varied as their orbits. Among individual comets there are strong similarities and yet many variations. When viewed through a telescope, a comet will be seen to have a central *nucleus* which is quite bright and small. Surrounding the nucleus is a fainter and more nebulous structure called the *coma;* and this together with the nucleus composes the *head* of the comet (Figure 7-11*b*).

Extending from the head is the *tail* of the comet, although unlike most tails in nature it is sometimes in front of the comet. The tail (and not all comets have tails) points away from the sun, so as the comet recedes from the sun the tail will

*As a rule of thumb, the velocity of a comet is comparable to the velocity of a planet at the same distance from the sun. Actually, since a comet's orbit is generally more elliptical than that of a planet, its velocity will be somewhat greater near perihelion and less near aphelion than that of a planet at the same distance.

lead the way. This seems contrary to what we might expect, but we must remember that in the space between the planets there is little that would cause the tail to trail behind as the resistance of the air in our atmosphere causes the smoke from a diesel truck to trail.

As the comet approaches the sun the coma and tail grow larger. (Their growth is shown schematically in Figure 7-12 and pictorially in Figure 7-13.) The obvious cause of this effect is the sun.

The streams of particles (electrons, protons, and nuclei of other light atoms) comprising the solar wind are the cause of both the coma and the tail of a comet. Gases are given off by the warming nucleus of the comet, are caused to emit their own light, and are driven away from the sun all by the solar wind. But studies of Comet Ikeya-Seki in 1965 indicate that only a thin outer shell of the comet is heated, even during a close approach to the sun. At perihelion, this comet was only 290,000 miles from the surface of the sun — and it survived the intense radiation. There have been a number of "sungrazing" comets as they are called.

If we analyze the light from a comet with a spectrograph we are able to learn much more about it. The nucleus gives a continuous spectrum which is apparently reflected sunlight. The coma and tail yield a bright-line spectrum composed of both the bright lines caused by atoms and bright *bands* from molecules. (Molecules, because they are several atoms together, emit a great many lines in groups which are called bands.) These molecules absorb some of the energy of the sunlight and reradiate it. Many of these molecules detected in comets, however, are not the ones we are likely to find on the Earth or any other planet. Instead, we find such strange molecules as C_2 (carbon), CN (cyanogen), NH (nitrogen hydride), CH (methylidyne), and OH (hydroxyl). In addition, there are a few common molecules, such as CO (carbon monoxide) and N_2 (nitrogen). Some of these have become ionized, that is, they have lost one electron (for example, CH^+, CO^+, OH^+, and N_2^+).

The strange molecules are incomplete and are called free radicals. If any one of them were placed in the Earth's atmosphere it would quickly grab one or more atoms and by doing so become a stable molecule. Under very low

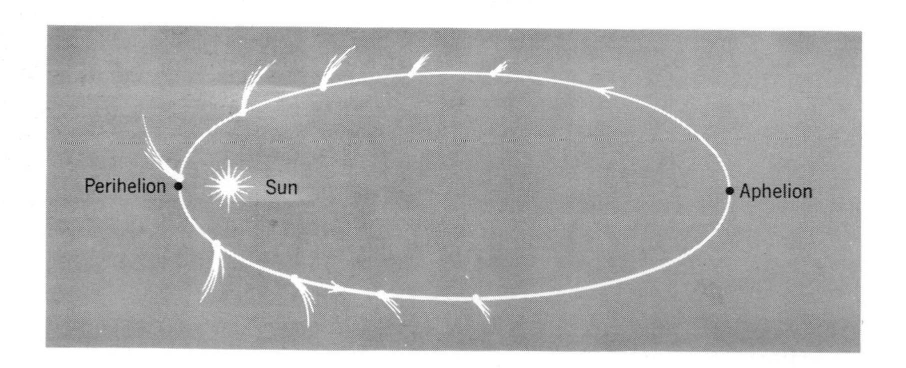

Figure 7-12
The comet's tail always points directly away from the sun.

pressure, however, atoms are too far apart for an incomplete molecule to capture free atoms. The fact that these strange molecules are found in the gases composing the coma and tail of a comet, then, leads to the conclusion that these gases are in what would be considered a vacuum by terrestrial standards.

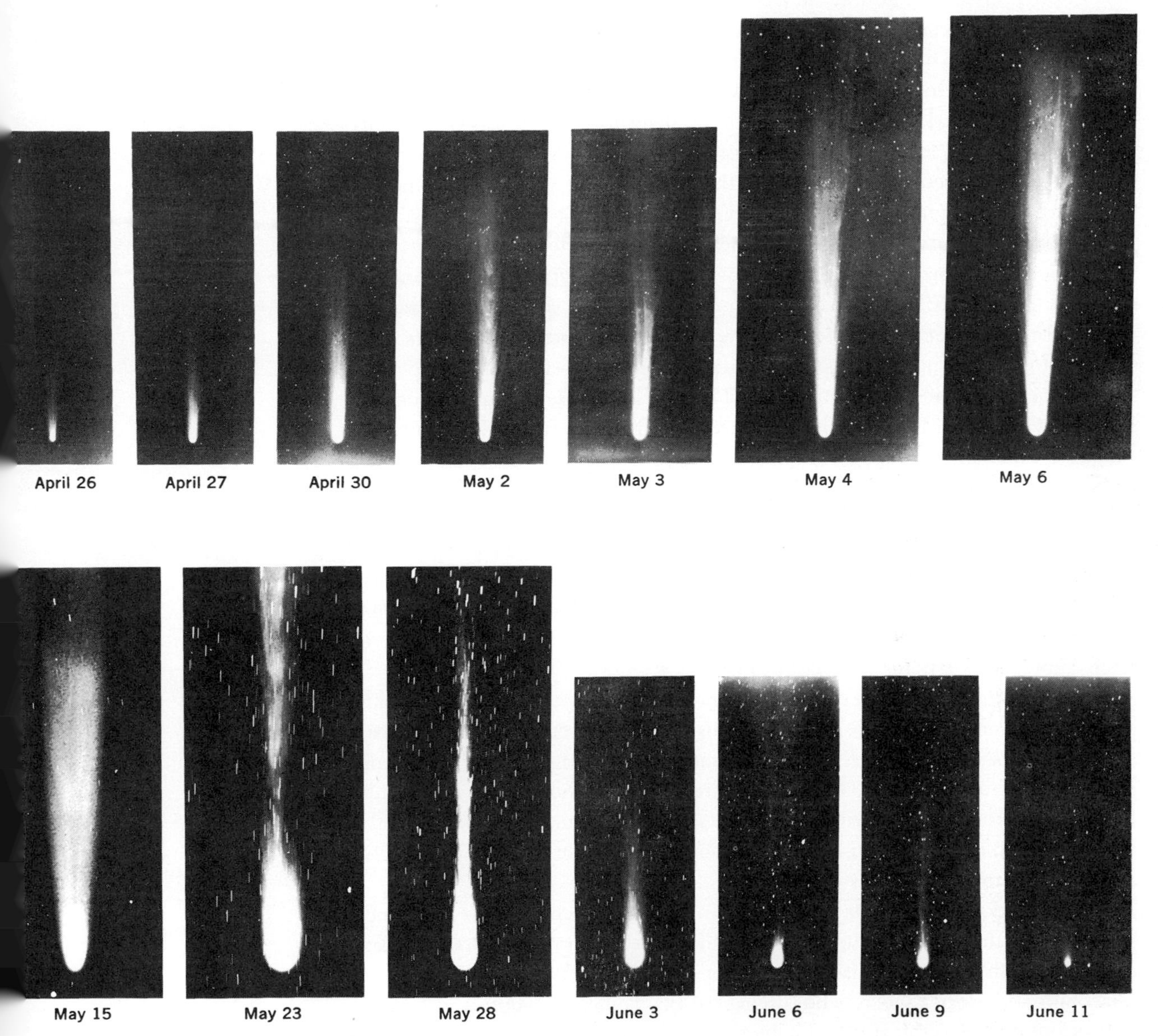

Figure 7-13
As Halley's comet approached the sun in 1910, its coma and tail grew in size. As the comet receded from the sun both the tail and the coma grew smaller. (Photographs from the Hale Observatories.)

C. THE NUCLEUS OF A COMET It is of interest to consider the nucleus of a comet, since both coma and tail originate in this small and less conspicuous part. From the nucleus we obtain a spectrum that is reflected sunlight, and we therefore assume that it is denser than the gaseous portions. Beyond this we are left with little to go on. One clue to the nature of the nucleus was obtained by Dr. Elizabeth Roemer, now at the University of Arizona. It was noticed in 1957 that Comet Wirtanen, 1956c (the *c* indicates that it was the third comet discovered in 1956) had split into two parts (Figure 7-14). By continued observation over 2 years, Dr. Roemer was able to determine the velocity with which the two nuclei separated. Making the assumption that this velocity was the velocity of escape of the nuclei, she was able to estimate that the total mass of the comet was about 10^{14} kilograms. On the Earth this would weigh about 10^{15} pounds. A rock with that weight on Earth would have a diameter of about 2 miles.

We have yet another clue to the structure of the nucleus of a comet. Not infrequently the Earth passes through the orbit of a comet. This occurrence is generally marked by a display of many meteors, called a *meteor shower*. It thus appears that a comet scatters debris in its orbit; comets litter their orbits.

A photograph of meteors that constitute a meteor shower makes it appear as though they originated in the same part of the sky like spokes radiating from the hub of a wheel. This is the result of perspective. The point from which the trails radiate is called the *radiant*. The radiant can exist only if these meteors had been traveling in the same direction in space before they collided with the Earth's atmosphere.

To find the actual speed and direction of travel of meteors in space, we observe their speed and direction of fall; from these we can subtract the velocity of the Earth and the effect of the Earth's gravitational field. It has been found that shower meteors travel around the sun in elongated ellipses with periods of from 1.6 to 415 years for different groups.

The Earth travels through one such stream about August 12 of each year, and since the radiant of this shower is in the constellation Perseus, these meteors are called the Perseids (see Figure 7-5). Their rate of fall averages about 50 meteors per hour. On October 9, 1933, there appeared a meteor shower (the Draconids) in which approximately 350 meteors were seen each minute at the peak of the shower. Again on October 9, 1946, the same meteor swarm was intercepted and another brilliant shower was observed.

The Perseid shower results from the debris left by the comet 1862 III, a long period comet that should return near the end of this century. The Draconids are associated with the comet Giacobini-Zinner, named after the two men, Signore Giacobini and Herr Zinner, who discovered this comet in 1900.

It is significant to the study of comets that during a meteor shower not one chunk of material has ever been observed to survive the burning in the atmosphere. We can only conclude, therefore, that the debris left by the comet in its orbit is composed of very small particles.

Figure 7-14
Comet Wirtanen, 1956c, showed a double nucleus in 1957 and 1958. This photograph was taken on April 27, 1958, at the U. S. Naval Observatory, Flagstaff, Arizona. (Official U.S. Navy photograph.)

So the question arises: How can the nucleus of a comet several miles in diameter leave tiny particles of debris behind in its orbit? In addition, how can that nucleus form a coma and a tail as it nears the sun?

These questions, and others as well, can be answered by the model of a comet proposed by Fred Whipple of Harvard University. Dr. Whipple suggested that the nucleus of a comet is a ball of frozen gases such as water, ammonia, methane, and others. If those frozen gases have a liberal sprinkling of grains of sand throughout, we can better understand the comet; it may be something like a huge dirty snowball. From the estimated mass of a comet, the diameter of such a "dirty snowball" would be 3 to 4 miles.

As the comet nears the sun, some of the frozen gases vaporize by the warmth of the sun. This process forms the coma; the solar wind causes some of the gas to blow away from the sun forming the tail. As the frozen gases vaporize, they can no longer hold the tiny grains of material, which are then left behind in the comet's orbit. This debris causes the meteor shower.

Since the gases that leave the comet by way of the tail never return (Figure 7-15), and since their supply is not inexhaustible, the comet becomes less spectacular after repeated trips near the sun. The bright "naked eye" comets that appear unexpectedly from time to time have apparently made only a few trips around the sun, whereas the short-period comets that spend more of their time close to the sun have very short tails or none at all.

This model of a comet seems to fit the observations better than any other. The one aspect it does not answer for sure is where the comets come from. Were

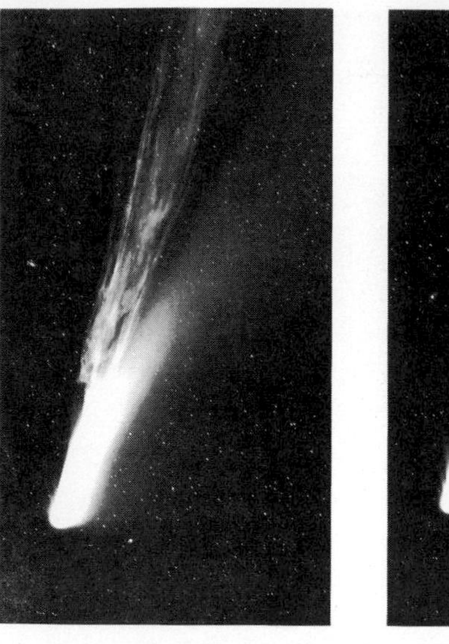

August 22

August 24

August 26

August 27

1957

Figure 7-15

Four views of Comet Mrkos taken with the 48-in. Schmidt telescope on Mount Palomar. (Photographs from the Hale Observatories.)

they formed as part of the solar system or do they permeate interstellar spaces and exhibit themselves only when the sun comes close enough to capture one?

The constituents of the solar system are many and varied: the planets orbiting the central sun, some attended by satellites; the minor planets in the gap between Mars and Jupiter, some of which we find as meteorites; the chondrules formed so early in the solar system's life; the comets; and there is even a modest sprinkling of gas and dust throughout it all.

It is the central sun, however, that dominates the system and has from the very beginning. It is the sun, a modest star, that not only permits and influences life here on Earth, but whose study takes us eventually beyond the confines of the solar system.

BASIC VOCABULARY FOR SUBSEQUENT READING

Achondrite
Asteroid
Astrobleme
Carbonaceous chondrite
Chondrite
Chondrule
Comet
Fireball
Meteor
Meteor shower
Meteorite
Minor planet

QUESTIONS AND PROBLEMS

1. Describe how the height of a meteor can be determined.

2. Explain one observation which indicates that meteors are not much larger than a grain of sand.

3. What evidence indicates that meteorites may be small minor planets swept up by the Earth?

4. What evidence indicates that many meteorites may have belonged to larger bodies?

5. Describe the appearance of a comet as it approaches the sun from beyond the orbit of Neptune, passes close to the sun, and returns to aphelion again.

6. What information leads astronomers to suspect that meteor showers are related to comets?

FOR FURTHER READING

Glasstone, S., *Sourcebook on the Space Sciences,* Chapter 7, Van Nostrand Co., Princeton, N. J., 1965.

Mason, B., *Meteorites,* John Wiley and Sons, New York, 1962.

Page, T., and L. W. Page, ed., *Neighbors of the Earth,* The Macmillan Co., New York, 1965.

Wood, J. A., *Meteorites and the Origin of Planets,* Chapters 1, 2, 3, and 4, McGraw-Hill Book Co., paperback, New York, 1968.

Anders, E., "Diamonds in Meteorites," *Scientific American,* p. 26, October 1965.

Barringer, D. M., "The Meteorite Search," *Natural History,* p. 56, May 1964.

Chapman, C. R., and David Morrison, "The Minor Planets: Sizes and Mineralogy," *Sky and Telescope,* p. 92, February 1974.

Fireman, E. L., "Freshly Fallen Meteorites from Portugal and Mexico," *Sky and Telescope,* p. 272, May 1969.

Florensky, K. P., "Did a Comet Collide with the Earth in 1908?" *Sky and Telescope,* p. 268, November 1963.

Futrell, D., "Some Notes on Tektites," *Sky and Telescope,* p. 272, May 1967.

Hartmann, W. K., "Craters — A Tale of Three Planets," *Natural History,* p. 58, May 1968.

Jacchia, L. G., "The Brightness of Comets," *Sky and Telescope,* p. 216, April 1974.

Jacchia, L. G., "A Meteorite That Missed the Earth," *Sky and Telescope,* p. 4, July 1974.

Lawless, J. G., et al, "Organic Matter in Meteorites," *Scientific American,* p. 38, June 1972.

Matthews, M. S., "The Asteroid Conference in Tucson," *Sky and Telescope,* p. 22, July 1971.

McCrosky, R. E., "The Lost City Meteorite Fall," *Sky and Telescope,* p. 154, March 1970.

O'Keefe, J. A., "Tektites and Impact Fragments from the Moon," *Scientific American,* p. 50, February 1964.

Whipple, F. L., "The Nature of Comets," *Scientific American,* p. 48, February 1974.

Wood, J. A., "Chondrites and Chondrules," *Scientific American,* p. 65, October 1963.

"Bosomtwe: An African Meteorite Crater?" *Sky and Telescope,* p. 15, July 1965.

"Formaldehyde in Allende," *Sky and Telescope,* p. 352, June 1972.

"The Great Comet of 1970," *Sky and Telescope,* p. 351, June 1970.

"Origin of Some Iron Meteorites," *Sky and Telescope,* p. 282, May 1965.

"Protostuff," *Scientific American,* p. 46, August 1971.

"Shower Meteorites Collected by Balloon," *Sky and Telescope,* p. 276, November 1965.

"Two More Ancient Canadian Meteorite Craters," *Sky and Telescope,* p. 198, October 1963.

"Woher der Gegenschein?" *Scientific American,* p. 47, August 1971.

CHAPTER CONTENTS

LEARNING OBJECTIVES

BE ABLE TO:

1. GIVE IDENTIFYING FEATURES OF THE SUN'S: CORE, PHOTOSPHERE, CHROMOSPHERE, AND CORONA.
2. DESCRIBE THE THREE TYPES OF SOLAR RADIATION: RADIANT ENERGY, SOLAR WIND, COSMIC RAYS.
3. DEFINE SPECTRAL LINE INTENSITY.
4. EXPLAIN HOW THE TEMPERATURE OF THE SUN IS DETERMINED FROM OBSERVATIONS OF SPECTRAL LINE INTENSITIES.
5. GIVE A NONMATHEMATICAL STATEMENT OF BOTH THE STEFAN-BOLTZMANN LAW AND WIEN'S LAW.
6. EXPLAIN HOW THE TEMPERATURE OF THE SUN IS DETERMINED USING THE STEFAN-BOLTZMANN LAW.
7. EXPLAIN HOW THE SUN'S TEMPERATURE IS DETERMINED USING WIEN'S LAW.
8. EXPLAIN POSSIBLE USES OF SOLAR RADIANT ENERGY.
9. DESCRIBE THE SUN'S SOURCE OF ENERGY.
10. DESCRIBE THE PHOTOSPHERE AND CAUSES OF ITS ACTIVITY.
11. DESCRIBE THE SUNSPOTS AND THEIR CYCLIC NATURE.
12. DESCRIBE SOLAR FLARES AND THEIR EFFECT ON THE EARTH.
13. DESCRIBE THE CHROMOSPHERE AND ITS PROMINENCES.
14. DESCRIBE THE CORONA AND ITS RELATION TO THE SOLAR WIND.

Chapter Eight

THE SUN

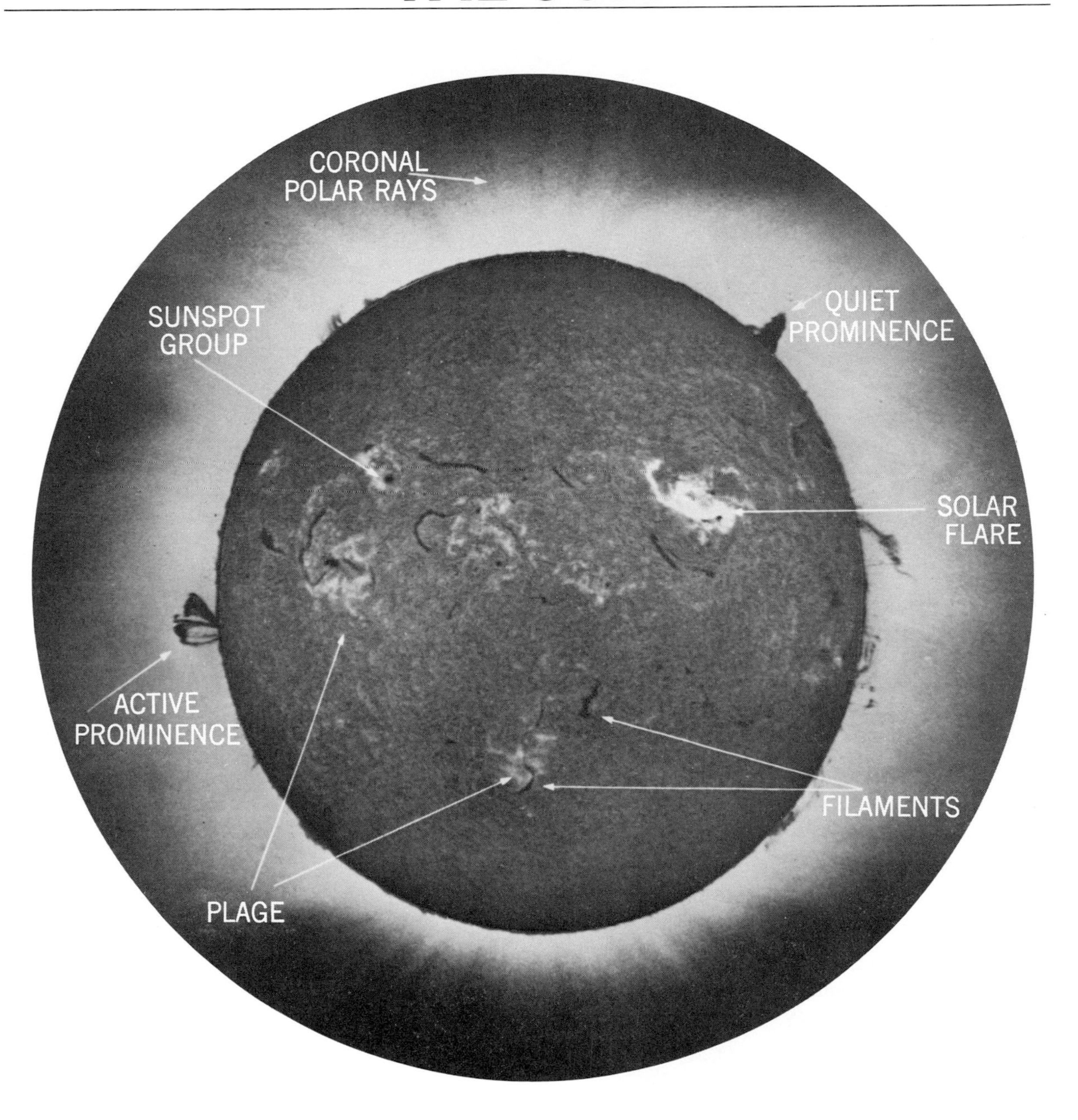

At the center of the solar system resides its largest and most important member, a modest star which we call the sun. The fact that it is relatively close makes possible fairly detailed observations that can tell us something about the other more distant stars. But of more immediate importance to us is the sun's influence on the Earth. Not only is it our source of heat and light, but it has other effects on the Earth that, although less noticeable, are interesting none the less.

8.1 GENERAL FEATURES

The sun is a huge ball of gas* 864,000 miles in diameter, with a mass 333,000 times that of the Earth. But since its volume is 1,300,000 times that of the Earth, its density, 1.41 times that of water, is less than the Earth's density of 5.5, about the same as that of the Jovian planets. The force of gravity operating on its surface is 28 times that on the surface of the Earth. Therefore, if a man weighing 180 lb on the Earth could stand on its surface, he would weigh 5040 lb. Such a large force of gravity also means that the escape velocity on the sun is very high, about 386 miles per second as compared with 7 miles per second for the Earth.

The portion of the sun that we see in the sky, its visible surface, is called the *photosphere* (sphere of light). This surface is gaseous rather than solid like the surface of a planet. The photosphere is the visible part of the sun because it is the outermost region to emit white light, that is, a continuous spectrum. The photosphere is, therefore, opaque and hides the solar interior from our view. There is no abrupt change as there is between the Earth's surface and its atmosphere, but there is a more gradual change in density and temperature.

Above the photosphere there are two clearly differentiated regions of gas. Immediately above the photosphere and extending for an estimated 5000 to 10,000 miles is the *chromosphere*. The chromosphere can be seen during a total solar eclipse just seconds before and after totality. During these fleeting moments, the moon covers the more intense photosphere and permits observation of the reddish chromosphere. In fact its reddish color is the origin of its name which means the "colored sphere." The gases of the chromosphere cause the absorption lines which show up in the solar spectrum by selectively absorbing

Chapter Opening Photo

Composite of two photographs of the sun, one of the disk in hydrogen light and the other of the limb with the disk masked out by an artificial eclipse. (Lockheed Aircraft Corporation.)

*This gas is quite unlike any found on Earth because its temperature, pressure, and density are so much higher than gases in our atmosphere. Yet the gases in the sun obey the gas laws met with on Earth.

Figure 8-1
The corona seen during the total solar eclipse on October 12, 1958. (Courtesy of F. L. de Romana, Peru.)

light emitted by the photosphere. Accordingly, the chromosphere must be at a lower pressure and temperature than the photosphere which emits the background continuous spectrum.

Above the chromosphere the *corona* extends for millions and millions of miles as a very tenuous gas (Figure 8-1). There is good evidence to conclude that many of the planets, including the Earth, actually revolve within the limits of the corona. The corona is best seen during a total solar eclipse, although both the chromosphere and the inner corona can be seen by optical means which simulate a solar eclipse.

8.2 SOLAR RADIATION

Observations of the sun take a much different form than observations of the planets. The main reasons for the differences are that the sun is so much hotter and brighter than the planets. Space probes have been landed on the moon, Venus, and Mars, and close approaches are possible with Jupiter or any of the planets, but none of these would be possible with the sun because of its very high surface temperature. Thus we must rely on the brightness of the sun, that is, we must study the energy it sends out into space, for this is the only way the sun communicates with us. What is the form of this energy? What can we learn about the sun by studying that small fraction of solar energy which reaches us here on Earth? Is this study of any value to us?

The fact that the sun is so big and so hot means that even at the great distance of 93,000,000 miles the relatively small Earth still receives a great deal of energy. It is certainly within the realm of possibility that in the future we shall be converting much more of that energy for our everyday use. Already solar

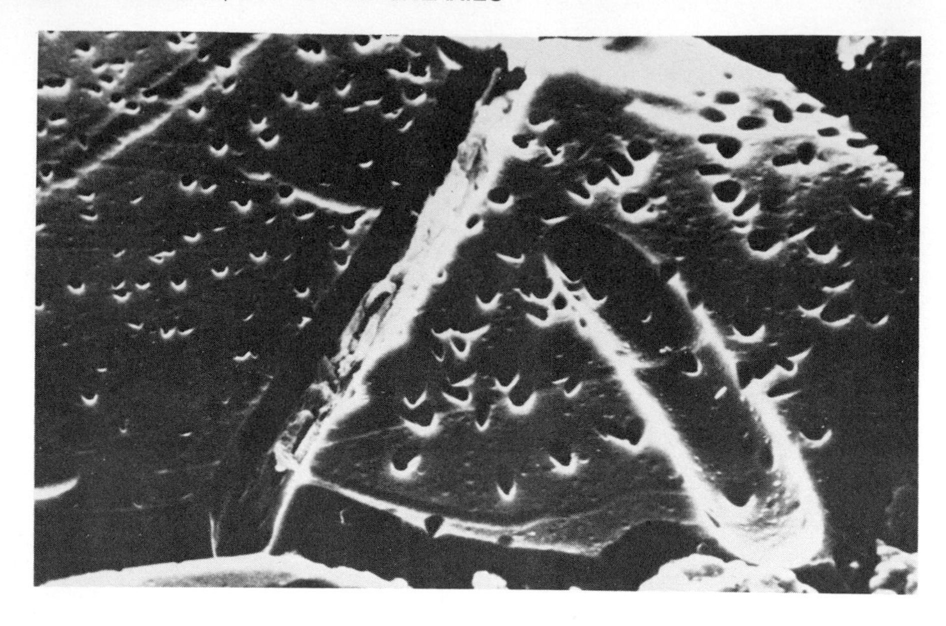

Figure 8-2
Scanning electron microscope photograph of cosmic ray tracks in a glass spherule brought back from the moon by the Apollo 11 crew. (NASA photograph.)

cells are being used on a limited basis as are solar heaters and solar furnaces. The more we know about the sun and the energy it radiates, the better able we shall be to convert this inexhaustible store of energy for our own use.

The energy that the sun radiates out into space can be grouped into three kinds: cosmic rays, the solar wind, and electromagnetic radiation. We shall discuss each of these in turn, saving the most important for the last.

A. COSMIC RAYS The term "ray" implies a beam of light, but don't be misled, for cosmic rays are nuclei of atoms traveling at tremendous speeds and with energies vastly greater than any of the particles the physicist can throw about in his big particle accelerators. About 90% of the cosmic ray particles are nuclei of hydrogen atoms (i.e., protons); helium nuclei account for some 9%; elements such as carbon, oxygen, nitrogen, neon, iron, and a few others make up the bulk of the remaining 1%. Evidence of long-time cosmic-ray bombardment of glass particles on the moon is shown in Figure 8-2.

Most of the cosmic rays originate, as the name implies, out in our Milky Way galaxy someplace, perhaps in exploding stars. But the sun produces a considerable amount of low energy cosmic rays. By studying these low energy solar cosmic rays it is hoped that we can learn more precisely what the sun is made of, that is, how much of the sun is hydrogen, how much is helium, carbon, etc. It is also hoped that solar cosmic rays will reveal information about the solar surface and interior, for they originate in hot spots on the surface and in eruptions called solar flares. The source of energy that causes these hot spots and solar flares must certainly be in the sun's interior.

B. THE SOLAR WIND In addition to the cosmic ray particles, the sun sends low energy protons and electrons out into space. The protons and electrons ejected by the sun and forming the solar wind are not bound to one another, they fly through space as individual particles. The electrons in the solar wind generally travel faster than the protons.

Studies of the solar wind are carried on mainly by space probes and give us information of the sun's upper atmosphere, which seems to be the origin of this stream of particles.

The solar wind must cause some of the erosion on the surfaces of the moon and Mercury, bodies that have neither a magnetic field nor an atmosphere. The Earth's magnetic field deflects these particles, trapping some into the Van Allen radiation belts. The trapped particles ultimately lose their energy in the north and south polar regions, causing the auroral displays as the particles enter the upper atmosphere.

It is the solar wind that forces the tails of comets to stream away from the sun. The solar wind, therefore, has a considerable influence on the inner part of the solar system.

C. RADIANT ENERGY (ELECTROMAGNETIC RADIATION) Our main source of information about the sun is neither cosmic rays nor the solar wind, however. It is the stream of electromagnetic radiation: gamma rays, X rays, ultraviolet, light, infrared, microwaves, and radio waves. Each of these is observed with appropriate instruments.

Observations of solar gamma rays, X rays, and most of the ultraviolet must be made from space probes above the Earth's atmosphere, because our atmosphere absorbs these energetic forms of radiation. It is a good thing, too, because gamma rays and X rays have energy enough to disrupt molecules in and, thus,the body chemistry of any living organism.

The sun is studied by radio telescopes and attempts are made to correlate the changes in the solar radio signal with changes in other forms of energy. For example, solar flares not only emit cosmic rays but also bursts of radio energy. The bursts of radio energy indicate that charged particles such as electrons and protons are being ejected explosively by the flare.

The solar spectrum is a fountain of information that has been studied for years and yet shows no signs of running dry. The various spectral lines give us information of the sun's chemical composition, for each element (such as hydrogen, helium, lithium, carbon, iron, etc.) has its own distinguishing sequence of spectral lines.

1. Spectral Line Intensities Not only is the spectrum of one element different from that of another by having lines of different wavelength (compare the spectra of iron and hydrogen in Figure 2-11), but the lines have different intensities.

By *intensity* we mean the amount of energy involved in any particular spectral line. For a bright spectral line, it is the amount of light energy in that line compared with other spectral lines. For a dark or absorption spectral line, it is

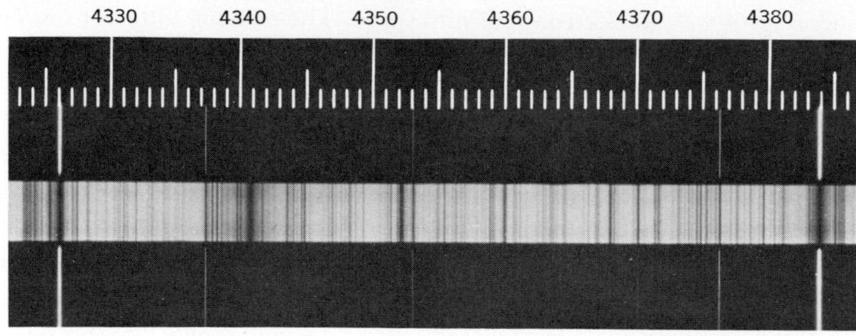

Figure 8-3
A portion of the solar spectrum with comparison lines on either side and a scale of Ångstrom units above. (Photograph from the Hale Observatories.)

the amount of light energy that the gas absorbs from the continuous spectrum. Recall that a dark-line spectrum is formed by white light passing through a gas that is cooler than the source of white light. The gas absorbs selectively from the white light forming the absorption lines in the otherwise pure continuous spectrum. If that gas absorbs more light at one particular wavelength than another, then that spectral line is more intense than the other. For example, the line 4326 Å in Figure 8-3 is more intense (brighter in the bright-line comparison spectrum and darker in the dark-line solar spectrum) than the line 4337 Å which, in turn, is more intense than the line 4353 Å.

Since the solar spectrum is an absorption spectrum, we know that the gases causing the absorption are cooler than the photosphere, which emits the white light that forms the continuous spectrum. But more than that, the spectral lines reveal the actual temperature, since some lines will form only at high temperatures (i.e., high energy), and others will form only at lower temperatures. If the high energy lines of an element are intense and the low energy lines are weak, the astronomer knows that the gas which caused those lines is at a high temperature. The actual temperature determines the intensity of each of the spectral lines. Numerical relationships between line intensities and temperatures have been worked out.

From studies of spectral line intensities, astronomers have learned that the gases in the lower atmosphere of the sun have a temperature of about 4700° K* (about 8000° F).

2. *Spectral Line Broadening* Spectral lines also reveal the pressure of the gas

*All of the countries in the world, except the United States and a few smaller ones, use a temperature scale called Celsius (centigrade) designed so that water freezes at 0° C and boils at 100° C. Another scale, with divisions the same size as the Celsius scale, is fundamentally important to science. It is called the Absolute or Kelvin scale. Zero on the Kelvin scale (−273° C or −460° F) is absolute zero, the temperature at which the energy of atoms is at a minimum. A temperature of 6000° K is equal to about 11,000° F.

forming those lines. Spectral lines formed in a higher pressure gas are broader than lines formed in a gas at low pressure. Studies of the amount of *pressure broadening* (as it is called) yield information about that layer of gas which causes the spectral lines. The pressure of the gas in that layer of the sun is close to 0.01 times the atmospheric pressure here on the surface of the Earth.

3. The Continuous Spectrum If the spectral lines yield information such as the temperature, pressure, and chemical composition of the gases in the lower atmosphere of the sun, how can we learn anything about the photosphere? The photosphere produces the white light that is spread out into a continuous spectrum by the spectrograph. Without the atmosphere, it would be without absorption lines. How can we decipher any information from a continuous spectrum?

Think of a toaster, a light bulb, and some molten metal being poured from a crucible in an iron foundry. The filaments in a toaster glow a deep red color, they are relatively cool; the filament of a light bulb glows a bright yellowish, it is hotter than the toaster filament; the molten metal in the foundry is the hottest of the three and seems to glow nearly white. If the light of each of these substances were sent through a spectroscope you would see a continuous spectrum for each, but with some differences.

The toaster-filament spectrum would be faint with the red region predominating, the violet would be very faint indeed. The light-bulb spectrum would be brighter with the red, orange, and yellow region the brightest, but the green, blue and violet would still be obvious. The molten-metal spectrum would be the brightest of all, it is from the hottest source. All of the colors would be bright, but the red, orange, and yellow would still be the brightest part of the spectrum.

These three examples permit us to draw a few generalizations and, since these statements are generalizations, we shall expand on the examples a bit and substitute the words radiant energy for light so as to include the entire radiant energy spectrum. First, the amount of radiant energy emitted by a source depends on the temperature of that source. *The hotter the source the more radiant energy it emits.* This first generalization is a nonmathematical statement of the Stefan-Boltzmann law.

The second generalization concerns the distribution of light within the continuous spectrum. By distribution we mean the amount of light in each part of the spectrum. Is the red region brighter than the green? Or is the yellow region the brightest of all? A more precise way of saying that the yellow region is the brightest of all is simply that the wavelength of maximum intensity is in the yellow.

Our three examples would lead us to conclude that cool sources (i.e., relatively cool), such as a toaster filament, emit mostly red light with very little violet, blue, and green light. Very hot sources emit more energy in the blue and violet region making their light appear white. This can be stated as our second generalization: *as sources of ever higher temperature are considered, the wavelength of maximum intensity decreases.* This generalization is a nonmathematical statement of Wien's law. If a single source could be heated from room temperature to a very high temperature, then when it became hot enough it would begin to glow first a dull red, then a bright red followed by a reddish-orange, orange, and finally yellowish. If it was heated to a high enough temperature, it would appear white, white-hot!

We can extend the second generalization by considering even hotter sources: the stars. The summer sky has good examples: Antares in the constellation Scorpio, Arcturus in the constellation Bootes, and Vega in the constellation Lyra (see Star charts). Antares is distinctly red, its surface temperature is about 3000° K; Arcturus is yellow, its surface temperature is about 4000° K; Vega is bluish, its surface temperature is about 10,000° K. Most of the light emitted by a cool star is in the red region of the spectrum; medium temperature stars emit more yellow light than any other colors; hot stars emit most of their light in the blue-violet. To extend this beyond the visible part of the radiant energy spectrum, we could add that very cool stars emit most of their energy in the infrared and very hot stars emit most of their energy in the ultraviolet.

By studying the background continuous spectrum of the sun we find that it emits more light in the blue than any other region. According to the mathematical statement of Wien's law, this yields a photosphere temperature of 6100 Å.

It would appear from the temperature of the sun's photosphere that it ought to appear blue, but the sun still emits a great deal of light in the green, yellow, orange, and red region, so its overall color is yellowish-white.

The temperatures of the photosphere and the gases in the lower atmosphere are fortunately consistent with what we know about the formation of a dark-line spectrum: the temperature of the lower atmosphere, where the absorption lines are formed, is 4700° K, and cooler than the temperature of the photosphere, 6100° K, the source of the continuous spectrum.

4. Uses of Solar Radiant Energy Of great consequence to us is the fact that the sun's light can and is being converted into electric energy by means of solar cells that are very similar to the light meter used by photographers. There are only two drawbacks to the conversion of solar radiant energy into electric energy: (1) it is not very concentrated, and (2) it is received for only 50% of the time in any one location.

Because the sun's light is not very concentrated, you and I can walk out in it without fear of being fried. There are exceptions, of course, but most people stay out of Death Valley in the summertime. Sunlight can be concentrated with a lens or concave mirror. Temperatures reached at the focus of a large solar mirror are in the thousands of degrees Fahrenheit. Large solar furnaces have been used in research for decades.

Another way of making use of the vast amounts of solar energy, even though it is not very concentrated, is to use very large collectors, that is, large banks of solar cells. The size of the collectors needed to produce the amount of energy generated by a typical power plant, however, run so large as to be out of the question. Collectors with an area of many square miles would be needed.

Yet a single house has a roof area large enough to collect enough solar energy to nearly meet the electrical demands of that house. During the day, when solar energy is being converted to electric energy, it can be stored in batteries, which would then supply the electrical energy needed at night. The house can be connected to the local electric power company so that during periods of excessive cloudiness, electrical energy would still be available.

If half the houses in this country were equipped with their own system of solar cells and batteries to convert radiant energy to electrical energy, the demand on the electric power companies would be materially reduced. Not only would this save fossil fuels, but in many regions of the country it would in turn reduce air pollution resulting from the burning of fossil fuels. Solar energy is available, it is nonpolluting, and it is inexhaustible.

The sun's surface loses an amount of energy equal to 5.8×10^{27} calories each minute, giving it a Detroit power rating of about 5×10^{23} horsepower! Yet the surface maintains the high temperature of 6100° K. The source of all this energy lies deep inside the sun.

8.3 THE SUN'S INTERIOR

The sun's interior is of utmost importance in understanding the more obvious solar features; it is the furnace that keeps the sun "burning." The interior, however, must be more than a furnace, for the sun is too hot to burn — fire as we know it cannot exist there because the sun is too hot for the common chemical reactions to take place. The source of light and heat in the sun is the energy derived from nuclear reactions. These nuclear reactions take place deep inside the core of the sun where the temperature is hot enough to force atoms together so fast that instead of rebounding they combine to form new atomic nuclei.

It is the simplest atoms of all, the hydrogen atoms, that combine the easiest, that is, at the lowest temperatures. Four hydrogen nuclei (protons) combine to form one helium nucleus. This one helium nucleus, however, has less mass than the original four protons. The mass which appears to have been lost in the reaction is converted into energy, both radiant energy and thermal energy (i.e., heat).

The temperature required for the hydrogen-helium conversion is much higher than the surface temperature. Deep in the interior of the sun where this reaction takes place the temperature is about 15,000,000° K. The gases that form the core of the sun are trapped at this high temperature by the pressure of the overlying layers of gas. The pressure in the core is about 3×10^{12} pounds per square inch, which is about 2×10^{11} times the atmospheric pressure here on the surface of the Earth. The density of these gases is about 110 times the density of water. By comparison, the density of lead is 11 times that of water. A chunk of lead the size of this book would weigh about 20 pounds; a similar sized chunk of material from the core of the sun would weigh about 200 pounds if it were brought to the surface of the Earth.

The radiant energy that is generated deep in the interior of the sun must make its way to the surface in order to radiate out into space. The mathematical expression of Wien's law indicates that the radiant energy formed at the high-core temperature is X rays. As these make their way outward, the temperature of the surrounding material decreases and the wavelength of the radiant energy increases until it becomes ultraviolet light. This ultraviolet light eventually strikes the *convective envelope*, the outer layer of which is the photosphere (Figure 8-4). The conditions of pressure and temperature in this envelope are such that the radiant energy is absorbed rather than transmitted through it. As they are absorbed, their energy is converted to thermal energy in the gases at the bottom of the convective envelope. You feel a similar conversion of energy when you stand in the hot summer sun.

The gases in the bottom of the convective envelope are heated in much the same manner as the air above a hot road in the summer time. You may have seen the convective currents of air rising above such a road. The gases heated at the bottom of the convective envelope rise in a similar way but on a much grander scale. Because such a vast amount of energy strikes the convective envelope from below, these currents of gas become turbulent; the result is the turmoil of the photosphere.

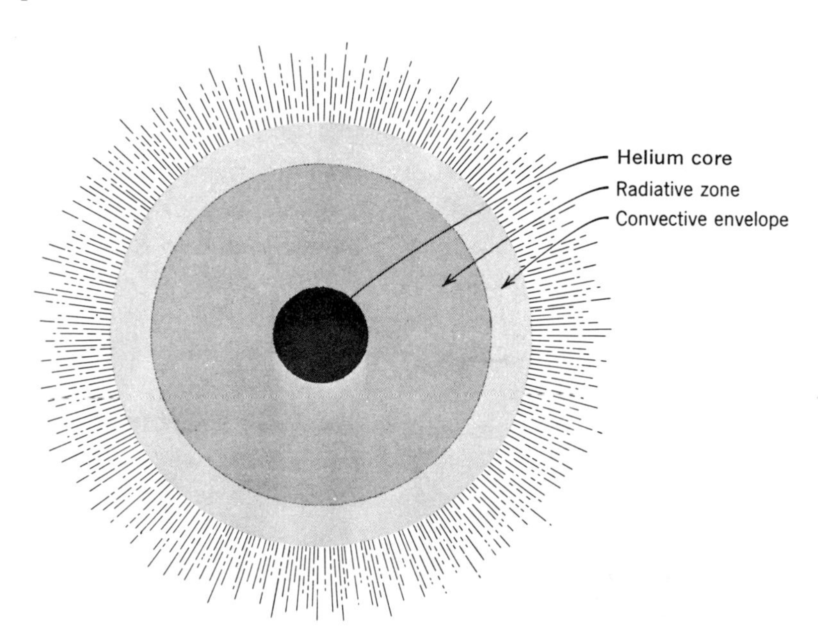

Figure 8-4
A cross section of the sun.

8.4 PHOTOSPHERIC ACTIVITY

A. GRANULES The photosphere has a mottled appearance with bright *granules* on a darker background (Figure 8-5). Since the photosphere is the top of the convective envelope and since hot gases will radiate more light than cooler gases (the Stefan-Boltzmann law), those bright granules must be the hot gases rising through the convective envelope. When they reach the surface, they radiate their energy out into space, cool, and sink back into the convective envelope to be reheated, etc. Spectroscopic observations confirm the idea that the bright granules are rising gases, for spectral lines of the granules are shifted slightly to the shorter wavelength.

B. SUNSPOTS The larger very dark blotches in Figure 8-5 are sunspots. These spots appear from time to time. Sometimes the sun has many of them (Figure 8-6); sometimes it has very few if any. Some of the sunspots are small and normally do not last very long. Others are very large by terrestrial standards and may last as long as several months. Some spots are single and travel with the rotating sun as a lone spot. It is more common, however, for the spots to appear

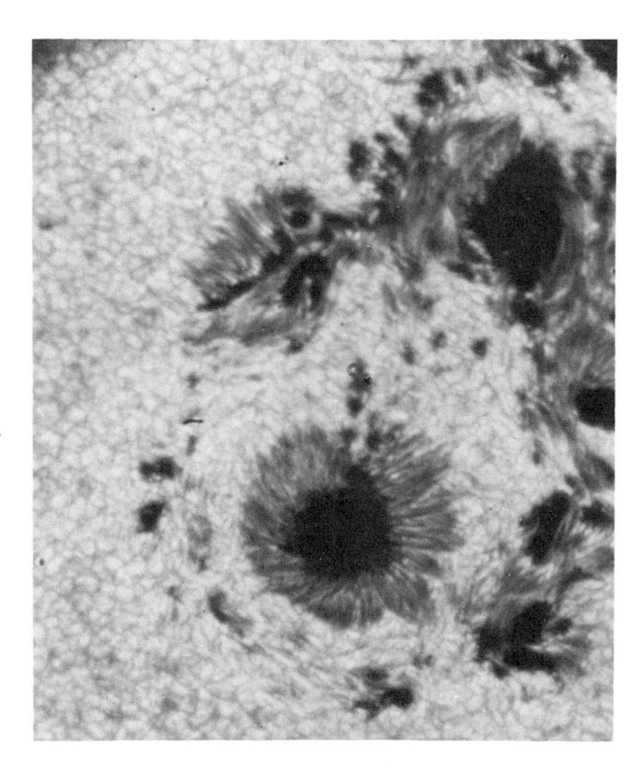

Figure 8-5
A sunspot group and solar granulation as photographed from a balloon at an altitude of 80,000 feet on August 17, 1959. This photograph was taken as part of Project Stratosphere of Princeton University, sponsored jointly by the Office of Naval Research and the National Science Foundation. (Courtesy of M. Schwarzschild.)

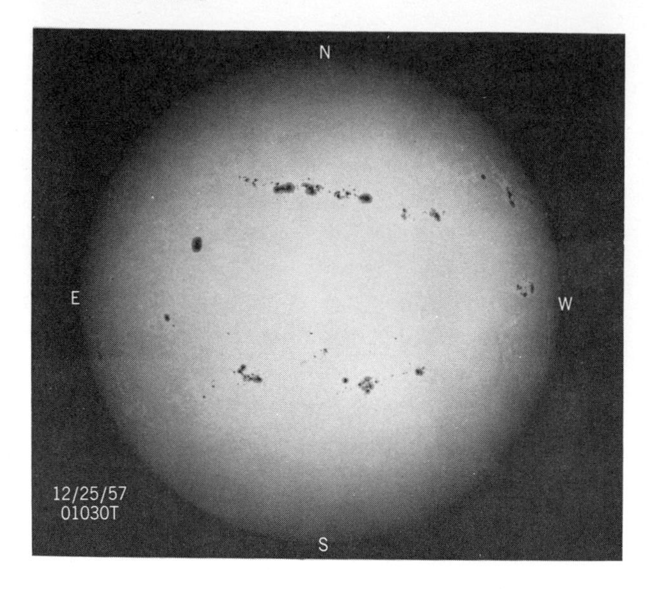

Figure 8-6
The sun with a large number of sunspots each some distance from the sun's equator. This photograph was taken on December 25, 1957. (Courtesy Hans Haber, Manila, Philippines.)

in groups (Figure 8-7). Each group is usually divided into two main parts — the preceding spot (preceding in the sense of solar rotation) and the one following.

The center of a sunspot, being dark, is called the *umbra.* It is surrounded by the *penumbra*, which is brighter than the umbra but still not as bright as the photosphere. The spot is dark when compared with the photosphere which is indicative of a lower temperature; this fact is verified by spectrograms made of sunspots. In these spectrograms we find lines that are typical of a gas at 4000° K. But a sunspot is not black in any sense of the word; if the rest of the sun were to be masked off we could see a large bright spot with the naked eye. It is dark only

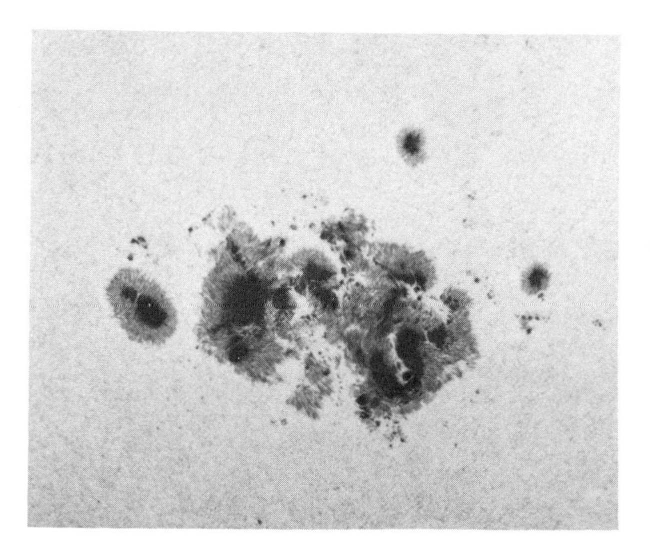

Figure 8-7
A large group of sunspots taken under exceptional seeing conditions, May 17, 1951. Notice the granulation on the photosphere. (Photograph from the Hale Observatories.)

when compared with the photosphere, because its temperature is about 2000° K less.

C. SUNSPOTS AND SOLAR ROTATION As seen from the Earth, the sunspots move across the disk of the sun in a regular manner. This motion was first observed in 1610 by Galileo, who suggested that the sun rotates with a period of about 26 days. It isn't all that simple, however. Studies more detailed than those made by Galileo indicate that the period of rotation of the sun's surface depends on latitude. For example, if a spot appears on the sun's surface near the equator, its period of rotation will be about 25.1 days; if it appears at 15° north or south of the equator, its period of rotation will be about 25.5 days; and if it appears at 30° latitude, its period of rotation will be about 26.5 days.

D. THE SUNSPOT CYCLE Since about 1750 the number of sunspots appearing each day has been recorded. By 1851 enough observations had been compiled to recognize that the number of spots varies over the years. Figure 8-8 shows that this variation, called the *sunspot cycle,* is not regular; neither all the maximum years nor all the minimum years have the same number of spots. In fact, even the period of 11 years is not the same.

The number of spots is not the only variation, however, for the latitude at which spots occur changes as the cycle progresses. A cycle begins during a minimum, at which time the spots in the new cycle appear at latitudes of about 35° north and south (see Figure 8-9). A given spot maintains approximately the same latitude, but as the cycle progresses the succeeding spots appear ever closer to the equator, until by the time minimum again occurs the spots in that cycle appear at about 5° north and south of the equator. Often two cycles overlap; a few spots belonging to the old cycle may appear near the equator,

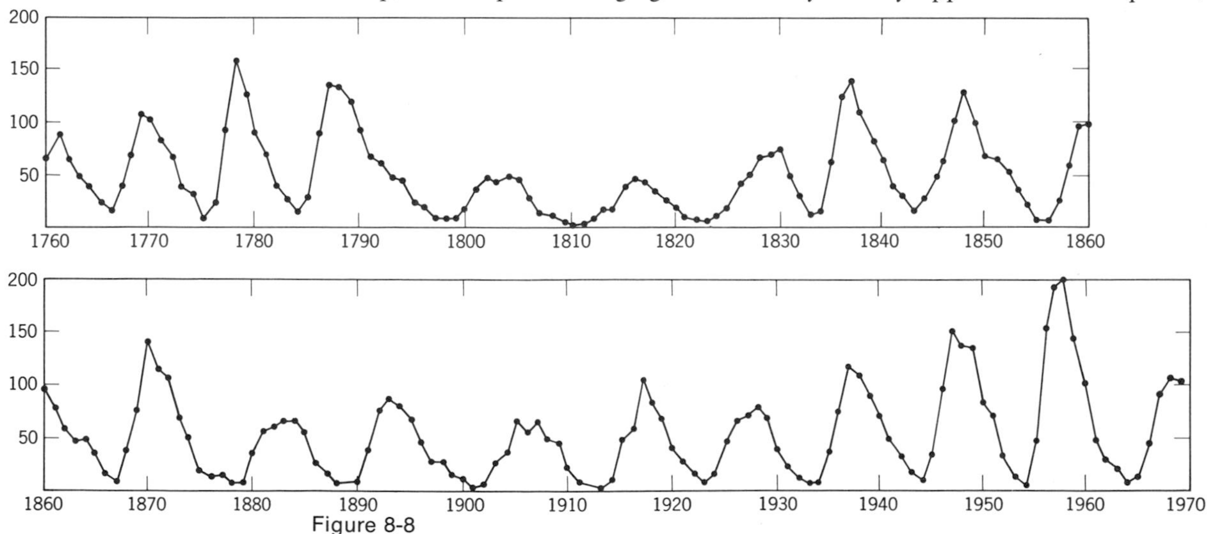

Figure 8-8

The sunspot cycle. The variations in the number of sunspots since 1760. The records before 1860 are not very reliable because they were not kept systematically.

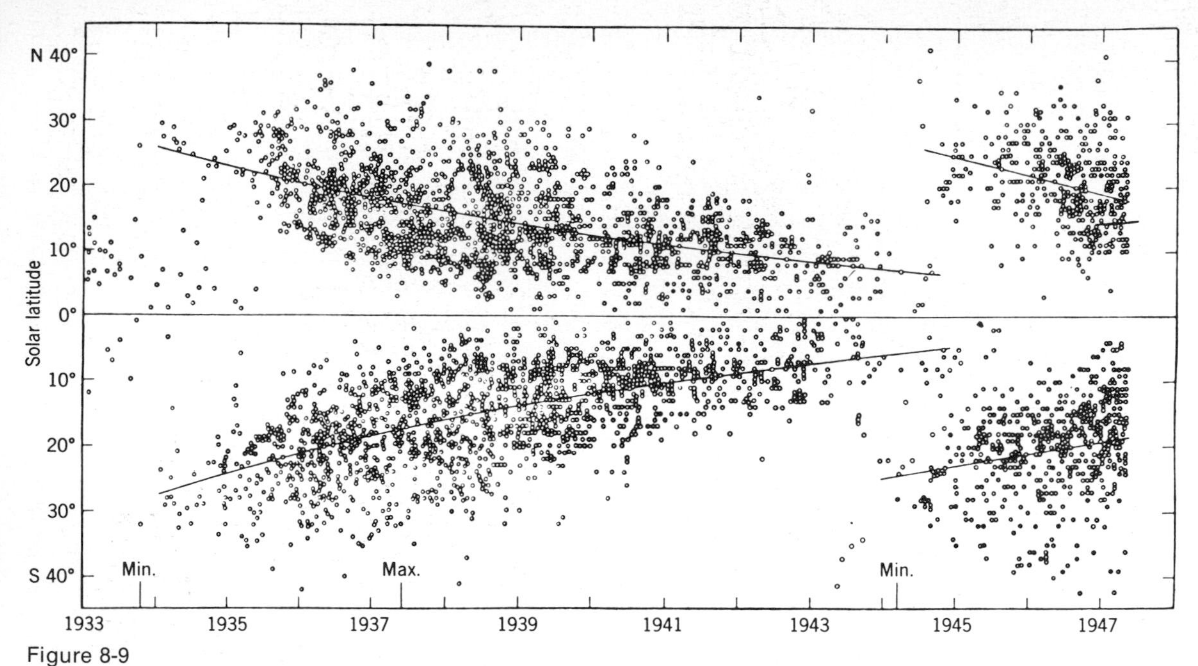

Figure 8-9

The distribution of sunspots during the 11-year cycle. At the beginning of each cycle the sunspots form at some distance from the equator. As the cycle progresses they form ever nearer the equator. (Courtesy of Giorgio Abetti, *The Sun,* The Macmillan Co., 1957.)

while the spots of the next cycle are appearing at about 35° from the equator. Spots are rarely seen more than 40° from the equator, and only occasionally do they reach the equator. Figure 8-9 shows a distribution of sunspots according to latitude and year.

During the 11-year cycle, then, both the number of spots and their latitude vary. But there is yet another variation, magnetic in nature. Associated with each spot is a magnetic field that behaves as if a long bar magnet had been inserted into the spot with one end sticking out. The evidence for this is spectrographic: when a spectral line is formed in a strong magnetic field it will be split into several components (the Zeeman effect). The characteristics of this splitting depend on the strength of the magnetic field and the direction in which the light leaves it.

Studies of the magnetic fields associated with sunspots reveal that if a preceding spot in a group has the north pole of the magnetic field pointing out from the sun, the following spot will have the south pole sticking out. The spots in the other hemisphere (northern or southern) will have the opposite magnetic polarity; the preceding spot will be a south magnetic pole and the following spot will be a north pole. During the next sunspot cycle, however, the polarity is reversed in both hemispheres. Any two succeeding cycles, then, are different in their magnetic aspects; consequently the cycle does not repeat itself in every respect until about 22 years have elapsed.

Star Plate 1
The Earth from outer space (NASA photograph).

Star Plate 2

The Crab nebula (copyright by the California Institute of Technology and Carnegie Institution of Washington; reproduced by permission from the Hale Observatories).

Star Plate 3

The Rosette nebula (copyright by the California Institute of Technology and Carnegie Institution of Washington; reproduced by permission from the Hale Observatories).

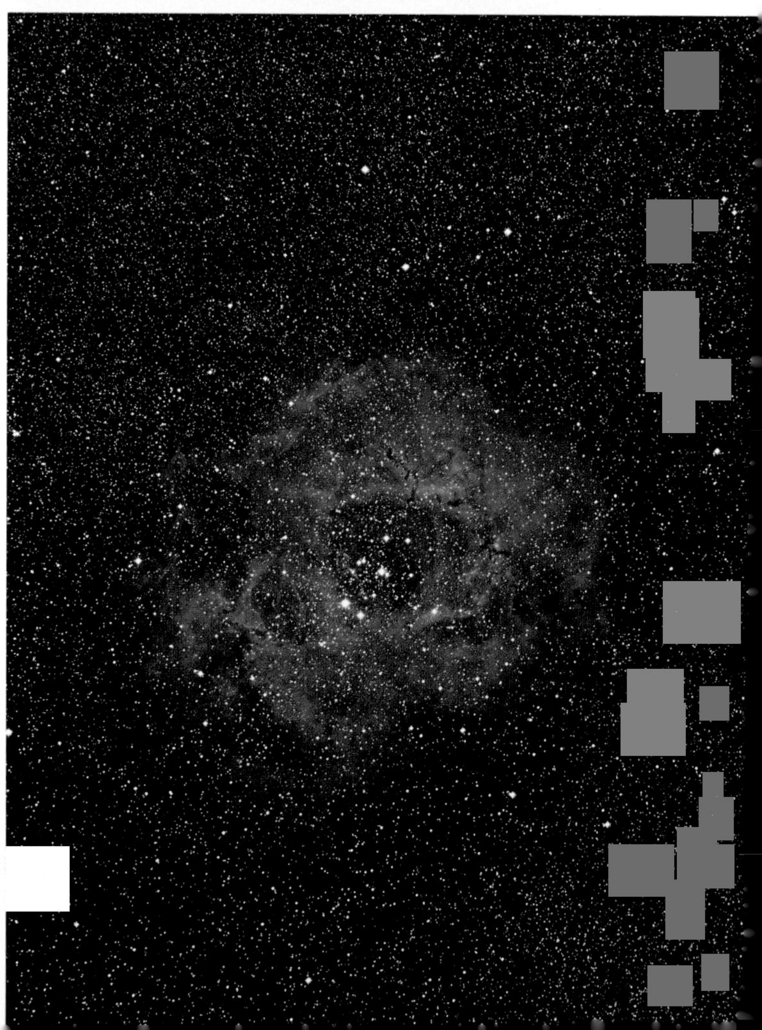

Star Plate 4

The Horsehead nebula (copyright by the California Institution of Technology and Carnegie Institution of Washington; reproduced by permission from the Hale Observatories).

Star Plate 5

The Andromeda galaxy (copyright by the California Institution of Technology and Carnegie Institution of Washington; reproduced by permission from the Hale Observatories).

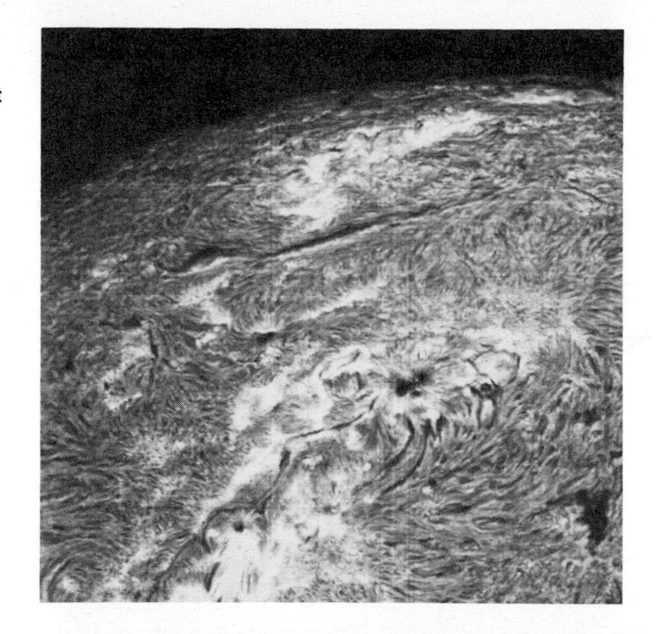

Figure 8-10
(a) The sun photographed on May 4, 1958, in the light of the H-Alpha line of hydrogen. (Photograph from the Hale Observatories.)

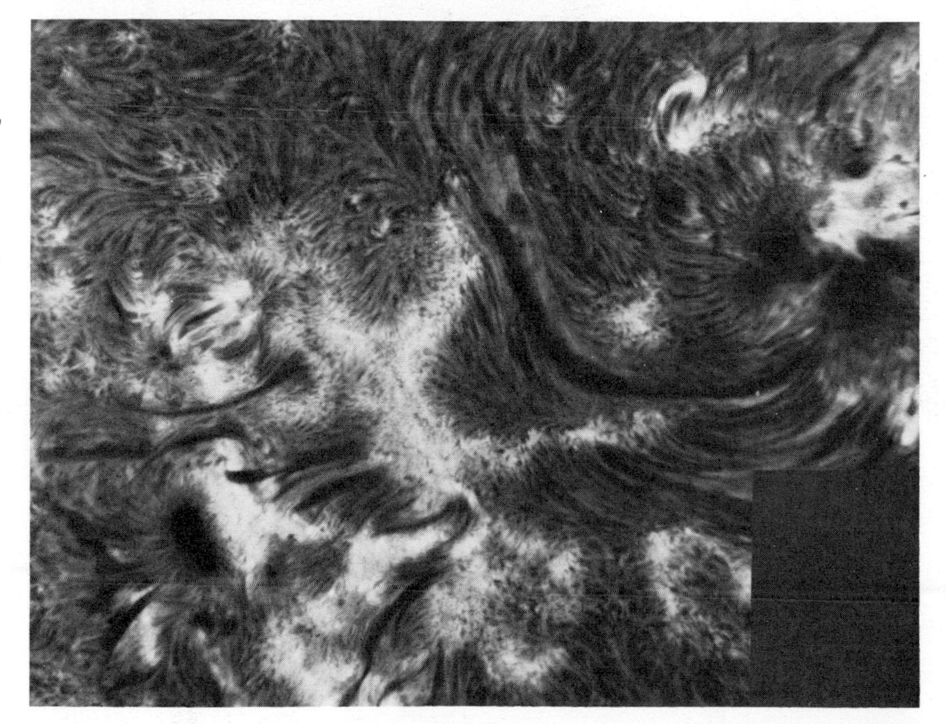

(b) Detail of two closely spaced sunspot groups is seen in this photograph taken on October 26, 1969, in the H-alpha light of hydrogen. The bright regions are hydrogen in emission and, thus, are hotter regions. The darker (cooler) filaments follow the magnetic field. (Big Bear Solar Observatory.)

The reason for this magnetic variation is quite unknown; in fact, we do not know with certainty why sunspots have magnetic fields associated with them at all. But it is suspected that the magnetic field causes the spot, not the spot the magnetic field. The main problem is one of refrigeration! How can the magnetic field maintain the temperature 2000° cooler than the surrounding photosphere?

E. THE PHOTOSPHERE AND THE SPECTROHELIOGRAM We could learn more about the photosphere in general and sunspots in particular if we knew where the various spectral lines originated in the sunspots and in the chromosphere. Spectral lines are, after all, indicative of the conditions under which they form. Since some of the lines in the solar spectrum are rather broad and quite intense, it is possible to take a picture of the sun by the light of any one of these broad lines alone. (It should be pointed out that even absorption lines are not black but contain some light that has been absorbed and re-emitted by the gas.) Special filters, both mechanical and optical, have been devised that eliminate all the solar spectrum except that spectral line of special interest. The resulting photograph, called a *spectroheliogram*, reveals the distribution of that particular gas whose spectral line has been allowed to pass through the filter.

Figure 8-10 shows two photographs of the sun taken in the light of hydrogen alone. The gases in the brighter regions are hotter than those in the darker. The dark streaks follow magnetic lines of force recalling to mind the iron filings about a magnet. The larger dark streaks in Figures 8-10*a* and 8-11*a*, called *dark filaments*, are relatively cool clouds of hydrogen hovering above the photosphere. They appear dark because they are cooler than the gases below them. Being cooler and being composed of hydrogen gas, they absorb the light emitted by the hotter hydrogen gas below them. They are not visible in Figure 8-11*b* because that photograph was made with the light in a line in the violet region of the calcium spectrum. The hydrogen gas does not absorb light at the wave-

Figure 8-11
The sun photographed simultaneously in (a) hydrogen H-alpha light, and (b) ionized calcium light. The plages (bright areas) and hydrogen dark filaments are clearly seen. (Sacramento Peak Observatory, Air Force Cambridge Research Laboratories.)

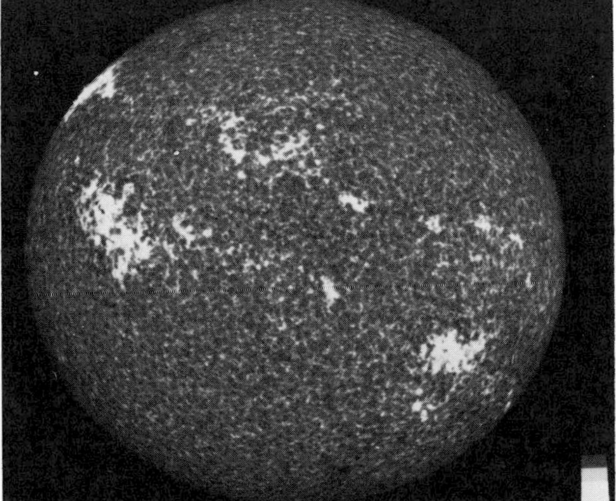

length of this calcium spectral line. The brighter regions are called *plages*. Each is evidence of some disturbance on the sun. Sunspots invariably appear in plages.

F. SOLAR FLARES Plages are the seat of not only the sunspots but also of the *solar flares* (Figure 8-12). One of the first signs of a flare is a local increase in brightness and the emission of short bursts of radio energy. After 15 to 30 minutes, the explosive phase begins, and for 15 minutes or more the flare emits short-wavelength X rays, ultraviolet radiation, visible light, and radio energy in

Figure 8-12
The eruption of a solar flare as seen in hydrogen light; total time lapse is about 1 hour 35 minutes. (Lockheed Aircraft Corporation.)

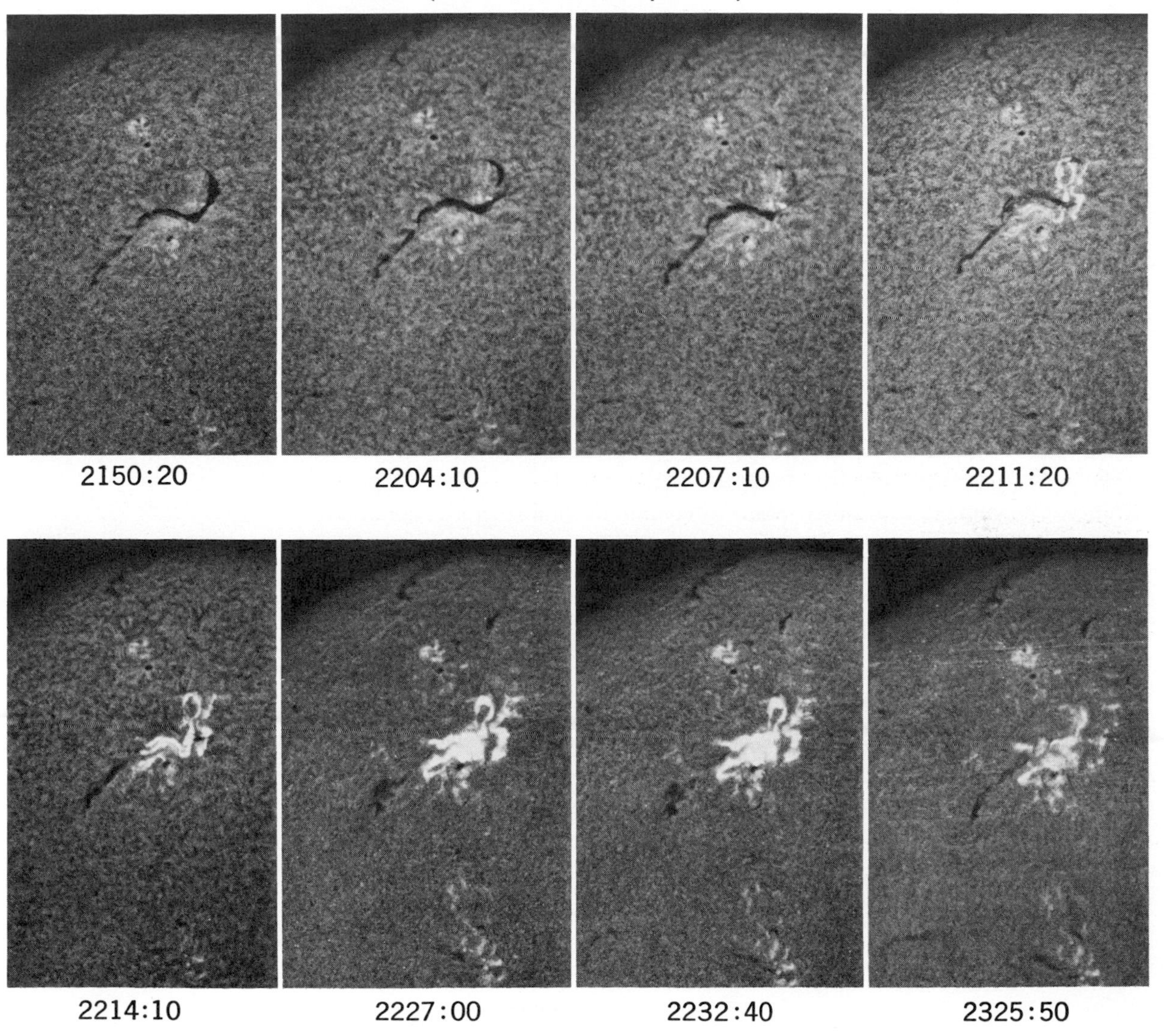

the meter, decameter, and centimeter range. Gases are ejected from the region at velocities in excess of 600 miles per second. Shock waves are propagated over the solar surface at velocities of 1000 miles per second. Vast amounts of energy are emitted during a flare. Clearly, the Earth does not escape the effects of such an explosion.

The gases, composed mostly of ionized hydrogen, that is, protons and electrons, reach the Earth in a very short time. Some of the most energetic particles travel at nearly the speed of light and reach the Earth in just over 8 minutes. Other less energetic particles take longer. Some of the very energetic protons reach the Earth in an hour or so, indicating that they do not travel in straight lines. The path of these particles is presumably deviated by the sun's magnetic field. Upon arriving at the Earth, the low-energy particles are deviated by the Earth's magnetic field and enter the atmosphere in the polar regions. These particles are related in a manner not well understood with the aurora — the northern lights and the southern lights. The more energetic particles come right on through the magnetic field and strike the surface of the Earth. They have been known to play havoc with electrical equipment, such as causing circuit breakers for towns and cities to go out.

Ultraviolet and X ray radiation disrupt the ionized layers in the Earth's atmosphere, which by reflecting radio broadcasting waves ordinarily enable radio operators in New York to receive messages from London, Sydney, Caracas. The ionized layers themselves are a result of the sun's usual ultraviolet radiation, but the intense radiation associated with solar flares has a disrupting effect and causes a corresponding fading in long-distance radio communications. Since we rely so heavily on long-distance radio communication (especially for ships and airplanes), there is a very practical purpose in observing the sun continuously: we may predict disruptions in the radio service and permit advance arrangement with other means of communication such as a communication satellite.

8.5 CHROMOSPHERIC ACTIVITY

We have learned that above the photosphere lies the chromosphere extending about 500 to 10,000 miles. The thickness of the chromosphere varies somewhat with the sunspot cycle. During a sunspot minimum its thickness at the poles is about 10% greater than at the equator, and during a maximum its thickness is the same all around the sun.

A. TEMPERATURES Temperatures in the chromosphere can be determined from a study of the intensities of spectral lines present in the chromospheric spectrum. The range of temperatures is considerable. At a height of only 400 miles above the photosphere, the temperature of the chromosphere reaches a minimum value of 4200° K. At an altitude of 2500 miles, it is 6000° K, the same as that of the photosphere. At its rather indefinite top, where it merges with the lower corona, its temperature is 50,000° K.

The increase in temperature of the chromosphere with increased altitude above the photosphere indicates that there is a conversion of some form of energy into thermal energy. At the higher altitudes, the chromosphere is nearly transparent to electromagnetic radiations. Since they are not absorbed it cannot be the X rays, ultraviolet, light, etc., that are maintaining such high temperatures of 50,000° K. It is possible that shock waves set by the turbulence on the photosphere is the source of energy.

B. METHODS OF DIRECT OBSERVATION The chromosphere was first seen during total eclipses when its scarlet color is clear and startling. This color is a result of a combination of a bright red line in the hydrogen spectrum and two bright violet lines in the spectrum of calcium. But the astronomer gets impatient waiting for those few favorable solar eclipses every year or so to observe the chromosphere and its activities. In addition, such scattered observations can never reveal any systematic changes in the chromosphere. Consequently, instruments have been invented that simulate a solar eclipse.

The *coronagraph* is an instrument that permits observations not only of chromospheric activity but also (as the name implies) of the sun's inner corona. A disk properly placed in the telescope blocks out the bright light of the photosphere. Such instruments must be located on high mountains to reduce the scattering of light inside the telescope by air and dust particles.

Since the coronagraph is able to simulate a solar eclipse whenever the sun is visible, it is possible to take time-lapse movies in which each frame is taken at a specified time interval (for example, 1 per minute). When such a movie is run through a motion-picture projector at 16 frames each second, any motions of the gases are speeded up. In this way it is possible to determine not only how fast the gas moves but also along which paths it moves.

C. PROMINENCES By means of the coronagraph and time-lapse movies, many varieties of upheaval have been discovered in the chromosphere. These upheavals, generally called *prominences,* take different shapes and exhibit different motions.

1. Surge prominence Figure 8-13*a* shows a burst of gas upwelling from the photosphere like a huge column. Its top may reach 30,000 miles above the photosphere before it fades and falls back along the same path. This must result from some sort of explosion near the surface.

2. Falling gases Time-lapse photography made clear that many of the prominences result from gases falling back into the sun. In the *loop prominence* (Figure 8-13*b*) the gases fall into an associated sunspot along definite lines. The gases seem to enter the top of the loop, divide, and flow down. The gases are constrained to the loop by magnetic fields.

The *quiescent* prominence (Figure 8-13*c*) is a dark filament seen from the side. It may remain for weeks or even months before breaking up by either vanishing or exploding. As with the loop prominence, the motions of gas within the quiescent prominence are down and constrained by magnetic fields.

An *arch prominence* is also a dark filament seen from the side. The motion of the prominence as seen in Figure 8-14 may result from an outward motion of a magnetic field rather than from the gases themselves moving out.

If the gases fall into the sun to form these prominences, why are the gases seen in only restricted regions? Why aren't the falling gases generally visible in the outer chromosphere and inner corona?

This gas is composed largely of hydrogen. As indicated earlier, the temperature in the outer chromosphere reaches as high as 50,000° K. At these temperatures, the hydrogen atoms become completely ionized, that is, the nuclei

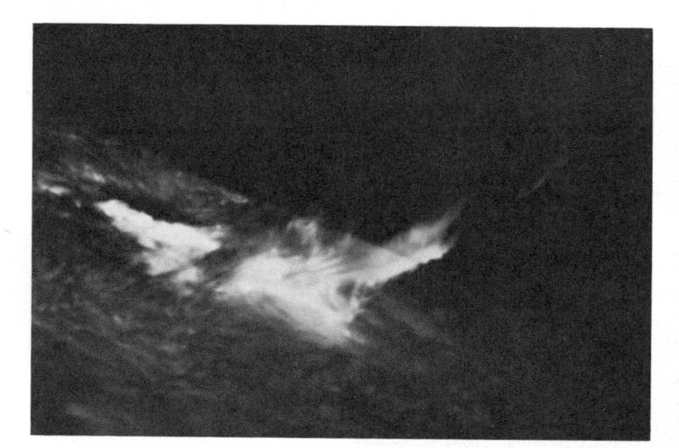

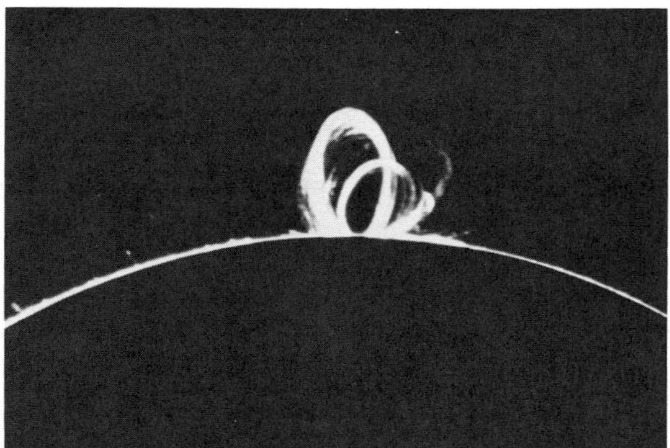

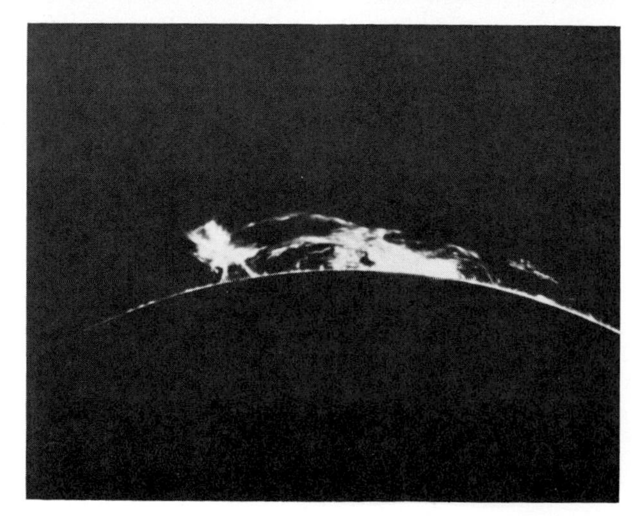

Figure 8-13
(a) Surge prominence associated with a flare. (b) Loop prominence. (c) Quiescent prominence. (Photographs from Sacramento Peak Observatory, Air Force Cambridge Research Laboratories.)

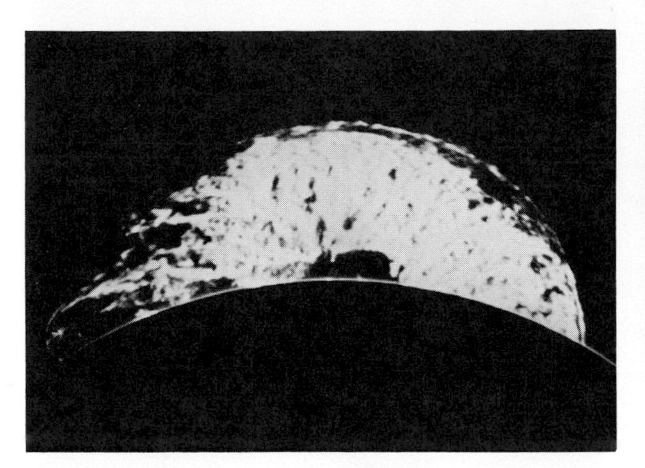
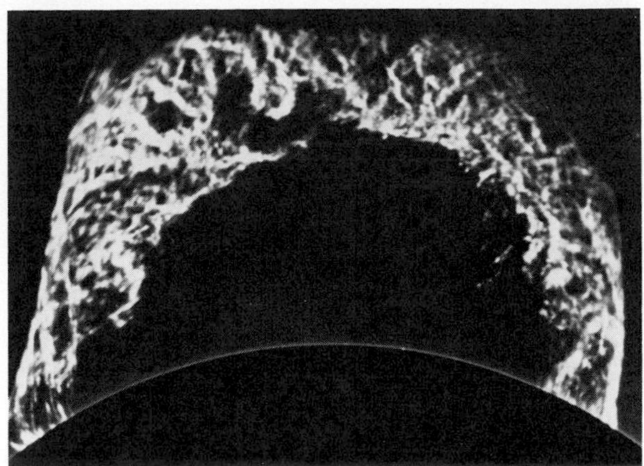

Figure 8-14
The great arch prominence of June 4, 1946. The two exposures were taken only 15 minutes apart. The diameter of the sun, roughly 864,000 miles, gives some idea of how rapidly this arch left the sun. (Sacramento Peak Observatory, Air Force Cambridge Research Laboratories.)

(protons) and the electrons are forced apart by the violent motions of particles in a high temperature gas. The gas is, therefore, not strictly hydrogen, but one of electrons and protons. As such it is not able to emit light; atoms can emit light, bare nuclei and free electrons do not.

Apparently magnetic fields cool the gases enough so that the protons and electrons recombine to form hydrogen atoms. In this recombining process, light is emitted, and continues to be emitted as long as the gas is hot enough to emit light but is not so hot as to ionize the hydrogen atoms.

8.6 THE CORONA

A. APPEARANCE Outward from the chromosphere there extends the pearly-white tenuous gas called the *corona* (meaning crown), which is seen so beautifully only during a total solar eclipse (Figure 8-15). Its height above the photosphere, as well as its shape, varies with the sunspot cycle. During a sunspot maximum it is fairly regular, but during a sunspot minimum it extends out much farther in huge petal-like configurations.

B. TEMPERATURE AND PRESSURE As with the photosphere and chromosphere, the temperature and pressure of the solar corona are determined by an analysis of the spectra, but these must be taken during total solar eclipses.

Figure 8-15
A specially designed filter was used for the March 1970 eclipse to bring out details in the bright inner corona as well as the much fainter outer corona. (Courtesy Gordon Newkirk, Jr., High Altitude Observatory.)

The outer corona produces a bright-line spectrum, but one with unusual spectral lines. After a thorough study it has been concluded that some of these strange lines are the result of iron ionized thirteen times.

The normal iron atom has 26 electrons about its nucleus, but in the outer corona, many of the iron nuclei have only 13 electrons. This can only result from a gas's being at very high temperatures and at very low pressures. At extremely high temperatures the atoms of a gas collide so hard that they knock one another's electrons off; the higher the temperature, the faster the atoms move, and the harder they collide.

Once ionized the ions (charged nuclei) and electrons can recombine only when they collide and the positively charged ion attracts and holds onto the negatively charged electron. If the gas is at a very low pressure, collisions and recombinations are less frequent than in a gas at higher pressure. Hence, in the low pressures of the outer visible corona, once an atom is ionized, it remains an ion for a long period of time, and indeed may be ionized a second, third or, as with iron, a thirteenth time. These studies indicate that the temperature of the outer corona is 1,000,000° K or more!

C. THE CORONA AND THE SOLAR WIND The very high temperature of the corona and the solar wind are related. If the shock waves set up by the turbulent photosphere heat the chromosphere to 50,000° K they must also be the source of energy that heats the corona to a million degrees or more. This continued increase in temperature with increasing distance from the photosphere means that the particles (ions and electrons) in the outer corona move very fast indeed. At a height of 6,000,000 miles (about 14 solar radii) from the solar surface, the corona is expanding at a speed in excess of 100 miles per second. This is faster than the speed of sound in the same gas, so it amounts to a supersonic wind. The outer solar corona turns into the solar wind that whizzes past the Earth at a velocity roughly 250 miles per second. The solar wind continues out beyond the orbit of Jupiter, but just how far remains to be seen by future research. The existence of the solar wind has been verified by measurements made by spacecraft on their way to the moon, Venus, and Mars. The solar wind at the Earth's distance from the sun is composed of about 5 protons per cubic centimeter, or perhaps more. A comet is truly a "solar wind-sock."

The sun has been and is now being studied by all available methods and with all the imagination and ingenuity that can be brought to bear on it. Very important to us here on the Earth is its 11-year cycle of activity with its sunspots and flares, prominence activity, varying coronal shape and size, and their corresponding effect on our planet. It has been found, for example, that the rate of growth of trees, which depends on the weather, varies in an 11-year cycle which corresponds to the cycle of the sun. In fact, a study and understanding of this growth as seen in tree rings has been used to date pieces of wood found by archeologists in their diggings.

The effect of solar activity on radio broadcasting reception has already been mentioned. Its effect on the weather of the Earth has led meteorologists, too, to become interested not only in the 11-year cycle but also in solar flares; there is reason to believe that those powerful eruptions affect our weather to an as yet undetermined degree. Furthermore, since large flares emit cosmic rays in such deadly amounts, they will also concern the space traveler who may find it healthier to venture out beyond our shielding atmosphere only during a sunspot minimum, when solar flares are quite rare.

The astronomer, however, studies the sun primarily because it is a star, the only one whose surface we can see and study in some detail. And stars, after all, are the astronomer's first calling, for they are the building stones of the universe.

BASIC VOCABULARY FOR SUBSEQUENT READING

Chromosphere
Convective envelope
Corona
Cosmic rays
Dark filament
Granules
Photosphere
Plage
Prominence
Solar flare
Solar wind
Spectroheliogram
Sunspot

QUESTIONS AND PROBLEMS

1. What are cosmic rays?

2. How does the solar wind differ from cosmic rays?

3. What is meant by the intensity of: (a) a bright spectral line? (b) a dark spectral line?

4. Which region of the sun produces the continuous portion of the solar spectrum?

5. Which region of the sun produces the dark absorption lines in the solar spectrum?

6. A low pressure gas is made to emit a bright-line spectrum. Gradually the pressure of the gas is increased. How can this change in pressure be detected by observing the bright-line spectrum?

7. What are the differences in the radiant energy emitted by a toaster filament and a blob of molten iron?

8. What are the differences in the radiant energy emitted by two stars of different temperature even if both are the same size?

9. How can the temperature of the photosphere be determined?

10. Why is the photosphere the visible portion of the sun?

11. What is the source of energy in the sun?

12. What are the solar granules?

13. What is a spectroheliogram?

14. Why does a loop prominence maintain such a distinctive shape?

15. What effect might a solar flare have on the Earth?

16. Give the differences in the following for each of these regions of the sun — core, photosphere, chromosphere, and corona:
(a) temperature, (b) pressure, (c) method of observation, (d) spectrum caused by each.

FOR FURTHER READING

Bray, R. J., and R. E. Loughheed, *Sunpots,* John Wiley and Sons, New York, 1965.

Glasstone, S., *Sourcebook on the Space Sciences,* Van Nostrand Co., Princeton, N.J., 1965, Chapter 6.

Struve, O., and V. Zebergs, *Astronomy of the 20th Century,* Crowell Collier and Macmillan, New York, 1962, Chapter VII.

Bahcall, J. N., "Neutrinos from the Sun," *Scientific American,* p. 29, July 1969.

Carroll, G. A., "The Star Telescope of Lockheed Solar Observatory," *Sky and Telescope,* p. 10, July 1970.

Dunn, R. B., "Sacramento Peak's New Solar Telescope," *Sky and Telescope,* p. 368, December 1969.

Eddy, J. A., "The Great Eclipse of 1878," *Sky and Telescope,* p. 340, June 1973.

Goldberg, L., "Ultraviolet Astronomy," *Scientific American,* p. 92, June 1969.

Howard, R., "The Rotation of the Sun," *Scientific American,* p. 106, April 1975.

Liebenberg, D. H., and M. M. Hoffman, "Swift as the Moon's Shadow," *Sky and Telescope,* p. 351, June 1973.

Livingston, W. C., "Measuring Solar Photospheric Magnetic Fields," *Sky and Telescope,* p. 344, June 1972.

Malitson, H. H., "The Solar Energy Spectrum," *Sky and Telescope,* p. 162, March 1965.

Maxwell, Alan, "Radio Emissions from Solar Flares," *Sky and Telescope,* p. 4, July 1973.

Maran, S. P., and R. J. Thomas, "OSO-7 Year of Discovery," *Sky and Telescope,* p. 4, January 1973.

McIntosh, P. S., "August Solar Activity and Its Geophysical Effects," *Sky and Telescope,* p. 214, October 1972.

Ney, E. P., "Balloon Observations During July Eclipse," *Sky and Telescope,* p. 251, November 1963.

Parker, E. N., "The Solar Wind," *Scientific American,* p. 66, April 1964.

Parker, E. N., "The Sun," *Scientific American,* p. 42, September 1975.

Pasachoff, J. M., "The Solar Corona," *Scientific American,* p. 68, October 1973.

Pneuman, G. W., "The Chromosphere-Corona Transition Region," *Sky and Telescope,* p. 148, March 1970.

Zirin, H., "The Big Bear Solar Observatory," *Sky and Telescope,* p. 215, April 1970.

"Acoustical Waves on the Sun," *Sky and Telescope,* p. 351, June 1972.

"The Moon's Wake in the Solar Wind," *Sky and Telescope,* p. 93, February 1968.

"OSO 4 Ultraviolet Solar Observations," *Sky and Telescope,* p. 362, December 1967.

"Round-the-Clock Solar Movies," *Sky and Telescope,* p. 296, May 1969.

"Solar Flare Symposium," *Sky and Telescope,* p. 89, February 1965.

"Some Results from Skylab's Solar Experiments," *Sky and Telescope,* p. 11, July 1974.

"The Temperature Minimum in the Sun," *Sky and Telescope,* p. 362, June 1969.

CHAPTER CONTENTS

LEARNING OBJECTIVES

BE ABLE TO:

1. DESCRIBE THE METHOD AND LIMITATIONS OF HELIOCENTRIC PARALLAX TO DETERMINE STELLAR DISTANCES.
2. EXPLAIN HOW PROPER MOTION OF A STAR IS DETERMINED.
3. EXPLAIN HOW A STAR'S RADIAL VELOCITY IS DETERMINED.
4. EXPLAIN HOW A STAR'S RELATIVE VELOCITY IS DETERMINED.
5. EXPLAIN HOW THE STEFAN-BOLTZMANN LAW CAN BE USED TO DETERMINE A STAR'S RADIUS.
6. DESCRIBE HOW STELLAR TEMPERATURES ARE DETERMINED.
7. GIVE A GENERAL DESCRIPTION OF THE STELLAR SEQUENCE INCLUDING IDENTIFYING SPECTRAL FEATURES.
8. DESCRIBE HOW A STAR'S BRIGHTNESS IS MEASURED.
9. DESCRIBE HOW A STAR'S LUMINOSITY CAN BE DETERMINED.
10. EXPLAIN THE DIFFERENCE BETWEEN BRIGHTNESS AND LUMINOSITY.
11. DESCRIBE THE H-R DIAGRAM.
12. EXPLAIN HOW THE H-R DIAGRAM IS USED TO DETERMINE A STAR'S DISTANCE.
13. DESCRIBE THE MASS-LUMINOSITY RELATION.
14. DESCRIBE THE LUMINOSITY FUNCTION.
15. GIVE THE PRINCIPAL CHEMICALS OUT OF WHICH STARS ARE MADE.

Chapter Nine
STABLE STARS

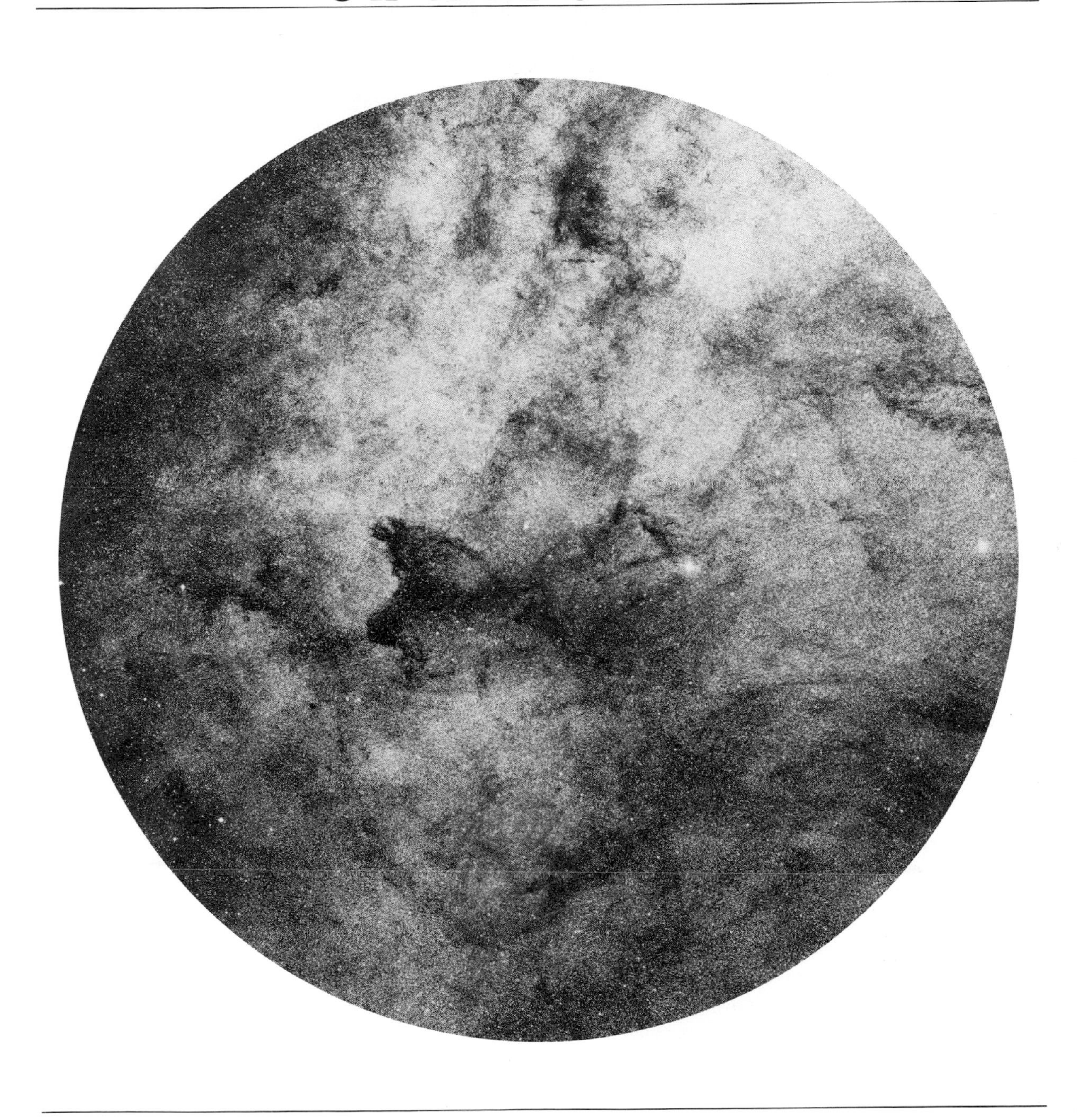

Studying the stars is like putting together a jigsaw puzzle for which we do not have all the pieces. What is more, the pieces that are available to us do not simply lie around waiting to be picked up and fitted into place; they are often hard to find and when found their shapes are sometimes indistinct. The pieces at hand are the observed physical characteristics of the stars. When fitted together they leave some gaps in our picture. We then deduce what ought to fill those gaps so that what we see may make sense. The astronomer hopes thus to obtain knowledge of the stars' present state, of their origin and past, and of their future. He wants to understand how stars are born, what stages they undergo, and how they end.

The prodigious amount of energy that a star emits each second is evidence that it is not a dead and static body, but that it undergoes an evolutionary process. A basic assumption of science, the principle of conservation of energy, leads us to conclude that any emission of energy from a star must have a source that is consumed during the star's lifetime. As this source of energy is consumed the star must change.

9.1 GENERAL FEATURES

In order to understand and predict stellar changes we must be able to describe a star. Such a description should include a knowledge of each of the star's:

1. Distance from the sun or, more exactly, the position of the star in our galaxy
2. Velocity
3. Size
4. Temperature
5. Luminosity
6. Mass
7. Density
8. Chemical composition
9. Age

9.2 STELLAR DISTANCES

The distance of a star from the sun is not only one of the more difficult characteristics to determine but also one of the more important. All the life processes of the star are determined by means of the amount and types of energy radiated. But the amount of energy a star radiates out into space cannot be known until its distance is known. Consequently, there has been a great deal

Chapter Opening Photo
Star cloud in the region of Sagittarius. The dark streaks are intervening gases that obscure the stars behind. (Photograph from the Hale Observatories.)

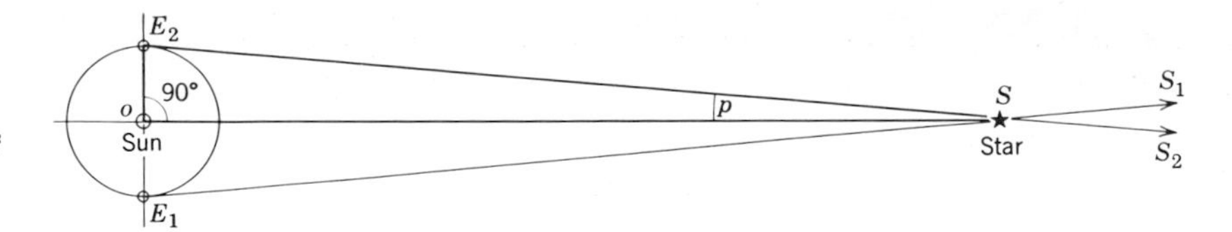

Figure 9-1
The angle of parallax is actually the angle subtended by the radius of the Earth's orbit as seen from a nearby star.

of work done on the distance determination of stars, some of which has proved successful and some of which at one time was thought valid but was later proved incorrect. Hence, there is still much thought and work applied to this problem.

A. HELIOCENTRIC PARALLAX The first method used to determine a star's distance from the sun is the most fundamental, for it is an actual measurement of distance after the manner of the surveyor. It is the method of heliocentric parallax discussed on p. 89. If the nearer stars appear to oscillate back and forth in front of the background stars as the result of the Earth's revolution about the sun, we need only measure the *angle of parallax p* (Figure 9-1) through which a star appears to move. This angle when considered trigonometrically with the radius of the Earth's orbit as a base line yields the distance *oS*.

The need for an astronomical unit of measure to fit stellar distances becomes apparent when we consider that the nearest star is something like 24 trillion (24×10^{12}) miles from the sun. The method of heliocentric parallax has given rise to a unit of distance called the *parsec*. One parsec is equal to about 19×10^{12} miles and is the distance at which a star must be in order to have a *par*allax of 1 *sec*ond of arc. The closest star must have the largest angle of parallax; it is called Proxima Centauri and its angle of parallax is 0.76 arc seconds.

Another unit of measurement employed in stellar distances is the *light year*, or the distance light travels in 1 year. Light travels 186,000 miles every second or about 6×10^{12} miles in 1 year. A light year is, therefore, shorter than the parsec; 1 parsec equals about 3.26 light years. Proxima Centauri is about 4.2 light years from the sun, or 1.3 parsecs.

B. LIMITATIONS OF HELIOCENTRIC PARALLAX As a method of determining distances, however, heliocentric parallax has inherent limitations. The farther the star is from the sun, the smaller the angle that must be measured. Since this angle must be measured by mechanical means (usually by the location of a star's image on a photographic plate) the measurement of the very small angles involved not only becomes difficult but the error in making the measurement soon becomes as large as the angle itself. Hence, only the distances of the nearer stars can be determined by heliocentric parallax. Distances for some 6000 stars have been so determined, but only for the nearest 700 have they been determined with an uncertainty of less than 10%. The limiting distance is about 100 light years.

Astronomers also need to be able to determine distances for many of the billions of stars beyond the limit of heliocentric parallax. As we see in this and

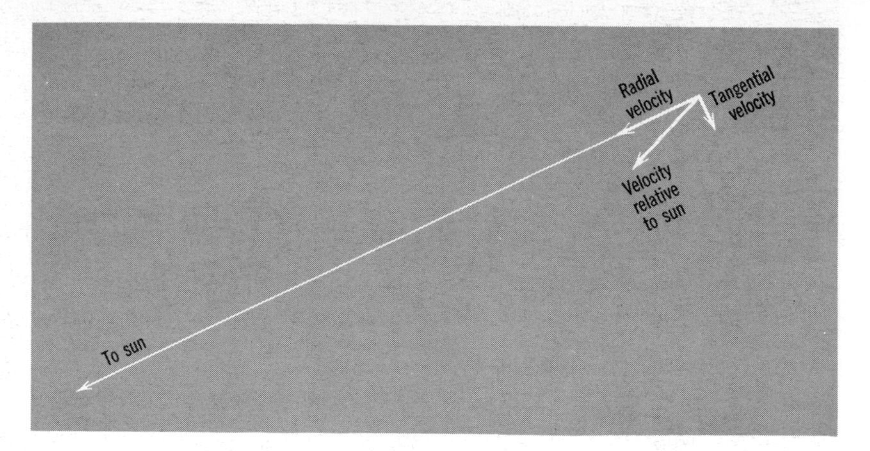

Figure 9-2
The velocity of a star with respect to the sun is broken up into two components — radial velocity and tangential velocity, for only these component velocities can be conveniently measured.

succeeding chapters there are many roundabout ways of determining stellar distances, all less accurate than the methods of heliocentric parallax applied to the nearest stars. Yet each contributes its bit to our knowledge of the size of our galaxy and ultimately of the size of our universe.

9.3 STELLAR VELOCITIES

We have seen (Chapter Two) that the Doppler shift of the spectrum can be used to determine the velocity of a star in the line of sight, that is, its *radial velocity*. This gives the velocity in only one direction; but certainly stars do not only travel directly away from or toward the sun. An airplane, for example, that is flying northwest may be due south of an observer at one moment; shortly thereafter it will not only be closer to the observer but to the west of due south. The observer could say, therefore, that the plane has two simultaneous velocities: one to the north, approaching him, and the other due west. These two velocities are called *components* of the actual velocity. When we observe the radial velocity of a star, we observe only one of the components of its motion, that one equivalent to the airplane approaching the observer. The other component, the one equivalent to the westward motion of the airplane, is called *tangential velocity* and is seen as the star moves perpendicular to the line of sight. This motion is revealed by the star's motion against the background stars, which are too far away to reveal any tangential motion of their own. When tangential velocity is expressed as an angular motion per unit of time, for example, seconds of arc per year, it is called *proper motion.*

Figure 9-2 shows the radial velocity component, the tangential velocity component, and the relative velocity of a star with respect to the sun. Figure 9-3 shows the proper motion of a star that has moved against the background stars during the time interval between the three photographs. The spectrum of this star reveals its radial velocity. Its velocity relative to the sun is obtained by the Pythagorean theorem from its radial and tangential velocities.

The fact that stars do have proper motion interferes to a certain extent with the determination of their parallax, for in determining a star's position from year to year its proper motion must be untangled from its parallactic motion. Since parallactic motion is periodic and proper motion is cumulative, the separation of the two is not an impossible task. The proper motion of some 200,000 stars has already been determined.

The difficulties of measuring parallax and proper motion are great because the stars are so very far apart in relation to their sizes and velocities. Although the velocity of a neighboring high-velocity star may be as much as 100 miles per second, its motion as seen from the Earth is very small indeed. Similarly, although some of the nearby stars are much larger than the sun, they appear only as points of light in our largest telescopes. To illustrate, let us imagine the sun shrunk to the size of an apple, and the nearest stars as well as the distances between each star shrunk in proportion. We would then have a group of apples, each about 1500 miles apart, with the fastest of them moving at a rate of about ½ in. per hour.

9.4 STELLAR SIZES

To measure the distances and velocities of stars is in itself a difficult task. But how about measuring the sizes of the stars? An apple at 1500 miles does not look very large. Nevertheless the astronomer has set himself the task of determining the sizes of the stars; how does he go about it?

The angular diameter of a star can be determined by the use of an interferometer. In 1920, two smaller mirrors were mounted on a 20-ft beam (it is called a *beam interferometer*) atop the 100-in. telescope on Mount Wilson. The angular diameters of 12 stars were measured by this instrument. Recently, however, an interferometer consisting of two separate telescopes each on a track has been installed at the Narrabri Observatory in Australia. Reports from work in progress indicate that the angular diameter of Sirius, for example, is 0.0062 arc seconds. A 45-ft crater on the nearside of the moon appears about as big as Sirius does in our sky. At a distance of 2.7 parsecs (8.7 light years) Sirius must have a diameter of 1.52×10^5 miles, 1.76 times that of the sun.

The diameters of some of the largest stars are astounding (Figure 9-4). Mira, in the constellation Cetus, has a diameter 460 times that of the sun. Were Mira to replace the sun, its boundaries would include the orbit of the planet Mars! The star Betelgeuse is a pulsating star and has a diameter which varies from 700 to 1000 times that of the sun! The star α Hercules has a diameter 800 times that of the sun.

Figure 9-3
(a) A composite of three photographs of Barnard's star (lower left). The three photographs have been superimposed so that the background stars (upper right) each show only one image. Motion to the north (down) is proper motion, motion east and west is the result of annual heliocentric parallax. (Courtesy Peter van de Kamp, Sproul Observatory.)

The diameters of smaller stars cannot be measured with interferometers. They are simply too far away and too small. In some instances, however, double stars revolve about their common center of mass in such a way as to eclipse one another as seen from the Earth. If the motions of the stars are well known, their diameters can be determined by the length of the eclipse.

A star's diameters can also be estimated by the Stefan-Boltzmann law. The amount of energy a star radiates depends on two factors: the star's surface temperature and surface area. Of two stars the same size, the hotter one will radiate more energy. Of two stars the same temperature, the larger one will radiate more energy. The mathematical expression of the Stefan-Boltzmann law permits the astronomer to determine the surface area (and thus the radius, since nearly all stars are spherical) once he knows the star's surface temperature and the amount of energy it radiates. To determine the amount of energy, however, he must know the distance of the star.

Large stars have diameters close to 1000 times that of the sun. The smallest stars have diameters smaller than that of the moon, and some special kinds of stars have diameters as small as 15 miles! Thus there is a considerable range of diameters: the larger stars have diameters something like 10^8 times larger than the smaller ones.

(b) A drawing indicating the location of Barnard's star at the specific dates shown in Figure 9-3*a*. When proper motion is taken out, the star appears to move in the little ellipse shown to the left. This elliptical motion is heliocentric parallax.

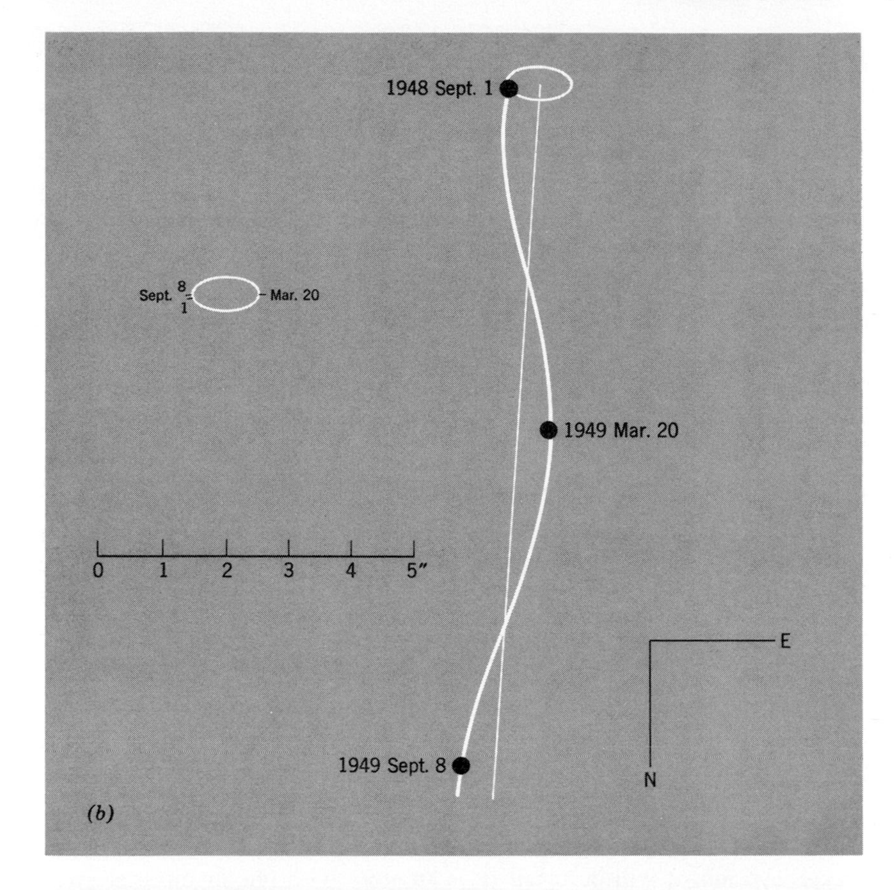

Figure 9-4
The relative sizes of some giant stars, the sun, and the orbits of the four terrestrial planets.

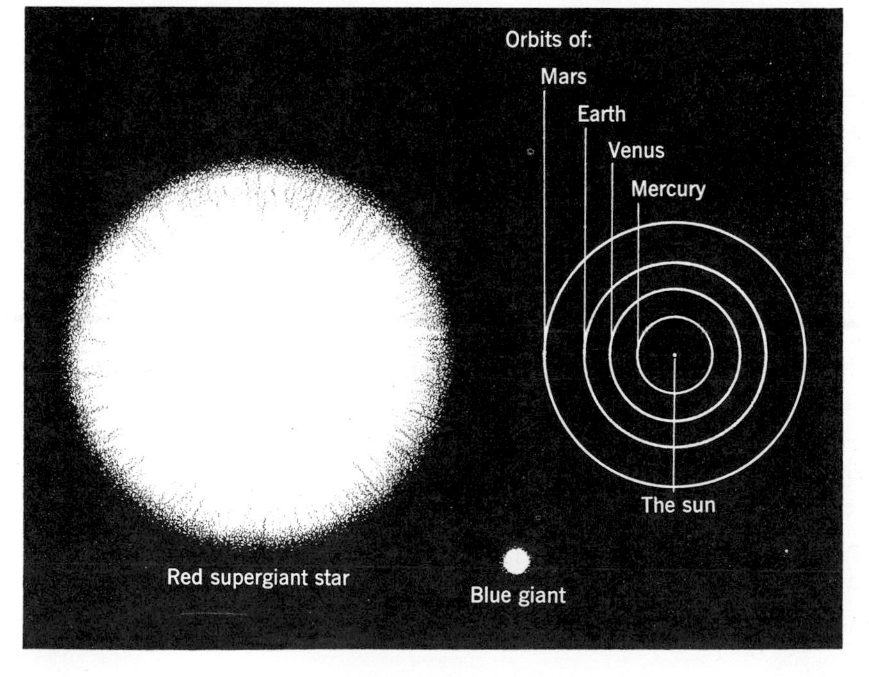

9.5 STELLAR TEMPERATURES

Stellar temperatures are fundamental in the study of the stars, for when a star's temperature is considered along with its size, we can find how much energy it radiates off into space and, hence, for a stable star, how much energy is generated in its core. Fortunately, stellar temperatures are one of the more easily observable characteristics. The range of temperatures is less than that of diameters; the hottest known star, at about 100,000° K, is only about 70 times as hot as the coolest one (about 1500° K). Since the temperature of the sun is about 6000° K, it may be classified as one of the cooler stars.

The surface temperature of a star can be determined by its color using Wien's law: cool stars are red, hot stars are white, and very hot stars are blue. It is simpler, however, to determine the temperature of the lower chromosphere by the intensity of the various spectral lines present in the stellar spectrum. If high energy lines are intense and low energy lines are weak, the star is a hot star. The star's spectrum, therefore, is the key to learning its temperature.

In cool stars, low energy lines are more intense and high energy lines are absent. These low energy lines are produced not only by atoms but by a number of molecules, that is, atoms joined together electrically. The spectral lines caused by molecules are many and grouped in what are called *bands*.

As stars of higher temperature are considered, the intensity of the molecular bands decreases, and the intensity of some of the atomic spectral lines increases.

As stars of still higher temperature are considered, the molecular bands disappear, atomic lines increase in intensity, and the spectral lines of ions begin to increase. An ion, it will be recalled, is an atom with one or more electrons removed. These electrons are removed by atomic collisions in a hot gas. The calcium atom is one of the more easily ionized atoms; two of its lines are intense in the solar spectrum, a star of modest temperature.

In stars of very high temperature, the spectral lines of atoms are less intense because more of the atoms are ionized. Therefore, the spectral lines of the ions become more intense.

Generally speaking, therefore, the molecular bands predominate in the spectra of cool stars; atomic lines are most intense in the spectra of stars of moderate temperature; while spectral lines of ions are strongest in the spectra of very hot stars.

That a hot star does have a very different spectrum from a cool star can be readily seen in Figure 9-5, where spectra are arranged according to the temperature of the emitting star. On the left-hand side of the figure, each spectrum has a letter which is used to designate that particular type. An O-type star is the hottest and the M-type is the coolest.

A. THE SPECTRAL SEQUENCE The letters used to designate this sequence of spectra do not really make any sense: O B A F G K M. It would almost appear as though the astronomer responsible for such a sequence of letters either did

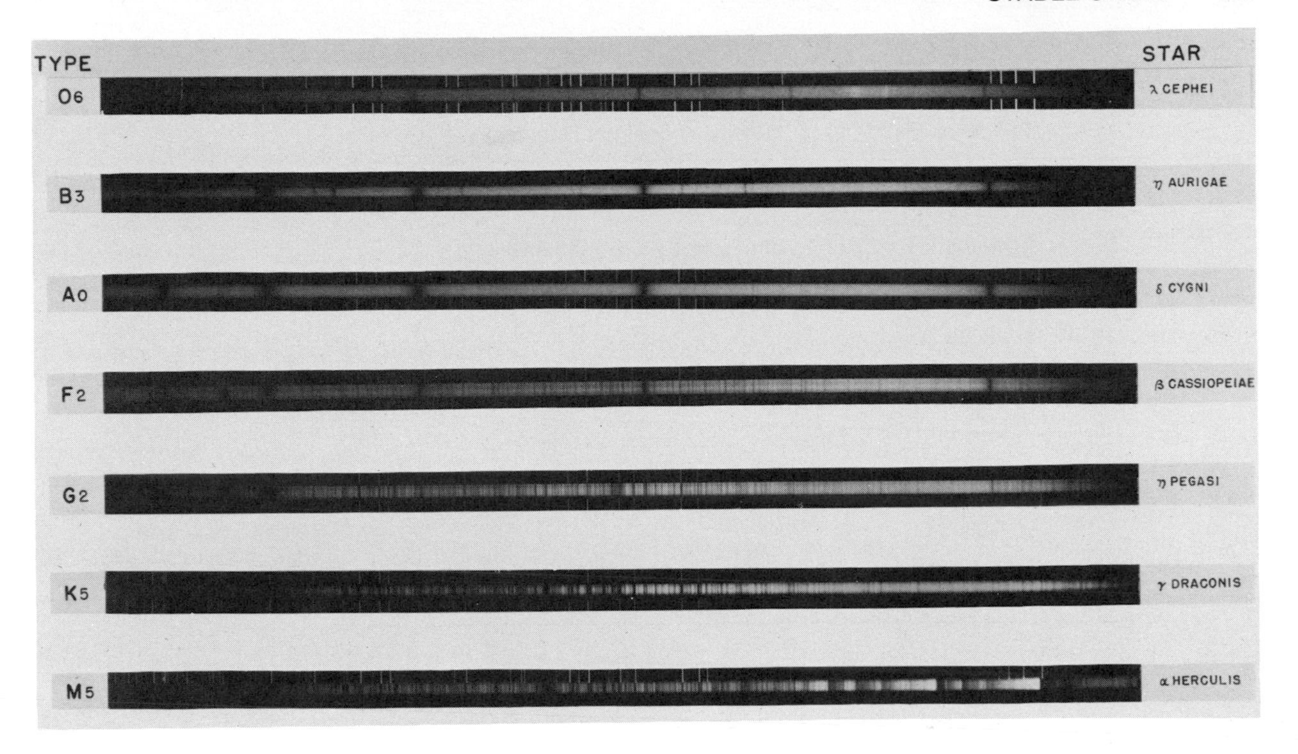

Figure 9-5
Principal types of the spectral sequence. (Photograph from the Hale Observatories.)

not know the alphabet or was simply not concerned with order. Fortunately, neither is the case. The history behind the formation of this sequence of letters helps explain the matter; it also shows that astronomers, like anyone else, have a tendency to become bound by an established system of nomenclature.

When the study of stellar spectra received the impetus of photography it became feasible to try to arrange the different spectra in some kind of order. Since hydrogen lines are present in so many of the spectra it was felt that these lines might be used as criterion. Accordingly, the stars with the strongest hydrogen lines were called A-type stars, those with slightly weaker hydrogen lines B-type stars, etc. At the time this scale was established the reason for the differences in stellar spectra was not well understood, and only later did it become apparent that they represented differences in stellar temperatures. Further study revealed that an A-type star is not the hottest; a B-type star is hotter, and an O-type star is the hottest of all. So the order of the spectra was rearranged according to temperature, although the letters designating the types were retained except for a few that had to be discarded. Thus, we see that the spectral sequence is basically a temperature sequence.

Since the spectral sequence is so important in astronomy a scheme has been devised to help the beginning student remember it. If we let the letters in the

Table 9-1
Spectral class characteristics

Spectral Class	Range of Temperatures	Identifying Spectral Features
O	Greater than 25,000° K	Lines of ionized helium present
B	11,000 to 25,000° K	Lines of neutral helium and ionized oxygen present
A	7,500 to 11,000° K	Lines of hydrogen at their strongest
F	6,000 to 7,500° K	Lines of both neutral and ionized iron present
G	5,000 to 6,000° K	Lines of ionized calcium at their strongest
K	3,500 to 5,000° K	Lines of neutral metals predominate
M	Less than 3,500° K	Molecular bands predominate

sequence, O B A F G K M (R N S), be the first letters of each word in a well-chosen sentence we get the following: *Oh Be a Fine Girl Kiss Me (Right Now, Smack)!* It is a rare student who forgets this, after even a cursory glance, and therefore it has so far survived the vicissitudes of expression in the English language.

Actually the R-, N-, and S-type stars are a branch of the spectral sequence. These stars have chemical compositions slightly different from the rest and are relatively cool.

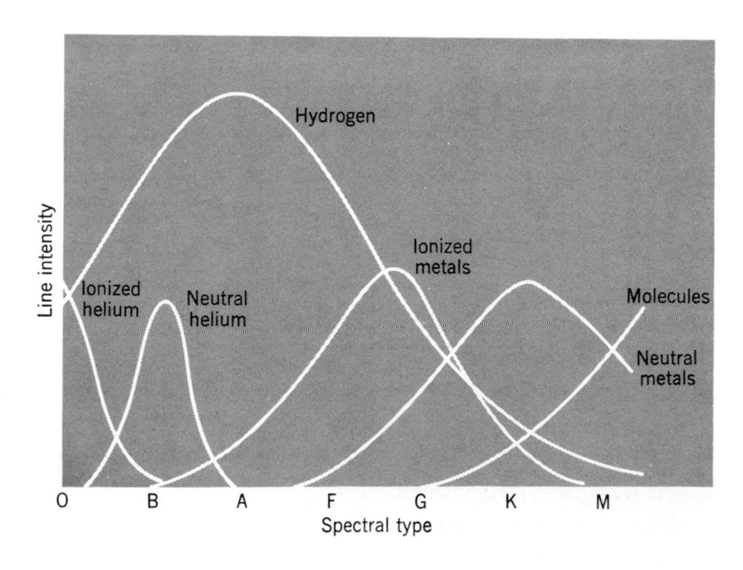

Figure 9-6
The intensity of a given stellar spectral line depends on the temperature of the star. Lines resulting from ionized atoms of a given chemical element will be more intense in a hotter star, and the lines of the neutral atoms will be more intense in the cooler star.

B. SPECTRAL LINES AND SPECTRAL TYPES The relationship between stellar temperatures and spectral line intensities is very detailed; consequently, it is best to summarize it in Table 9-1 and in a graph (Figure 9-6). It will be seen in both the table and the graph how the spectra change when considering stars of different temperatures.

The range of temperatures within each spectral class gives rise to a subdivision of the spectral classes by number. For example, there are stars that are classified as A0, other as A1, A2, A3, etc. An A3 star is cooler than an A2. Such subdivisions within each spectral type are recognized by differences in line intensities, the hydrogen lines being strongest in A0. The sun is a G2 star.

A peculiar terminology has arisen along with the spectral classes. Very often astronomers want to speak not of one particular spectral class but of a group of spectral classes or subclasses. It was once thought that the O-type stars were the youngest and that as stars grew older they cooled progressively into M-type stars. Therefore the stars classed as O, B, and A are loosely spoken of as *early-type* stars, and the K and M stars are referred to as *late-type* stars. An F0 or F1 star is sometimes called an early F star, but an F8 may be referred to as a late F star. This terminology is one of convenience only and has no strict definition or hidden meaning. It will be seen in Chapter Eleven that an O or B star is indeed likely to be younger that a type M.

Stars of very low temperature have recently been discovered and more are expected to be found as current researches continue at a number of observatories. These stars radiate mostly in the infrared region of the spectrum. Special equipment has been set up at Mount Wilson Observatory to survey the sky in search of *infrared stars*. Of those found so far, temperatures as low as 1000° K have been measured. What sort of stars are these? Why are they so cool? These are questions being asked by astronomers.

9.6 STELLAR LUMINOSITIES

The temperature of a star, according to the Stefan-Boltzmann law, determines how much energy is emitted per unit area of the star's surface. Of two stars the same size, the hotter will radiate more energy; of two stars the same temperature, the larger one will radiate more energy. The *luminosity* (intrinsic brightness) of a star, then, depends on two factors: its temperature and its size.

A. BRIGHTNESS AND APPARENT MAGNITUDE A star's *brightness*, however, is simply its apparent brightness as seen in the sky and depends not only on its luminosity but also on the distance between it and the sun. If we had three stars of the same luminosity at distances of 1, 2, and 3 parsecs, their brightnesses would progressively decrease with increasing distance according to the *inverse square law*. That is, if the brightness of the first star were 1 (on an arbitrary scale), the brightness of the second star would be ¼ which equals $(\frac{1}{2})^2$, and that of the third star would be $^1/_9$ which equals $(\frac{1}{3})^2$. If a star of the same luminosity as these three were 100 parsecs away, its brightness, on the same scale, would be $^1/_{10,000} = (^1/_{100})^2$. If the luminosity of any one of these were twice as great, its brightness would be twice as great.

The first astronomers concerned enough to make a record of the brightnesses of stars as they saw them were the Alexandrian Greeks. In the absence of a means for determining a star's distance from the sun they considered its brightness only. It was Hipparchus (about 150 B. C.) who first established a scale, called *apparent magnitude,* for comparing the brightnesses of stars. He did this in the course of compiling the first known star catalog, which contained the positions and brightnesses of just under 1000 stars. The brightest stars he arbitrarily called of the *first magnitude*; the faintest stars visible to the naked eye were of the *sixth magnitude,* and all other stars were of intervening magnitudes according to their brightnesses. The system worked quite well, for it is fairly easy to estimate the brightness of a visible star on this scale. But with the advent of the telescope and of photography, the scale has had to be extended to ever fainter stars, until the *faintest* stars now photographed are of about the twenty-third magnitude.

Hipparchus misjudged the magnitudes of some of the brighter stars, however. When the magnitude scale was extended and expressed by a mathematical formula, it developed that the brighter stars are brighter than those of the first magnitude; indeed they are even brighter than those of zero magnitude. The only way to express these hitherto unsuspected magnitudes and yet retain the old scale (thereby avoiding a wholesale revision of stellar catalogs) was to adopt a few negative magnitudes. As a result, Sirius, the brightest star in the sky aside from the sun, has an apparent magnitude of -1.4 and the star 61 Cygni A has an apparent magnitude of $+5.4$ and is just visible to the naked eye on a clear moonless night when viewed from a place far from the lights, smoke, and smog of our modern cities. The magnitudes of six of the stars in Figure 9-7 are given. Since the brightest has an apparent magnitude of 8.2, none of these stars is visible without the aid of a telescope.

There are stars over the entire sky which have been chosen as standard stars for magnitude determination. The magnitude of any star can be determined by comparing it with one of the standard stars which appear in the same part of the sky as the star being studied. The magnitudes are determined by comparing the electric currents set up when the star's light strikes a photoelectric cell. The brighter the star the greater the electric current produced in the cell. If a star whose apparent magnitude is to be determined produces an electric current equal to one of the standard stars, it has the same apparent magnitude as that standard star.

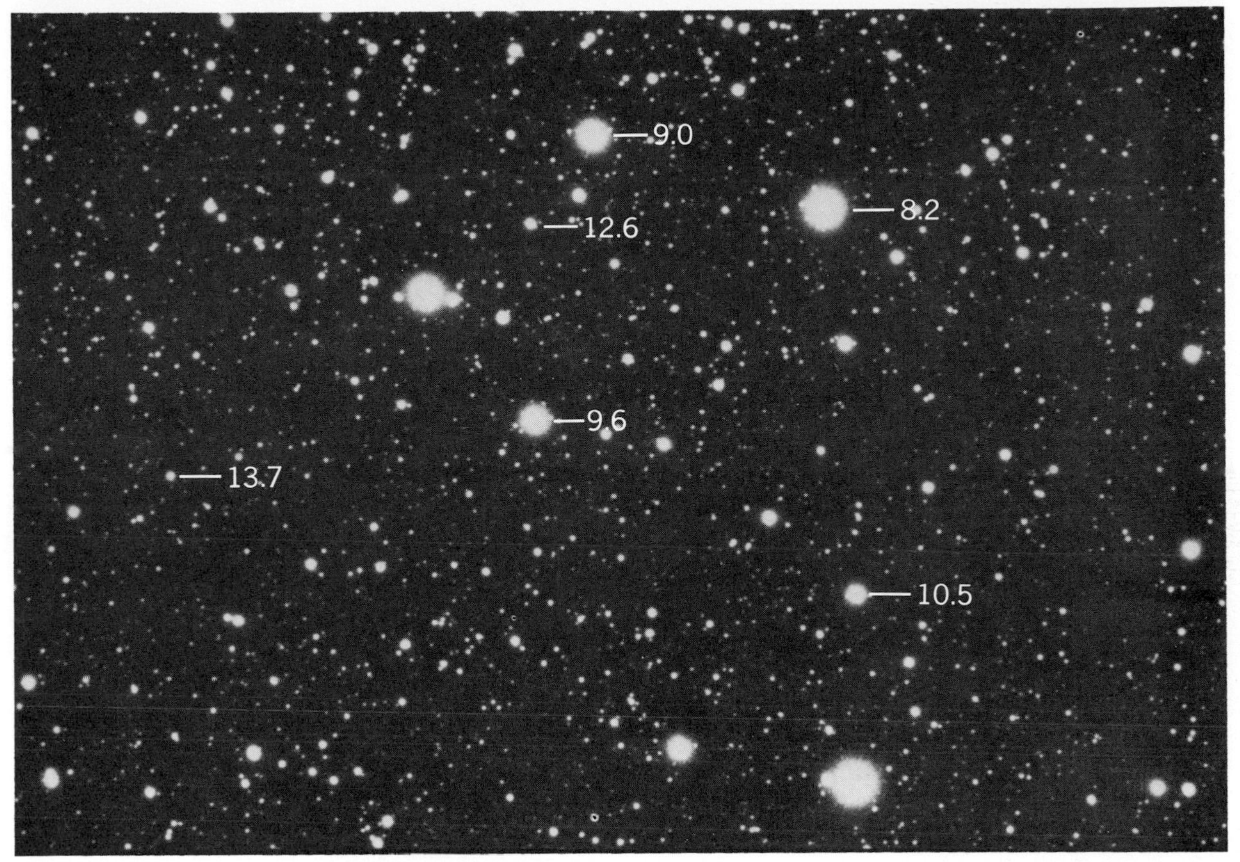

Figure 9-7
The magnitudes of some stars are marked in this photograph. The brightest stars leave the largest disks, since they expose the film more. (Courtesy Merle Walker, Lick Observatory.)

B. LUMINOSITY AND ABSOLUTE MAGNITUDE Before we can determine a star's *luminosity* (as distinguished from its brightness), its distance must be known. The scales used to measure luminosity are as arbitrary as the scale of apparent magnitude, and indeed one of the scales of luminosity, called *absolute magnitude*, is based on the scale of apparent magnitudes. It is an indication of the total amount of light radiated by the star and not the amount we receive here on the Earth. Unlike apparent magnitudes, absolute magnitudes are independent of the star's actual distance. By definition *the absolute magnitude of a star is the apparent magnitude the star would have, were the star 10 parsecs from the sun.* On this scale, then, all stars are compared from the same standard distance. If a star is closer than 10 parsecs, its apparent magnitude is numerically less (that is, it is brighter) than its absolute magnitude. If a star is more than 10 parsecs distant, its apparent magnitude is numerically greater (fainter) than its absolute magnitude.

Sirius, for example, is 2.7 parsecs from the sun. Its apparent magnitude is −1.4 and its absolute magnitude is +1.5. This means that if Sirius were moved away to 10 parsecs from the sun its apparent magnitude would be +1.5. Pollux, with an apparent magnitude of +1.1, is just barely more than 10 parsecs from the sun and thus its absolute magnitude is +1.0. On the other hand, Rigel, with an apparent magnitude of +0.14 is 200 parsecs from the sun so its absolute magnitude is −6.4. Since Rigel has an absolute magnitude of about 8 less than Sirius, it is about 1,600 times more luminous. The sun is so close that it has an apparent magnitude of −26.7, even though its absolute magnitude is only +4.9. Rigel's absolute magnitude is therefore 11.3 magnitudes less than that of the sun, that is, it is 33,000 times as luminous.

Paralleling the scale of absolute magnitude is another scale whose reference point is the sun, which is given an arbitrary luminosity of 1.00. A star with a luminosity of 10 would be 10 times more luminous than the sun; one with a luminosity of $^{1}/_{100}$ is $^{1}/_{100}$ as luminous as the sun.

The range in the luminosities of the stars is staggering. The most luminous star is about 10^{11} times more luminous than the least luminous star. The most luminous stars are about 100,000 times more luminous than the sun, whereas the least luminous stars known have a luminosity about $^{1}/_{1,000,000}$ that of the sun. Barnard's star (see Figure 9-3) has a luminosity of only 0.00044. With an absolute magnitude of +13.2 and at a distance of only 5.9 light years, its apparent magnitude is +9.5.

For the study of *intrinsic* characteristics of the stars, luminosity (absolute magnitude) has a great deal of meaning, whereas brightness (apparent magnitude) has essentially no meaning because of the different distances involved.

9.7 THE H-R DIAGRAM

It is of interest to graph the stars by plotting their luminosities against their temperatures (Figure 9-8). Since a star's luminosity depends on both its temperature and size (see p. 189), a graph plotting temperature against luminosity separates the stars according to size. If two stars have the same temperature but different diameters the smaller will be less luminous, since it has a smaller surface area and thus is located lower on the graph. Conversely, if two stars have the same luminosity but different temperatures the cooler must be the larger. Since it is cooler it is located farther to the right on the graph.

Such a graph is called an *H-R diagram* in honor of Hertzsprung and Russell, the two astronomers instrumental in compiling the first such diagram. Because of its tremendous significance in astronomy we shall refer to the H-R diagram frequently.

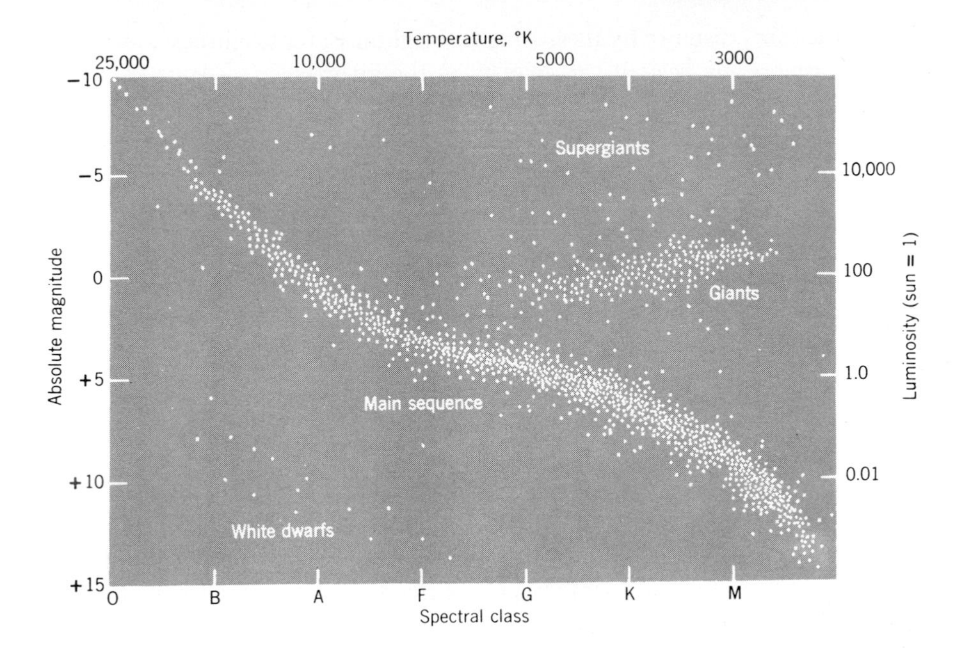

Figure 9-8
The H-R diagram.

We notice that there is a great number of stars forming a sequence from the upper left to the lower right of the diagram. This band is called the *main sequence*, because it includes the majority of the stars observed. Above the main sequence we see a rather large group of stars classified as *giants* and *supergiants*. The supergiants, as the name indicates, are more luminous and thus larger than the giants, which are in turn larger than the main-sequence stars of the same temperature. Below and to the left of the main sequence, we see a group of stars that must be smaller than the main-sequence stars of the same temperature and thus are called *white dwarfs*. The late main-sequence stars (the sun included) are often called dwarfs because of their smallness when compared with giants of the same temperature. Nevertheless they are larger than the white dwarfs and are not white in color.

That such a distinctive grouping of stars should exist when they are placed on a graph according to their temperatures and luminosities cannot be mere coincidence; there must be some meaning to this arrangement. It is well established that these groupings represent the stages that a star goes through in its evolution from birth to death. We shall examine this process in Chapter Eleven.

9.8 SPECTROSCOPIC PARALLAX

A. THE INVERSE SQUARE LAW OF BRIGHTNESS One of the more beneficial results of the H-R diagram is that it can be employed to find the distances of stars too far away for heliocentric parallax determination. This is possible because it gives us the luminosity of a star, and once the luminosity is known we can find the star's distance by measuring its brightness; for brightness, luminosity, and distance are related very nicely by the inverse square law. By using luminosity units (as against the system of magnitudes) the brightness B of a star in the sky is proportional to the star's luminosity L and inversely proportional to the square of its distance d from the sun:

$$B \propto \frac{L}{d^2}$$

This proportionality can be converted into an equality by inserting a constant of proportionality K:

$$B = K\frac{L}{d^2}$$

The numerical value of K depends upon the choice of units for the brightness B and L. If L is based on the sun's luminosity, B can be based on a group of stars just as the apparent magnitude is.* This equation, expressed in units of luminosity and brightness, permits us to see clearly how the brightness of a star depends upon its luminosity and distance. If both the brightness and luminosity of a star are known, its distance can be found by algebraic manipulation of the equation

$$d = \sqrt{\frac{K \cdot L}{B}}$$

But the problem of determining the luminosity of a star without first knowing its distance remains. To do this we must find some way of determining, independently of its luminosity, whether the star is a dwarf (late main-sequence), a giant, or a supergiant. If this can be determined, a star's luminosity can be found by correlating its spectral type with its size. For example, if a K0 star is estimated to be a dwarf, its luminosity must be about 0.4; if a K0 star is a giant, its luminosity will be a little less than 100 (see Figure 9-9). But how is it possible to determine the size of a star from a study of its spectrum?

B. PRESSURE BROADENING AND LUMINOSITY A supergiant star may have a diameter of 1000 times that of the sun, but its mass is likely to be only 20 times the sun's mass. (Stellar mass is discussed in Chapter Twelve.) Its tremendous size, then, does not correspond to its more meager mass. Is it possible that giant and supergiant stars are simply expanded versions of the more numerous main-sequence stars? We will proceed on the assumption that they are, and then let observations test this assumption.

*An equivalent expression used by astronomers which relates the apparent magnitude m of a star to its absolute magnitude M and its distance d (in parsecs), is logarithmic in nature: $M = m + 5 - 5 \log d$.

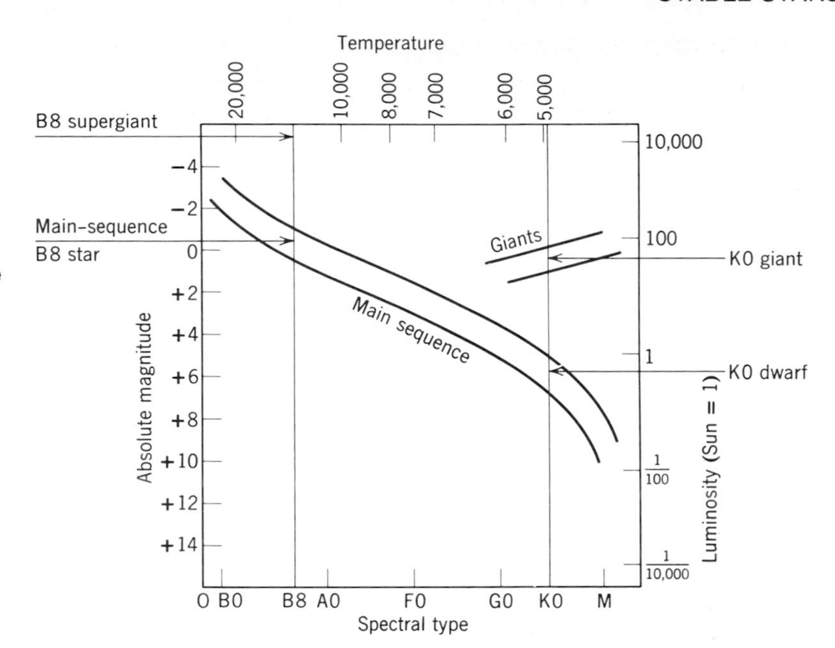

Figure 9-9
Spectroscopic parallax depends on the ability of the astronomer to determine from a star's spectrum alone whether a given star is a main-sequence star, a giant, or a supergiant. Once this is determined, the H-R diagram yields the star's luminosity. The temperatures refer to the main sequence.

If a giant star is an expanded main-sequence star, then the density and pressure of the gas composing a giant must be less than that in the main-sequence star. Is there an observation that might permit us to determine the pressure of the gas in the atmosphere of a star?

Our main source of information of stars is their spectra, namely, their spectral lines. The spectral lines are caused by atoms of the gas; but the movement of those same atoms causes the pressure of the gas. As those atoms move, they collide with one another resulting in gas pressure. These collisions are relatively infrequent in a gas at low pressure, so they do not affect the atoms much. But in a gas at high pressure, the collisions between atoms become so frequent that the energy levels of the atoms become distorted. A photon that results from transitions in an atom with distorted energy levels does not have a precise amount of energy; its energy for any transition can vary within the limits of the distortion. The greater the pressure, the greater the distortion, and the greater the resulting variance in the energies of the photons emitted.

It is the energy of the photons that determines their wavelength. If the energies of the photons resulting from a particular transition are all the same, the resulting spectral line will be of a single wavelength, that is, it will be very narrow. But photons emitted by a gas at high pressure will have a variation of energies, and thus a variation in wavelengths. Therefore, the spectral lines will not be narrow, but broadened.

This line of logic, then, would permit us to predict that giant stars, being composed of gas at low pressure, have very narrow lines in their spectra. On the other hand, main-sequence stars, being composed of gas at higher pressures,

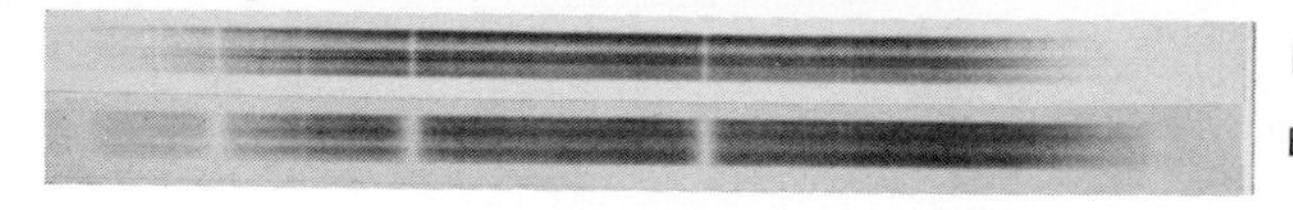

Figure 9-10
The spectral lines in a B8 main-sequence star are noticeably broader than those in a B8 supergiant. The most obvious lines in these spectra are those of hydrogen. The spectra are photographic negatives, so the normally bright continuous spectrum is dark and the normally dark absorption lines are bright. (Yerkes Observatory.)

have slightly broadened spectral lines. The spectra in Figure 9-10 are the observations used to test our initial assumption and the ensuing logic. The spectral lines of the supergiant B8 star are definitely narrower than those same lines of the B8 main-sequence star. Our initial assumption and the logic are acceptable. The broadening of the spectral lines resulting from pressure is, logically enough, called *pressure broadening*.

According to the H-R diagram (see Figure 9-9) a B8 supergiant star has a luminosity of about 30,000 times that of the sun, while the luminosity of the B8 main-sequence star is only about 100. These luminosities, then, permit us to distinguish between a dwarf, a giant, and a supergiant.

Once the luminosity of a star has been determined, we must obtain its brightness, before determining its distance. Its brightness can be determined either from a catalog of stars that lists the brightness or apparent magnitude, or by making direct observations of the star with either a photocell or photographic plate. Then, knowing both the luminosity and the brightness of the star, we can calculate its distance by means of the inverse square law given on p. 194.

Since this method of distance determination is based on a study of the spectrum, it is called the method of *spectroscopic parallax*. It has proved a very powerful tool in the astronomer's hands, although it involves certain difficulties. One difficulty is the determination of luminosity from the spectral lines. A slight error in judgment of line width and intensity means a large error in luminosity. Another difficulty lies in the fact that the inverse square law assumes that no light has been lost in transit from the star. But this is not always true, as will be seen in Chapter Thirteen.

9.9 THE MASS-LUMINOSITY RELATION

One of the most important and yet one of the most difficult characteristics to determine is the mass of a star. The only opportunity for making direct measurements of stellar mass is afforded by a star that is a member of a

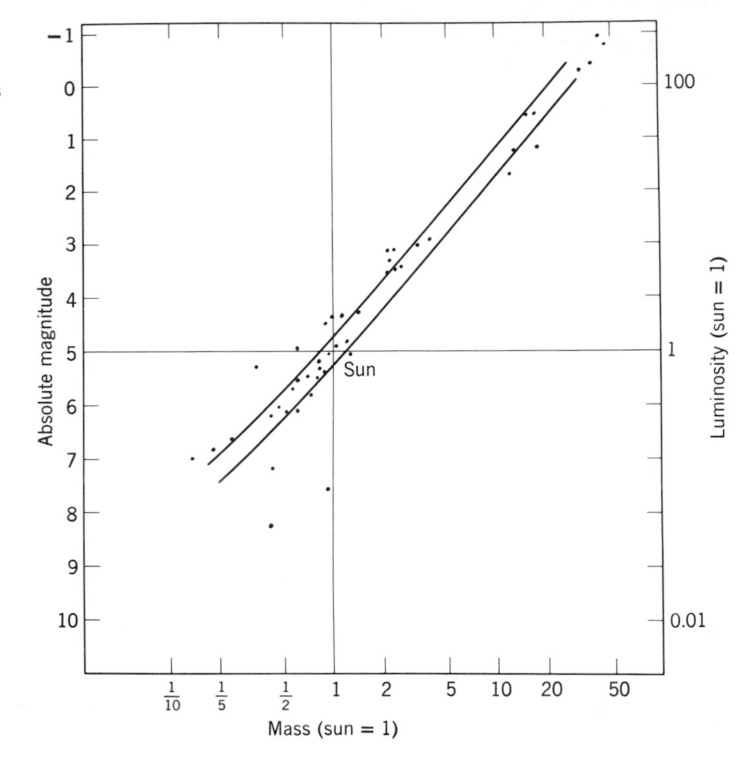

Figure 9-11
The mass-luminosity relation indicates that stars of increasing mass will be more luminous.

double-star system. Since the gravitational field about a star depends solely on its mass and since the motions of the two stars in a double-star system depend on their mutual gravitational attraction, their masses can be determined if we can describe their orbital motions. This process is discussed briefly in Chapter Twelve; we may, however, discuss a closely related point at this time.

After the masses of a number of stars had been determined a close correlation between mass and luminosity was recognized. The more massive a star, in general, the more luminous it is. This *mass-luminosity relation* gives us a rough method for determining the mass of an isolated star. The mass thus determined is only an approximation, for we can see in Figure 9-11 that the stars plotted scatter about a mean curve that must be used to determine the mass. Since there is considerable deviation from this mean curve, the mass of any one star is always in doubt. When a large number of stars is used, however, the method can be employed very effectively to determine the average mass of a particular *group* of stars. For example, the masses of the main-sequence stars of each spectral type have been determined by this method and are given in Table 9-2, from which it appears that the masses vary less than the other characteristics.

Since direct determination of mass can be made of only those stars that are members of binary systems, the mass-luminosity relation is based on such stars. If single stars differ from stars that are members of binary systems, this empirical mass-luminosity relation has limited value. It is known that some types of stars (for example, white dwarfs) do not follow the mass-luminosity

Table 9-2
The masses of main-sequence stars

Spectral Type	Mass	Spectral Type	Mass
B0	16	G0	1.0
B5	6	G8	0.9
A0	4	K0	0.8
A5	2	K5	0.6
F0	1.5	M0	0.5
F5	1.3	M5	0.2

relation and there are even special classes of binary stars that violate it. It is generally held, however, that most stars do follow this relation, which has become an important part of stellar astronomy.

The most massive star known is an O8 star which is a member of a double-star system. The name of that star is HD 47129 and each star in this system has a mass close to 50 times that of the sun. The least massive star known is less than 0.1 solar mass.

9.10 THE LUMINOSITY FUNCTION (OR STELLAR CENSUS)

One of the goals of the astronomer is to understand the entire life history of a star including its formation and ultimate death. To achieve this goal, he must know how many of each kind of star there are in our galaxy. Which kind of stars are the most numerous, the very massive ones or the less massive ones? The mass of single stars, however, cannot be determined and thus a study of masses of a large number of stars must be performed by studying their luminosities.

The number of stars of each luminosity have been counted, and a graph showing their distribution has been prepared. This graph, called the *luminosity function*, tells us directly what proportion of the stars have a particular luminosity and thus indirectly what proportion have a particular mass. It can be seen from Figure 9-12 that the low luminosity stars are much more numerous than the high ones. Thus, the less massive stars are much more numerous.

Does this mean that less massive stars form more readily? Or does it mean that they have a longer life? Houseflies are very abundant because they are born in quantities too numerous for comfort, despite the fact that they live for only a few weeks. Human beings, on the other hand, are numerous because they have a life expectancy of 65 to 70 years, even though their birth rate is much lower than that of the fly. We shall discuss the answer to the question "Why are less massive stars so numerous?" in Chapter Eleven.

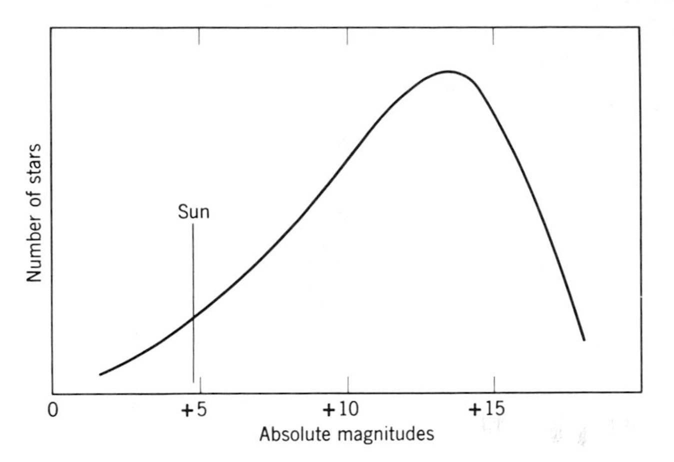

Figure 9-12
The luminosity function (or stellar census) indicates that the less luminous stars are far more common than the very luminous ones.

9.11 CHEMICAL COMPOSITION

The last characteristic, chemical composition, is not startling; it simply answers the question, "What are stars made of?" No matter what substances may compose it, we are fairly certain that every star is gaseous throughout, if by gaseous we include some rather exotic forms of gas such as are found in white dwarfs. It is possible, although difficult, to determine the chemical compositions of most of the stars' atmospheres from an analysis of their spectra. But if we want to go inside a star, we must, for obvious reasons, forego direct observation and use instead the fruits of theoretical studies.

These studies are in the forefront of astronomical thought. The conclusions reached indicate that stars begin life as balls of gas composed mostly of hydrogen and helium. As they generate energy by converting hydrogen into helium, the amount of hydrogen decreases and the amount of helium increases. As they age, the helium is converted into the heavier elements. Thus a star's chemical composition depends on its age.

There are some stars, however, that have in their atmospheres an unusual abundance, as yet unexplained, of one element or another. Examples are the carbon stars of the spectral classes R and N, and the S-type stars with their relatively high percentage of zirconium and technetium. These are only some of the many peculiar stars whose behaviors do not fall within the confines of this chapter's description of "normal" stars.

BASIC VOCABULARY FOR SUBSEQUENT READING

Absolute magnitude
Angle of parallax
Apparent magnitude
Early-type star
Giant star
Heliocentric parallax
Late-type star
Light year
Luminosity
Main sequence
Parsec
Pressure broadening
Proper motion
Radial velocity
Spectroscopic parallax
Supergiant star
Tangential velocity
White dwarf

QUESTIONS AND PROBLEMS

1. The angle of parallax p in arc seconds and the distance of a star d in parsecs are related by the equation $d = 1/p$. The smaller the angle of parallax, the farther away the star is. Given the angle of parallax, find the distance of the following stars in parsecs:

(a) Sigma Draconis $p = 0.17$ arc seconds
(b) Epsilon Eridani $p = 0.30$ arc seconds
(c) Barnard's star $p = 0.54$ arc seconds

2. One parsec equals 3.26 light years. Find the distance of each star in question 1 in light years.

3. A star has a proper motion of 0.5 arc seconds per year. How many years will it be before the position of that star on the celestial sphere changes by 1°. Recall that there are 60 arc seconds in one arc minute and 60 arc minutes in one degree.

4. What are the two primary factors that determine how much radiant energy a star emits?

5. Arrange the following types of spectral lines in order as they appear in stars of ever increasing temperature:
(a) neutral helium, (b) molecular bands, (c) neutral metals, (d) ionized helium, (e) ionized metals, (f) lines of hydrogen at their strongest.

6. By using their apparent magnitudes, list the following stars in order of decreasing brightness in our night sky (place the brightest first):

Star	Apparent Magnitude
Tau Ceti	+3.5
Arcturus	−0.1
Barnard's star	+9.5
Deneb	+1.3
Sirius	−1.4
Aldebaran	+0.9

7. By using their absolute magnitudes, list the following stars in order of decreasing luminosity (place the most luminous first):

Star	Apparent Magnitude
The sun	+ 4.8
Wolf 359	+16.8
Rigel	− 7.0
Ross 619	+12.9
Polaris	− 4.5
Aldebaran	− 0.8

8. By using absolute magnitude, list the following stars (all of which are at nearly the same distance) in order of decreasing brightness in our night sky (place the brightest first):

Star	Absolute Magnitude	Distance (parsecs)
Polaris	−4.5	200
Betelgeuse	−5.9	180
Epsilon Canis Majoris	−5.0	200
Lambda Velorum	−4.3	200
Beta Scorpii	−4.0	200

9. By using their distances, list the following stars (all of which have nearly the same absolute magnitude) in order of decreasing brightness in our night sky (place the brightest first):

Star	Absolute Magnitude	Distance (parsecs)
Alpha Persei	−4.1	150
Beta Scorpii	−4.0	200
Bellatrix	−4.1	140
Delta Scorpii	−4.0	180

10. (a) Using their apparent and absolute magnitudes, list the following stars in order of increasing distance (place the nearest star first):

Star	Apparent Magnitude	Absolute Magnitude
Ross 619	+12.9	+13.8
Wolf 359	+13.7	+16.8
Regulus	+ 1.3	− 0.8
Alpha Centauri	+ 0.0	+ 4.4
Ross 128	+11.1	+13.5
Deneb	+ 1.3	− 7.2
Pollux	+ 1.1	+ 1.0
AD Leo	+ 9.4	+11.0

(b) Which star in the above list is about 10 parsecs distant from the sun?

11. Describe how the distance of a star can be estimated from a study of its spectrum.

12. Describe how the following stellar characteristics are related:
(a) temperature, size, luminosity.
(b) luminosity, brightness in our sky, and distance from the sun.
(c) Mass, luminosity, and temperature (for main-sequence stars).
(d) Mass and diameter (for main-sequence stars).
(e) Distance from the sun and proper motion.

13. Given the spectral types (A1, B1, etc.) and the absolute magnitude ($M = +1.5$, etc.) form an H-R diagram for the following stars:

1. Sirius, A1, $M = +1.5$
2. Alpha Crucis, B1, $M = -3.7$
3. Ross 775, M4, $M = +11.3$
4. Beta Pegasi, M2, $M = -1.4$
5. 82 Eridani, G5, $M = +5.3$
6. Sirius B, A5, $M = +11.4$
7. Polaris, F8, $M = -4.5$
8. Rigel, B8, $M = -7.0$
9. Procyon, F5, $M = +2.7$
10. Canopuls, F0, $M = -3.1$
11. Epsilon Indi, K5, $M = +7.0$
12. Antares, M1, $M = -4$

FOR FURTHER READING

Kruse, W., and W. Dieckvoss, *The Stars,* The University of Michigan Press, Ann Arbor, Michigan, 1957.

Page, T., ed., *Stars and Galaxies,* Prentice-Hall paperback, Englewood Cliffs, N.J., 1962.

Struve, O., and V. Zebergs, *Astronomy of the 20th Century,* Crowell, Collier and Macmillan, New York, 1962, Chapters X-XIII.

Anderson, J. H., "The Stars of Very Large Proper Motion," *Sky and Telescope,* p. 76, August 1969.

Barton, R., "Some Effects of Stellar Proper Motions," *Sky and Telescope,* p. 4, January 1963.

Boyce, P. B., and W. M. Sinton, "Infrared Spectroscopy With an Interferometer," *Sky and Telescope,* p. 78, February 1965.

Brown, H. R., "The Stellar Interferometer at Narrabri Observatory," *Sky and Telescope,* p. 64, August 1964.

Gingerich, O., "Laboratory Exercises in Astronomy — Spectral Classifications," *Sky and Telescope,* p. 75, August 1970.

Hack, M., "The Hertzsprung-Russell Diagram Today," *Sky and Telescope,* part I, p. 260, May 1966; part II, p. 332, June 1966.

Kamperman, T. M. et al, "Results from the Utrecht Orbiting Spectrophotometer," *Sky and Telescope,* p. 85, February 1973.

Mumford, G. S., "Distance Modulus," *Sky and Telescope,* p. 274, May 1965.

Nielsen, A. V., "Ejnar Hertzsprung — Measurer of Stars," *Sky and Telescope,* p. 4, January 1968.

Sitterly, B. W., "Symposium on Abundances of Elements in Stars," *Sky and Telescope,* p. 11, January 1965.

Whitney, C. A., "New Trends in Spectral Classification," *Sky and Telescope,* p. 356, June 1968.

"Extremely Cool Stars," *Sky and Telescope,* p. 195, October 1965.

"Star Sizes Measured," *Sky and Telescope,* p. 139, March 1968.

CHAPTER CONTENTS

LEARNING OBJECTIVES

BE ABLE TO:

1. EXPLAIN THE CAUSES OF LIGHT VARIATION IN CEPHEID VARIABLES.
2. DESCRIBE THE PERIOD-LUMINOSITY RELATION.
3. EXPLAIN HOW CEPHEIDS AND RR LYRAE STARS ARE USED AS DISTANCE INDICATORS.
4. GIVE THE GENERAL NATURE OF A LONG PERIOD VARIABLE.
5. GIVE OBSERVATIONAL EVIDENCE, BOTH SPECTRAL AND DIRECT, INDICATING THAT A STELLAR EXPLOSION CAUSES A NOVA.
6. GIVE OBSERVATIONS AND REASONING INDICATING THAT THE CRAB NEBULA WAS A SUPERNOVA IN 1054.
7. GIVE THE ORIGIN OF SYNCHROTRON RADIATION.
8. DESCRIBE THE GENERAL NATURE OF A PULSAR.
9. DESCRIBE METHODS OF LOCATING SUPERNOVA REMNANTS.
10. DESCRIBE THE GENERAL NATURE OF PLANETARY NEBULAE.

Chapter Ten
NONSTABLE STARS

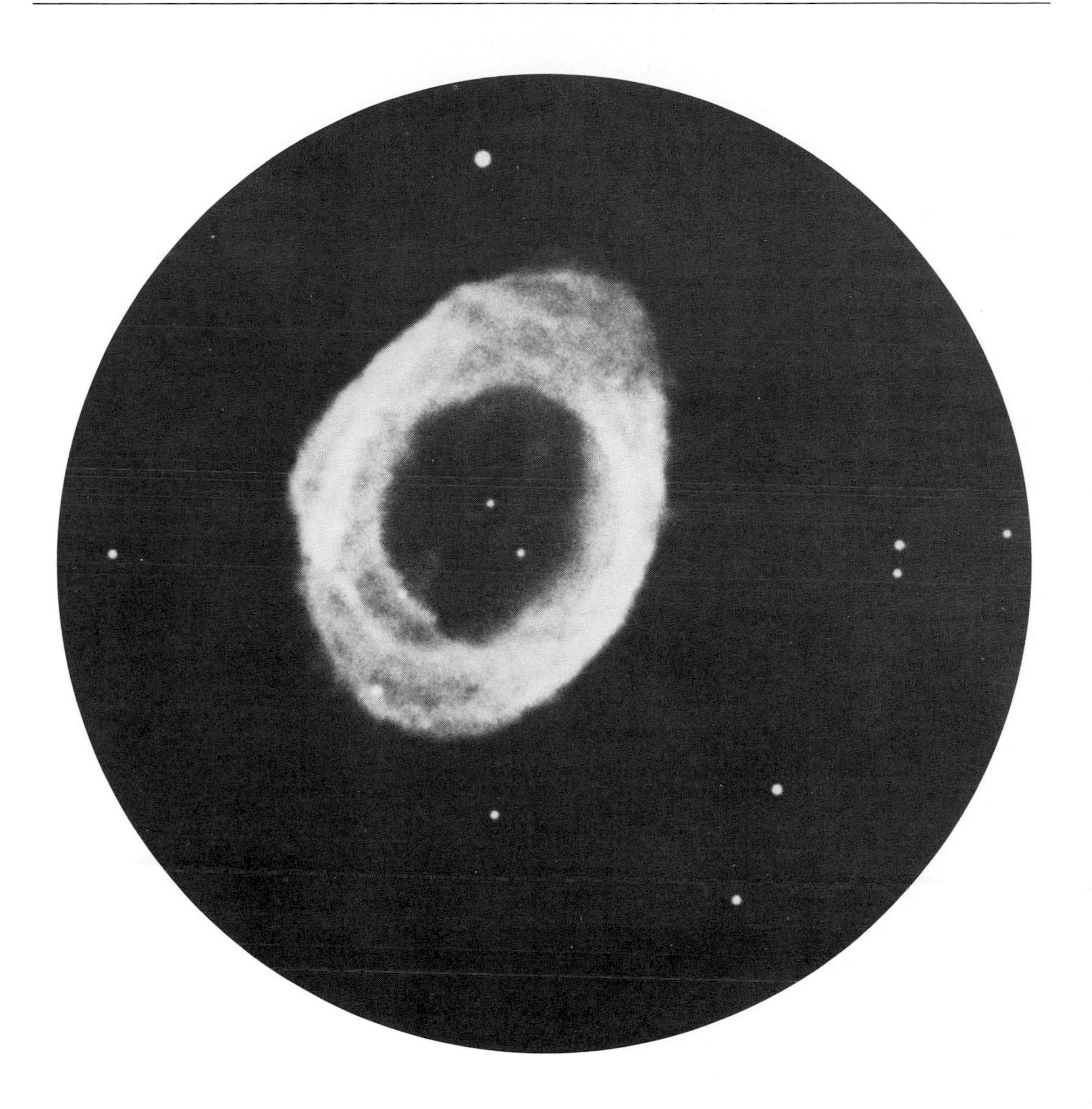

Nonstable stars, as their name implies, are stars that evidence a lack of stability or equilibrium. This evidence is a change in brightness which may indicate either a periodic pulsation, an irregular eruption, or a massive and nearly self-destroying explosion. It may be that many main-sequence stars will, by the normal process of stellar evolution, become unstable in a manner described in this chapter. For this reason, the nonstable stars are of particular interest to us; and, since their study is fraught with difficulties, their problems are a challenge.

10.1 CEPHEIDS

The most important class of these nonstable stars is called the *Cepheids*. The group derives its name from its most prominent member, Delta Cephei, a naked-eye star that misbehaves in that its luminosity varies in a periodic fashion. This variation of luminosity is detected by a variation in brightness or apparent magnitude. The apparent magnitude of Delta Cephei varies from about +4.3 at minimum brightness to about +3.6 at maximum. A variation of 0.7 in apparent magnitude amounts to a factor of almost 2 in the brightness. Thus, Delta Cephei at its maximum brightness is nearly twice as bright as at its minimum.

A. THE LIGHT CURVE The light variation of all the Cepheids is periodic in nature, repeating itself over a certain interval of time with only small deviations from the observed pattern. When the pattern of variation for a given Cepheid has been determined from repeated observations, it is possible to predict with considerable accuracy when it will be at its maximum brightness. Indeed, it is possible to predict when it will have any given magnitude within its range of variation. Such a prediction may be made from a graph on which the variation in brightness is plotted against time. The resulting graph is called the *light curve*.

The light curve for Delta Cephei is shown in Figure 10-1. The interval of time elapsing between two successive maxima or minima is called the *period*. Delta Cephei's light curve shows that its period is close to 5½ days. A light curve made from observations over a longer time interval enables us to measure its period at 5 days, 8 hours, 46 minutes, 38 seconds. But not all Cepheids exhibit this period of light variation. The shortest period for a Cepheid is just over a day and the longest is more than 50 days.

Since the periods are longer than 1 day, it is not possible to follow the variation of one entire period on any one night. The light curve, therefore, must be determined from many observations of brightness over many nights; all these

Chapter Opening Photo
An unstable star ejected gases that now form the Ring nebula in the constellation Lyra. (Lick Observatory photograph.)

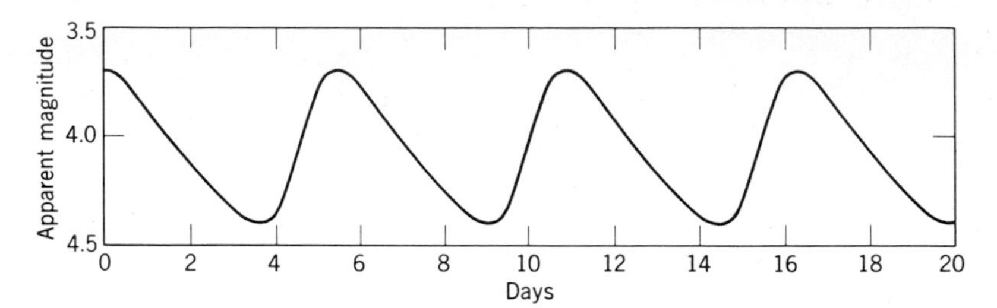

Figure 10-1
A schematic diagram of the light curve of Delta Cephei.

observations must then be fitted onto a curve that can account for the past variations and be used to predict future ones. Thus it may take many months to obtain the light curve of a Cepheid with a period of 45 days.

B. PULSATIONS The cause of light variations cannot be determined from the light curves alone; for this reason the spectra of the Cepheids have been investigated quite thoroughly and with startling results. There is not only a periodic change in the spectral lines but there is also a shifting of the lines; they oscillate back and forth across a mean position. Such a shifting of lines has been interpreted as a velocity or Doppler shift. If this interpretation is correct, and all indications are that it is, then there must be some periodic motion of the star; the star must either revolve in an orbit or pulsate. Both possibilities have been investigated, and astronomers have come to the conclusion that apparently the Cepheids are nonstable stars and pulsate like balloons that are partially deflated, then inflated, then partially deflated again, and so forth. Thus, the surface facing us periodically approaches and recedes from us, thereby causing the spectral lines to shift periodically to the blue and to the red.

C. THE PERIOD-LUMINOSITY RELATION The Cepheids, because of their light variation, constitute one of the classes of stars most important to the astronomer. Since astronomical distances are so great, and methods of measuring them are so difficult, astronomers are always looking for new and better ways to verify and extend our present knowledge of these distances. The Cepheids offer a unique method, for their luminosity and period are related so that once the period of a Cepheid is determined, its luminosity can easily be estimated. This relation, called the *period-luminosity relation*, was originally found by studying the Cepheids in the *Small Magellanic Cloud*, a large aggregation of stars visible as a faint patch of light in the southern hemisphere not too far from the South Celestial Pole (Figure 10-2).

In 1912, Henrietta S. Leavitt, of the Harvard College Observatory, plotted the periods of the Cepheids seen in the Small Magellanic Cloud against their brightness and found the relation shown in Figure 10-3: the longer the period, the brighter the star. It was realized that the cloud has a very small diameter compared to its distance from the sun and consequently all the stars in the cloud could be considered as being the same distance from the sun without introducing much error into measurements that compared the luminosities of these stars.

Figure 10-2
The two Magellanic Clouds are small galaxies that are satellites of our Milky Way galaxy. (Harvard College Observatory photograph.)

Thus the period-*brightness* curve discovered by Miss Leavitt should mean that a period-*luminosity* curve exists, and if the luminosity of even one Cepheid could be determined along with its period, the luminosity of each Cepheid in the Small Magellanic Cloud could be determined by reference to this standard Cepheid. That is, the scale on the left of Figure 10-3 could be converted from apparent magnitude (brightness) to absolute magnitude (luminosity). If the luminosity of each Cepheid were known, the distance of the cloud from the sun could be determined by the inverse square law, as could the distance of *any* Cepheid in the sky whose period was known.

Unfortunately, there is no Cepheid close enough to the sun for a heliocentric parallax determination. The problem of finding the luminosity of a Cepheid or group of Cepheids took on major interest for astronomers. The actual determination of a distance and thus a luminosity of Cepheids took several decades of

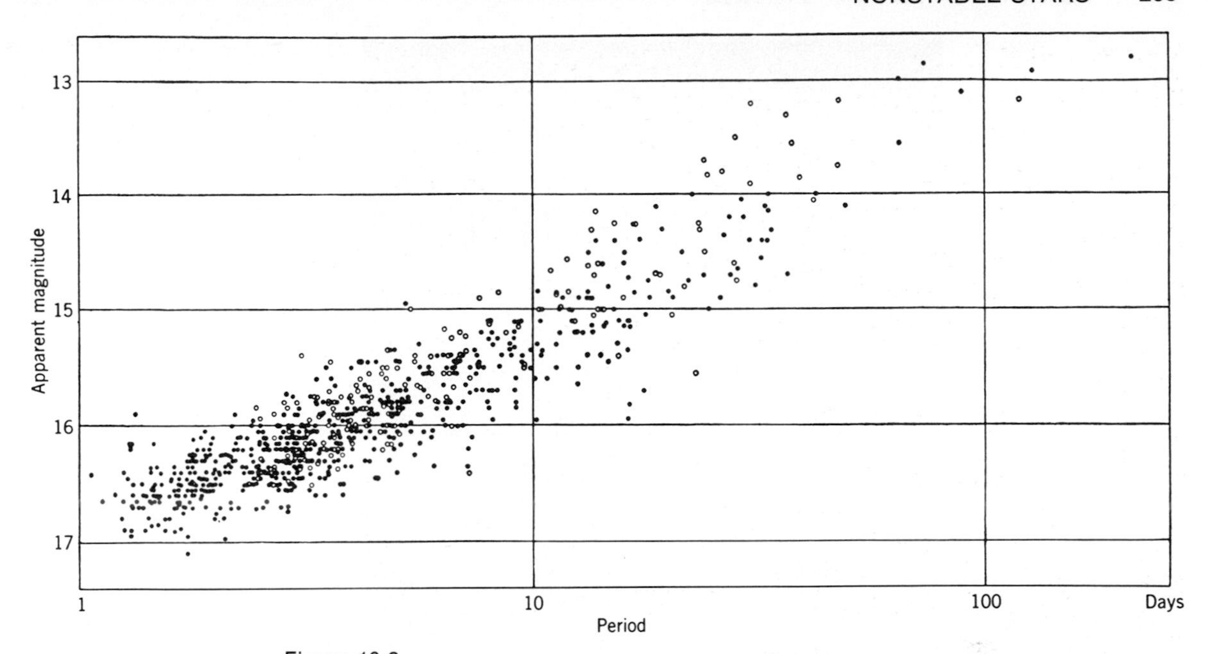

Figure 10-3

The period-luminosity relation for the Cepheids of the Small and the Large Magellanic Clouds. (Modified by permission from *Astrophysics,* edied by J. Hynek, Copyright 1951, McGraw-Hill Book Co.)

observations and study, and then had to be revised when it was realized that there is more than one kind of Cepheid. The kinds appear to differ in age; the younger ones are called Population I and the older Population II.*

Once the absolute magnitudes of a number of Cepheids were finally determined, the distance of every Cepheid seen in the sky could be calculated. It was mainly through this discovery that the astronomer began to realize that the universe is really much larger than he had previously thought, and he began to speak in terms of millions of light years. Before that time (in the mid-1920s) it was not known with any certainty whether the so-called "spiral nebulae" were a part of our galaxy or were outside and enormously farther away and larger.

D. CEPHEIDS AND THE H-R DIAGRAM Now that the luminosities of the Cepheids are known, it is instructive to plot them on the H-R diagram. (Since both their luminosities and their spectral types vary, the average of the extremes is used for such purposes.) Figure 10-4 shows the position of the Cepheids (along with some other variables to be discussed in this and later chapters) on the H-R diagram. The Cepheids' luminosity places them in the giant and supergiant class, which is fortunate because they are visible even at great distances from the sun.

*The nature of stellar populations will be discussed in more detail later.

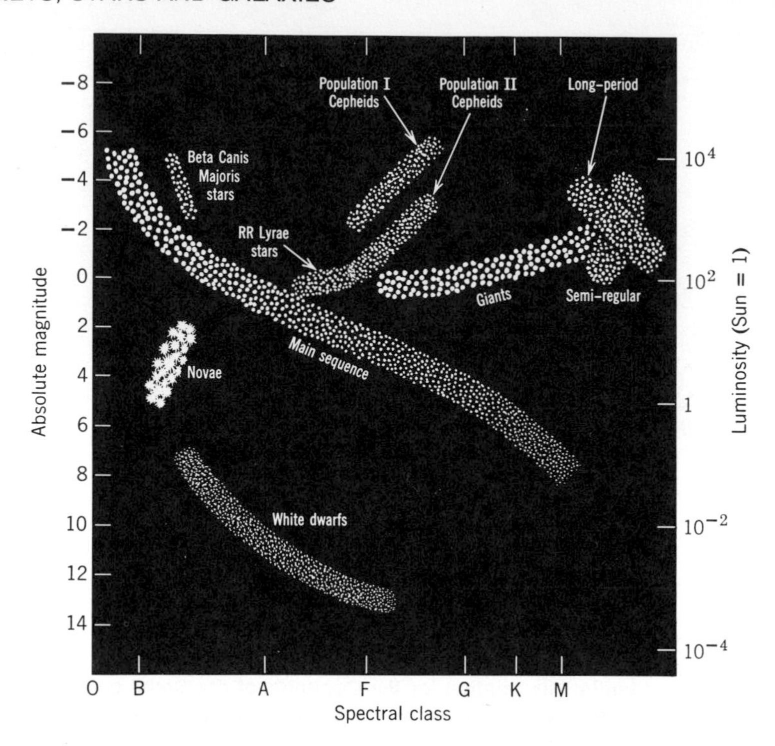

Figure 10-4
A schematic H-R diagram showing the location of many of the nonstable stars in relation to the main sequence and the giants.

The H-R diagram shows also that among the Cepheids there is a relationship between spectral type and luminosity; the later the spectral type, the more luminous the star. An F-type Cepheid has a luminosity of about 100, and a G-type has a luminosity of about 10,000. At first glance this may seem contrary to the Stefan-Boltzmann law until we recall that this law only refers to luminosity *per unit area.* Thus, the later-type Cepheids must be very much larger than those of early type.

There is, then, a very definite relationship between the period of the Cepheids and a number of their other characteristics. An increase in period is accompanied by: (1) an increase in luminosity; (2) an increase in diameter; (3) a decrease in temperature; (4) a larger range in temperature variation; and (5) a larger range in luminosity variation. That all of these are so related is significant and must depend on the masses and sizes of the various Cepheids.

10.2 THE RR LYRAE STARS

There are other stars related to the Cepheids in that they, too, appear to pulsate, and yet they differ enough to be classified separately. The first of these to be discussed are called the *RR Lyrae* stars after one of their members. Because they are often found in globular clusters (see p. 242) they are sometimes called cluster-type variables. These stars belong to Population II and have periods ranging from about 1½ hours to a little over 24 hours. The longer-period RR Lyrae stars blend into the type II Cepheids, since there are several stars that could fall into either group. The most luminous RR Lyrae stars have an absolute magnitude of 0.0 (close to 100 times the luminosity of the sun), and the least luminous have an absolute magnitude of +1.2 (Figure 10-4); the average is about +0.6. As with the Cepheids, the range in luminosity for a given period is attributed to differences in surface temperature.

The greater luminosity of the Cepheids results in their being seen at much greater distances than the RR Lyrae stars. The Cepheids can be used as distance indicators outside of our Milky Way Galaxy; RR Lyrae stars can be so used only inside our galaxy and its immediate environs. But, despite problems yet to be resolved, both groups of stars have been indispensable links in the measurement of distances within our galaxy and throughout the visible universe.

10.3 LONG-PERIOD VARIABLES

Long-period variables form another group of nonstable stars. The stars in this group are red giants and have periods ranging from about 90 to over 700 days, with the majority around 300 days. Their light curves do not repeat themselves as accurately as those of the Cepheids or the RR Lyrae stars, and there are irregularities not only in the shape of such a light curve but also in the period of the star. The reasons for these irregularities are not known, but it has been suggested that they might be connected with the large size and low density of these red giants, whose diameters are several hundred times that of the sun and whose average densities are only about $^1/_{1,000,000}$ the density of water.

The range of luminosity for the long-period variables is about 5 magnitudes, considerably larger than for the Cepheids. Thus a star at maximum will be about 100 times brighter, visually, than at minimum. We know of at least one star of this type whose luminosity varies by a factor of 4000.

Such a large change in the brightness of some stars accounts for the early discovery of this class of stars. In 1596 David Fabricius, a clergyman of East Friesland (now in The Netherlands), noticed a red star in the constellation of Cetus (the whale) that had not been seen before. In the next few months the star faded and finally disappeared from view. It was seen again a few years later, but no connection was made between these two isolated observations. It was not until 1667 that the regularity of its changes in brightness was observed. It has been observed with increasing interest since then. It was called *Mira* "The Wonderful" by the early observers, and even though it is known more technically as Omicron Ceti, the name Mira is still used.

Mira has a period of light variation of 331 days, or about 11 months, and is visible to the naked eye for about half of this period. At maximum it is a second magnitude star or brighter and consequently easily observable. It is a favorite of many amateur astronomers, and the changes in its spectrum make it a favorite of many professional astronomers as well.

Long-period variables have periods of more than 90 days and temperatures that range from more than 3000° K to less than 2000° K, although no one star has such a large temperature variation. Because their temperatures are so low, most of the energy emitted by these stars is in the infrared.

10.4 NOVAE

Some obviously nonstable stars erupt or burst with a display of energy so large it is difficult to imagine. Such outbursts have been noted in history when a star so increases in brightness that it becomes easily visible where no star was seen before; it is called a *nova* (new), even though we now know that novae are not new at all.

Many novae have been recorded in the history of Western civilization. One of the more notable, which occurred in 134 B.C., stimulated Hipparchus to compile the first catalog of the stars by position and apparent magnitude.

Bright novae which have occurred since the development of modern equipment have been observed as thoroughly as possible. Nova Aquilae (1918) was nearly the brightest star in the sky and is the first nova which had been recorded before it became a nova. Direct photographs show that Nova Aquilae was perhaps an A-type star of variable brightness before it became a "new" star. Thus, novae are really existing stars that increase in brightness many thousands of times. After such an outburst of energy the brightness of a nova will gradually decrease until it returns to its former relatively insignificant state.

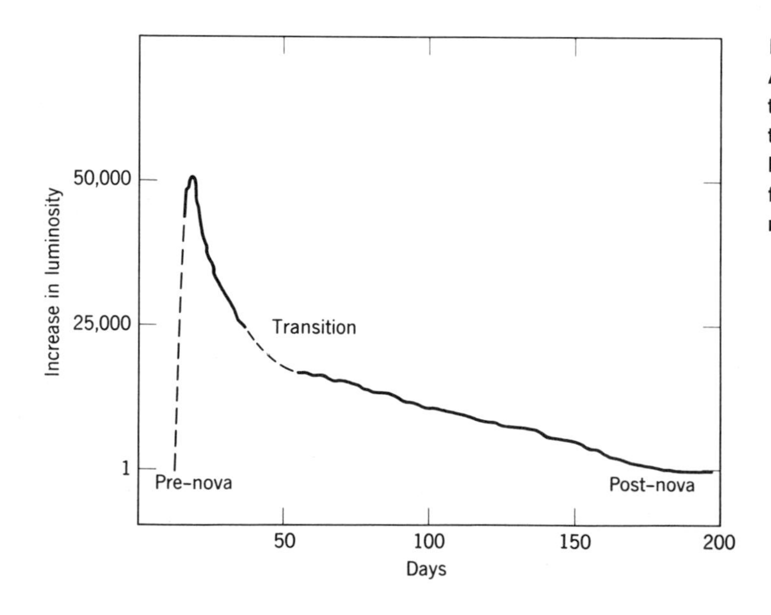

Figure 10-5
A schematic curve of a typical nova. During the transition phase, the brightness of the nova frequently varies as if the nova were pulsating.

A. THE LIGHT CURVE Figure 10-5 shows the light curve of a typical nova. The intensity of light may increase up to 160,000 (although the average is closer to 50,000) times its original brightness in a matter of hours. The exact time is difficult to determine, for a nova is not noticed until it has already achieved a brightness sufficient to distinguish it. After achieving this peak it begins to wane, at first rapidly and then, after a transition phase, more slowly, taking months or even years to return to its original brightness.

The transition phase between the initial rapid decline and the ensuing more gradual decline is very often observed as fluctuations of light, as if the star were pulsating. But a single and greater decrease in brightness has been observed during this phase in other novae, which then brighten again before declining gradually to their original magnitude.

B. SPECTRAL EVIDENCE FOR AN EXPLOSION That such an outburst of energy would be accompanied by changes in the spectrum of a nova seems a foregone conclusion. During the rapid increase in brightness the absorption lines in the spectrum are not only shifted to the violet, indicating that the side of the star facing us is approaching, but they are broadened indicating turbulence.

Since a nova continues to shine many years after the explosion with nearly the same luminosity as before, the explosion is apparently a surface phenomenon. The star literally "blows its top."

As the star's brightness decreases from its maximum, the spectrum contains broad bright lines indicating a hot shell of gas about the star. To the violet side of each broad bright line is a fine absorption line, caused by those gases directly between us and the nova. Since this absorption line is to the short wavelength side of each bright line, it indicates that the gases directly between us and the nova are approaching us; that is, the shell of gas is expanding.

C. OBSERVATIONS OF EJECTED GAS That novae do eject gases is confirmed by photographs of Nova Persei (1901); one of these, recently taken at the Palomar Observatory, is shown in Figure 10-6.

When the shell of expanding gas is visible it gives us a method for finding the distance of the nova. From the Doppler shift of the absorption lines we can determine the velocity with which the shell is expanding. Then, when after a number of years the shell becomes visible, we can compute its actual radius in miles because the radius must equal the velocity of the ejected material multiplied by the interval of time between the explosion and the appearance of the observed shell. Since we now know the linear diameter of the shell we can calculate its distance by simply measuring its angular diameter.

The shell of Nova Aquilae (1918) grew at the rate of 1 arc second per year; the radial velocity of the material ejected was about 1000 miles per second. For material to travel that fast and appear to move only 1 arc second (1/3600 of a degree) in one year, it has to be about 1200 light years away.

Figure 10-6
The expanding nebulosity about Nova Persei (1901). (Photograph of the Hale Observatories.)

At a distance of 1200 light years, the luminosity of Nova Aquilae must have been 40,000 times that of the sun. Such an increase in luminosity obviously required the sudden release of a tremendous amount of energy.

10.5 SUPERNOVAE

Novae, on the average, achieve an absolute magnitude of about −6 or −7 at maximum. That is, momentarily at least, very bright indeed, but there are other stars which in exploding achieve an absolute magnitude as great as −13 or even −16. In other words, an ordinary nova may achieve a luminosity of about 50,000 suns but the star that reaches an absolute magnitude of −16 achieves a luminosity of about 200 million. Stars that reach this latter, far greater magnitude, are called *supernovae*. A supernova indeed, when a star gives off as much light at maximum as 200 million suns! This amount of light may be similar to the amount given off by the entire galaxy in which the star resides, and represents a positively staggering expenditure of energy. Certainly an explosion of this sort must leave a star somewhat the worse for wear.

A. THE CRAB NEBULA The Crab nebula (Figure 10-7 and Star Plate 2) is the result of such an explosion observed in 1054. The connection between the Crab nebula of today and the supernova of more than 900 years ago is quite definite. We can observe and measure the rate of expansion of such a nebula by two methods. The first is to determine the radial velocity from the Doppler shift of a portion of the nebula that is approaching us. This gives us a linear velocity of expansion of 800 miles per second. We can also measure the angular rate of expansion by taking two pictures separated by a considerable interval of time. The combination of these two observations tells us that the nebular gas is about 6000 light years away, is about 3 light years in diameter, and is expanding at a rate such that it should have begun expanding about 900 years ago. The exact date was derived from the chronicles of ancient Chinese astronomers who recorded a "guest" star of the brightness of Venus in the celestial region now occupied by the Crab nebula. This guest star, which was also recorded by the Japanese, was visible during the daytime for 23 days and at night for more than a year.

Pictures of the Crab nebula, taken in different colors with color filters and photographic plates sensitive to the colors transmitted by each filter, reveal that the gaseous filaments are rather complex. The photograph taken in red light is centered on the red line in the spectrum of hydrogen, so it appears that the nebula's complex structure is hydrogen gas emitting light. The photograph taken in infrared reveals an amorphous structure in the center.

Other observations reveal the manner in which energy is distributed in the entire spectrum of the Crab nebula. It emits strongly in the radio region, less strongly in the microwave region, still less strongly in the infrared, and so on to the shorter wavelengths. This is shown in the graph in Figure 10-8.

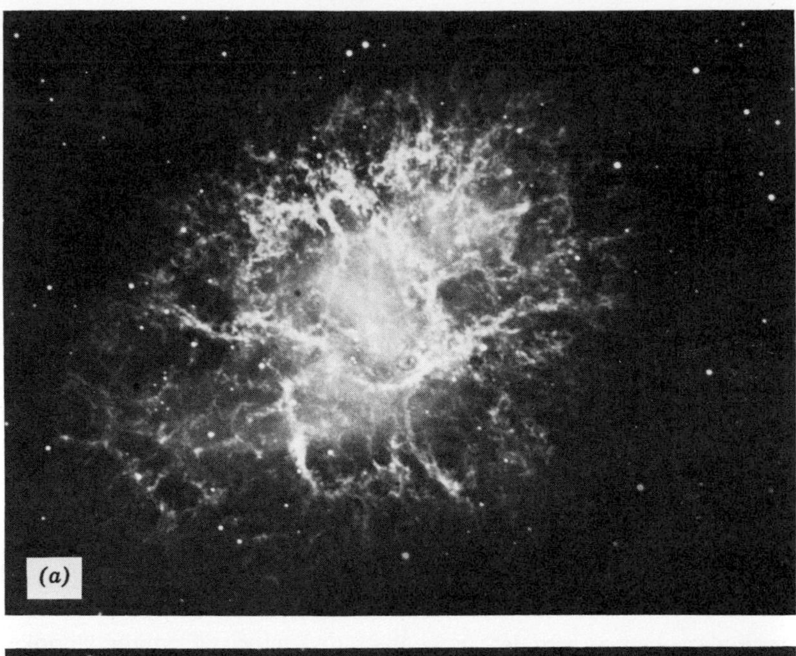

Figure 10-7
The Crab nebula as seen (a) in the red light of hydrogen, and (b) in infrared light. (Photograph of the Hale Observatories.)

The distribution of energy in the spectrum tells us a good deal about what is going on in this supernova remnant. The only known cause for the distribution of energy shown in Figure 10-8 is electrons traveling at speeds close to that of light and being constrained to curved paths by magnetic fields. This type of radiation was first observed in the physics laboratory in a huge particle ac-

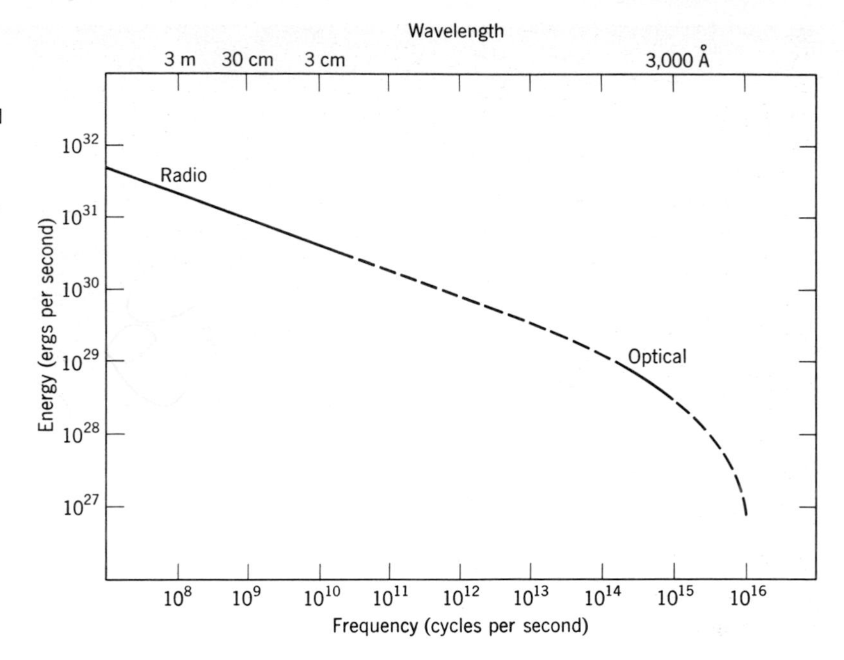

Figure 10-8
The continuous spectrum of the Crab nebula extending from the radio to the optical region. Solid lines represent observed regions. (After Steinberg, J. L., and J. Lequeux, *Radio Astronomy,* McGraw-Hill Book Co., New York.)

celerator called a synchrotron. Therefore, the radiation is called *synchrotron radiation*. In contrast, the energy emitted by a hot toaster filament or the photosphere of the sun is called *thermal radiation*, and has quite a different distribution of energy in the spectrum.

It is clear that a great deal of energy was released during the explosion some 900 years ago, but why there are still so many very high energy electrons remains a bit of a puzzle. Since these electrons do emit radiant energy, they should slow down with time, so apparently there is some other source that gives them energy. This source may be the star that survived the explosion.

B. PULSARS The star that remains near the center of all this gas appears to be a most fantastic object. It is one member of a class of objects called *pulsars*. These objects were discovered because their radio brightness varies, which in itself is not surprising, but their period of variation is surprising. Most pulsars have a period of variation of less than one second. The pulsar in the Crab nebula has a period of 0.033 seconds (see Figure 10-9). It flashes in radio, visible light, and X rays about 30 times each second. The period is amazingly consistent, so it can be determined very accurately as 0.033099522 second. The main difficulty is that the period for most pulsars is changing slightly: their periods are becoming longer. The pulsar in the Crab nebula is increasing by 3.652256×10^{-8} second each day! Even the rate of change of the period is itself changing.

The best ideas concerning the identity of these objects seem to center around a compact mass of neutrons. Apparently the core of the star is compressed during a supernova explosion to such an extent that the electrons and protons are

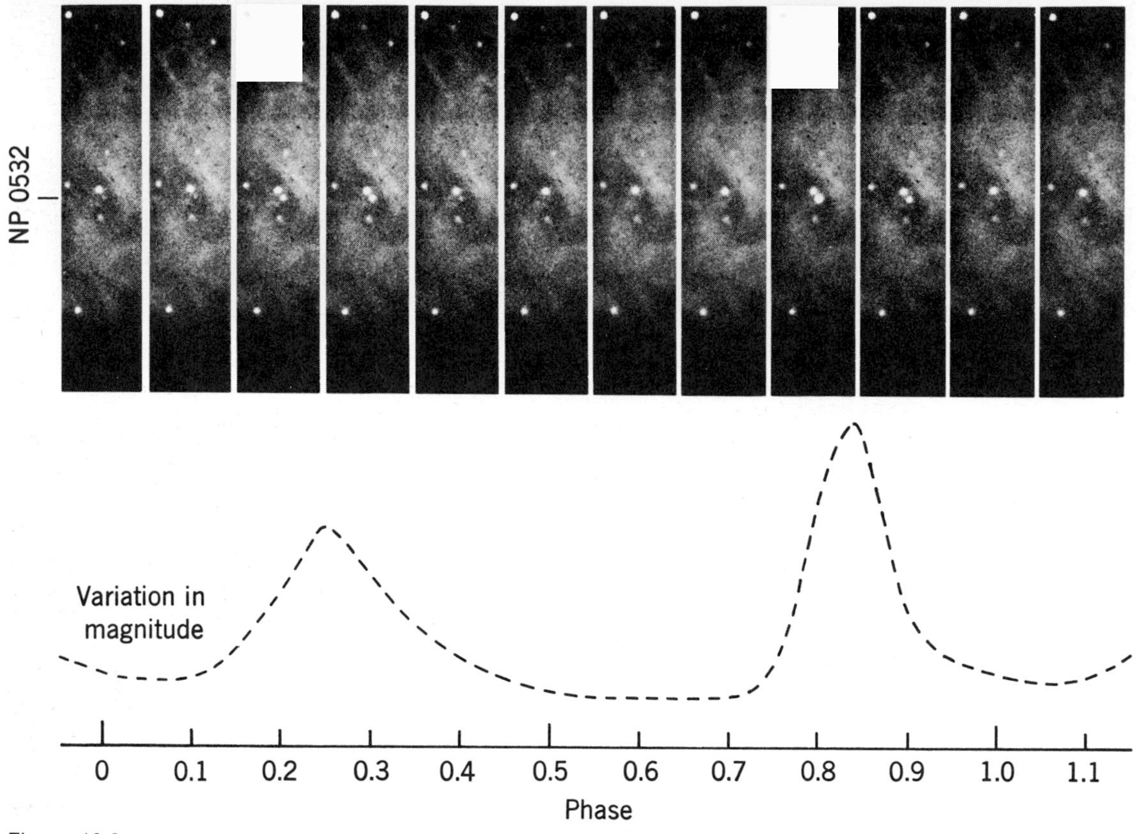

Figure 10-9

The variations in the brightness of the pulsar in the Crab nebula (NP 0532) have been photographed with the help of a stroboscopic device and an electronic image intensifier attached to the telescope. These attachments enabled the camera to make a 6-second exposure while viewing only a single phase of the pulsar's variation. One full period (phase 0 to 1.0) of the pulsar is 0.033 second. (Kitt Peak National Observatory, photographed by H. Y. Chiu, R. Lynds, and C. P. Maran.)

pushed into one another leaving only neutrons. This new form of matter has been called a *neutron star*. The mass of a neutron star must be close to that of the sun, but its diameter cannot be more than about 15 miles! Such an extremely compact blob of matter would rotate with the very short periods observed in pulsars.

Whether one portion of the star radiates more energy than another (which may be caused by a magnetic field), making it blink as it alternately exposes its brighter and then its darker side to us, or whether the star pulsates is not yet known with certainty. However, because of their recent discovery, in 1967, there is a great deal of work yet to be done on these fascinating objects. We may, in fact, find that pulsars do not pulsate at all. This is one problem astronomers meet: they must often apply names to objects about which they know very little, and these names often prove inappropriate at a later date.

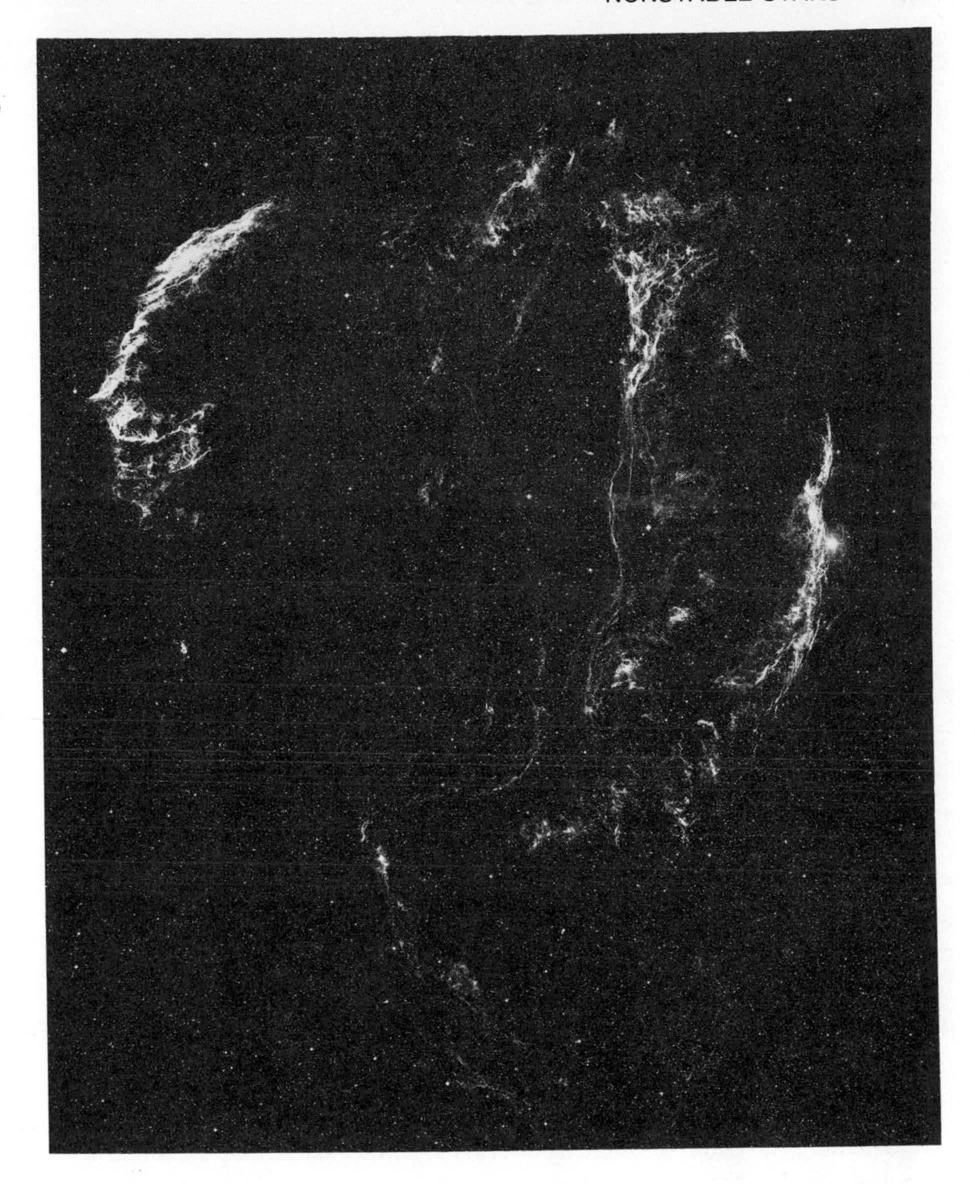

Figure 10-10
The Loop nebula in the constellation Cygnus is the result of a nova that exploded many thousands of years ago. (Photograph from the Hale Observatories.)

C. SUPERNOVA REMNANTS Bolstered by the successes of theories and observations of the Crab nebula, astronomers have searched the records and the skies for other supernova remnants. Their efforts have been rewarded because they have found the remnants of other supernovae. The most intense discrete radio source in the sky, Cassiopeia A, is now known to be a remnant of a supernova explosion that occurred about the year 1700. There is no reference in historical records to a nova at that time, and the records of those years are fairly adequate, but it is a distant object and behind a good deal of interstellar material, so it may not have been obvious in the nighttime sky. At present, fragments of this nebula

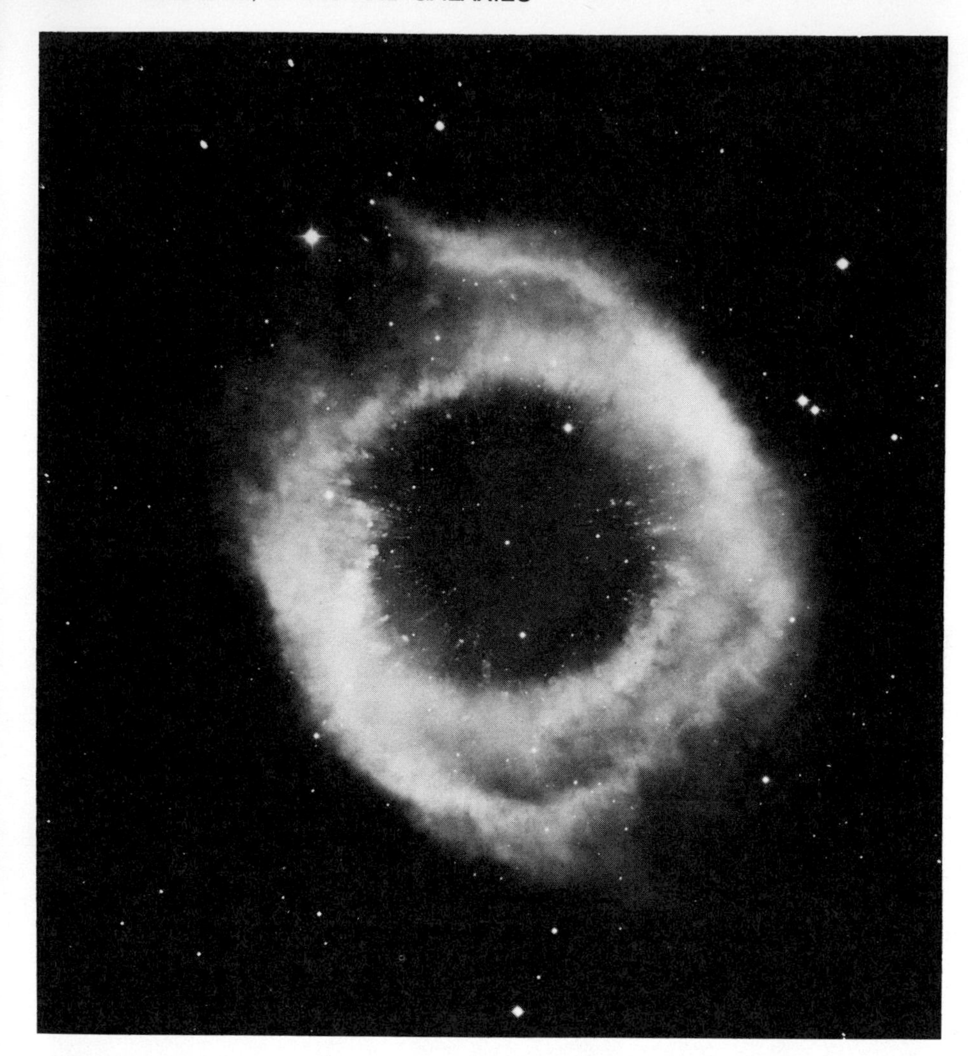

Figure 10-11
The planetary nebula NGC 7293. (Photograph from the Hale Observatories.)

are expanding at the staggering velocity of 4600 miles per second. Just why this velocity of expansion is so much greater than that of the Crab nebula is not known. At a distance of 1000 light years it is estimated that an amount of material equal to the sun's mass was ejected during the explosion.

The Loop (Veil) nebula in Cygnus is another nonthermal source of radio energy, and it, too, is a remnant of a supernova (Figure 10-10). Material in the Loop nebula is expanding at the rate of only 56 miles per second. Presumably, its expansion has been slowed by the interaction of the expanding gases with interstellar gases. The diameter of the Loop nebula is now about 120 light years, so this explosion must have occurred thousands of years ago. The center of the nebula is about 2300 light years from the sun.

10.6 PLANETARY NEBULAE

There are stars in the sky that have very obvious shells of gas about them (Figure 10-11). When these were first observed visually with smaller telescopes they gave the appearance of planetary disks and consequently were named *planetary nebulae*. (To be sure, this is a misnomer, for they have scarcely anything in common with the planets that revolve about the sun.) The gas in the shell is greenish in color because of several strong emission lines in the green and the lack of a continuous spectrum. The shells are observed to expand at the rate of 6 to 30 miles per second, considerably less than the gases ejected by novae or supernovae. The diameters of planetary nebulae range from ⅓ to 3 light years.

Photographing planetary nebulae with plates sensitive to blue and violet light discloses a central star that may be very faint, if not invisible, when viewed through a telescope directly with the eye. Since these stars are very hot (their temperatures range perhaps from 50,000° to 100,000° K), they emit most of their light in the blue and violet region where the eye is not very sensitive. There is much evidence to support the suggestion that the central stars are actually white dwarfs. This leads to the tentative conclusion that a planetary nebula is the product of a white dwarf in the making. Its shell eventually expands to invisibility and all that remains is the central white dwarf.

BASIC VOCABULARY FOR SUBSEQUENT READING

Cepheid
Light curve
Neutron star
Nova
Planetary nebula
Pulsar
RR Lyrae star
Supernova
Synchrotron radiation
Velocity curve
Thermal radiation

QUESTIONS AND PROBLEMS

1. What observations must be made before the distance of a Cepheid can be estimated?

2. What observations led astronomers to believe that Cepheids pulsate?

3. Why do Cepheids make good "yardsticks" for the astronomer?

4. Why can't the distance of the Andromeda galaxy, or any other galaxy, be determined by using RR Lyrae stars?

5. Explain how the astronomer detects an expanding shell of hot gas about a star. How can he determine its velocity of expansion?

6. What evidence indicates that the Crab nebula was a supernova in the year 1054?

7. Explain how the distance of a planetary nebula or supernova remnant can be determined if actual expansion can be measured in arc seconds.

FOR FURTHER READING

Struve, O., and V. Zebergs, *Astronomy of the 20th Century,* Crowell, Collier and Macmillan, New York, 1962, Chapters XV and XVI.

Burbidge, G., "Dissecting the Crab," *Natural History,* p. 66, October 1970.

Cox, A. N., and J. P. Cox, "Cepheid Pulsations," *Sky and Telescope,* p. 278, May 1967.

Gorgenstein, P., and W. Tucker, "Supernova Remnants," *Scientific American,* p. 74, July 1971.

Green, L. C., "Pulsars Today," *Sky and Telescope,* p. 260, November 1970; p. 357, December 1970.

Hack, M., "RU Camelopardalis — A Unique Cepheid Variable," *Sky and Telescope,* p. 350, June 1967.

Hewish, A., "Pulsars," *Scientific American,* p. 25, October 1968.

Kraft, R. P., "Pulsating Stars and Cosmic Distances," *Scientific American,* p. 48, July 1959.

Liller, W., and M. Liller, "Planetary Nebulae," *Scientific American,* p. 60, April 1963.

Mumford, G. S., "The Dwarf Novae," *Sky and Telescope,* part I, p. 71, February 1962; part II, p. 135, March 1962; part III, p. 190, October 1963.

Pacini, F., and M. J. Rees, "Rotation in High-Energy Astrophysics," *Scientific American,* p. 98, February 1973.

Percy, J. R., "Pulsating Stars," *Scientific American,* p. 66, June 1975.

Struve, O., "The Pulsating Star RR Lyrae," *Sky and Telescope,* p. 311, June 1962.

"A Rapidly Pulsating Radio Source," *Sky and Telescope,* p. 207, April 1968.

"Alcock's Nova in Delphinus," *Sky and Telescope,* p. 150, September 1967; p. 300, November 1967.

"Barnard's Loop Nebula," *Sky and Telescope,* p. 145, September 1967.

"Distances of Pulsars," *Sky and Telescope,* p. 378, December 1968.

"Expansion of the Crab Nebula," *Sky and Telescope,* p. 13, January 1969.

"Further Observations of Pulsars," *Sky and Telescope,* p. 339, June 1969.

"The Pulsar Industry," *Scientific American,* p. 43, June 1968.

"Visible Pulsar," *Scientific American,* p. 46, p. 49, March 1969.

CHAPTER CONTENTS

LEARNING OBJECTIVES

BE ABLE TO:

1. DESCRIBE THE ENERGY BALANCE AND PRESSURE BALANCE.
2. DESCRIBE HOW THE ENERGY BALANCE AND PRESSURE BALANCE OPERATE TO ESTABLISH EQUILIBRIUM IN A STAR.
3. DESCRIBE, IN GENERAL, THE THEORY OF STELLAR EVOLUTION.
4. DESCRIBE DEGENERATE MATTER AND THE WHITE DWARF.
5. DESCRIBE HOW THE EVOLUTIONARY TRACK OF A STAR ON THE H-R DIAGRAM DEPENDS ON THE STAR'S MASS.
6. GIVE OBSERVATIONS SUPPORTING THE THEORY OF STELLAR EVOLUTION, INCLUDING: HERBIG-HARO OBJECTS, T-TAURI STARS, STELLAR CLUSTERS, AND NEW WHITE DWARFS.
7. EXPLAIN HOW THE AGE OF A STAR CLUSTER IS DETERMINED.
8. GIVE THE GENERAL FEATURES OF THE SOLAR SYSTEM THAT MUST BE EXPLAINED BY ANY ACCEPTABLE THEORY OF ITS ORIGIN.
9. DESCRIBE THE THREE STAGES IN THE CONTRACTION (PROTOPLANET) HYPOTHESIS OF PLANETARY FORMATION.
10. DESCRIBE THE ACCRETION HYPOTHESIS OF PLANETARY FORMATION.
11. GIVE OBSERVATIONS IN SUPPORT OF THE CONTRACTION HYPOTHESIS.
12. GIVE OBSERVATIONS IN SUPPORT OF THE ACCRETION HYPOTHESIS.

Chapter Eleven

THE FORMATION AND EVOLUTION OF STARS

By converting hydrogen to helium deep in their interiors, stable stars transform nuclear energy into radiant energy. The radiant energy gradually makes its way to the surface where it is lost by radiation. All this is accomplished in a stable star with nothing more than a turbulent photosphere, a slight ruffle compared to the nonstable stars. How can a stable star remain so tranquil?

11.1 EQUILIBRIUM IN A STABLE STAR

The stable star is in equilibrium in a number of ways. First, the amount of energy generated equals the amount of energy radiated out into space. This equilibrium is called the *energy balance*. Second, the internal pressure of the very hot gases is balanced by the gravitational pressure tending to collapse the star. This equilibrium is called the *pressure balance*. If something happens to upset one of these balances, the other will shift to compensate.

For example, suppose that for some reason the star begins to generate more energy than it radiates out into space; what would happen? The energy not radiated would accumulate inside the star causing the interior temperature to increase. This increase in temperature would cause the gas pressure to increase and the star would expand. As the star expands, its internal temperature drops and a new equilibrium may be established.

On the other hand, suppose the amount of energy generated suddenly decreases. The internal temperature drops, the gas pressure drops, and the star collapses a little bit until a new equilibrium is established.

Any change in energy generation or energy radiation will upset the pressure balance and the star will either expand or contract to compensate.

11.2 THEORY OF STELLAR EVOLUTION

Since stars do lose energy by radiation into space and since they have only a limited amount of hydrogen as fuel, we assume that they have a limited lifetime. Therefore, stars must be born and they must die.

Astronomers further assume that any internal changes resulting from the depletion of hydrogen will result in some external changes. We want to observe these external changes and relate them to what is going on inside the star.

The theory of *stellar evolution* attempts to trace the life history of stars: How are they born? How long do they live? How do they die? Are stars being born even today?

Chapter Opening Photo
The Rosette nebula, NGC 2237. (Photograph from the Hale Observatories.)

A. A STAR IS BORN A star cannot be born from nothing so astronomers search our galaxy to find locations that have the ingredients out of which stars can form. The ingredients are large amounts of gas and dust.

A careful look at the night sky on a clear summer night in the dark of the moon and away from city lights will reveal dark lanes stretching down the middle of the Milky Way (see Figure 13-13). Those dark lanes are gases and dust that, like curtains, obscure the stars behind them. There are less obvious clouds of gas and dust in the winter sky in the constellations of Orion and Taurus.

A careful search of these parts of the sky with optical and radio telescopes

Figure 11-1
The Orion nebula. (Lick Observatory photograph.)

Figure 11-2
The Lagoon nebula (M8) with dark "globules" of gas and dust that could perhaps form new stars with planets about them. (Photograph from the Hale Observatories.)

reveals many clouds of gas and dust, some that glow brightly because stars are embedded in them. Wherever a cloud of gas and dust forms an obvious feature in the sky, we call it a *nebula*, for each is rather nebulous. Examples of bright nebulae are the Orion nebula (Figure 11-1), the Rosette nebula (Chapter Opener and Star Plate 3), and the Lagoon nebula (Figure 11-2). Other examples

Figure 11-3
Close association of bright and dark nebulae near the star Rho Ophiuchi. Two globular clusters can be seen just below the center of the bright nebulosity. (Yerkes Observatory photograph, University of Chicago.)

of nebulosity are the mixture of dark and bright nebulae near the star Rho Ophiuchi (Figure 11-3). The material for new stars is certainly there.

B. THE PROTOSTAR The process by which stars form from a nebula is not well known. The gases in nebulae are in motion, however (see the Lagoon nebula, Figure 11-2), and it has been suggested that a parcel of gas might, by some whirling motion, detach itself from the rest of the nebula. If the density of this parcel of gas becomes high enough, it will contract under its own gravitational influence. Such a whirling contracting parcel of gas is called a *protostar*, for it is not yet hot enough to start the hydrogen-helium conversion process. Until it radiates it own light, it will remain a protostar.

As the gas continues to contract the material in the center is heated by the contraction process. When its temperature is high enough hydrogen nuclei begin to combine to form helium nuclei. During this process more thermal energy is generated and the protostar becomes still hotter. Contraction stops when the core becomes hot enough for the internal gas pressure to balance the gravitational pressure.

If, in reaching the equilibrium pressure, the center is heated to a high enough temperature to sustain the hydrogen-helium conversion, the object radiates its own light into space; it becomes a star.

The more massive stars will require a higher central pressure than the less massive stars in order to balance the greater gravitational pressure. A higher central pressure means a higher central temperature, and the hydrogen-helium conversion proceeds more rapidly. More thermal and radiant energy are generated. The surface of more massive stars is hotter than that of less massive ones, and they radiate more energy out into space. That is, when a star first forms, the more massive it is the hotter and more luminous it is.

C. THE ZERO-AGE MAIN SEQUENCE The astronomer thinks of stars, in part, by their location on the H-R diagram. Where would a newly formed star fall on that diagram? Most of the stars on the H-R diagram are in the main sequence. The mass-luminosity relation, which was derived from observations of main-sequence stars, supports our reasoning above: the more massive a star, the hotter and more luminous it is. Perhaps newly formed stars are plotted along the main sequence.

Theoretical studies using the principles of physics have enabled astronomers to calculate what would happen to a parcel of gas as it contracts into a star. These calculations agree with the conclusion that newly formed stars are plotted along the main sequence.

In fact, astronomers have called the location of new born stars the *zero-age main sequence.* A star of 16 solar masses will become a B0 star. One with a mass about ⅓ solar masses becomes a late M-type star. Stars more massive than 16 or 20 solar masses are rare, and they are the O-type stars.

As a contracting parcel of gas becomes hot enough, it begins to radiate light, perhaps unsteady at first, with flare-ups from time to time, but certainly with changes. It surely begins to radiate energy before it settles down to its stable position on the main sequence. Therefore, after the time it begins to emit light and before its luminosity and temperature would place it on the main sequence, its changing temperature and luminosity means that it occupies successively different positions on the H-R diagram. If these different positions are all joined with a line, we have what is called the *evolutionary track* of a star on the H-R diagram. The evolutionary track of two stars during their *initial contraction* stage are shown in Figure 11-4.

Because of its stronger gravitational field, a more massive star will contract faster than a less massive star. Likewise, the more massive it is, the more luminous it is at each stage, so the higher its evolutionary track will appear on the H-R diagram.

A star with a mass of 0.2 solar masses is estimated to take as long as 600 million years to contract to the zero-age main sequence; it will be a late M-type star. A star with a mass equal to that of the sun requires 50 million years to

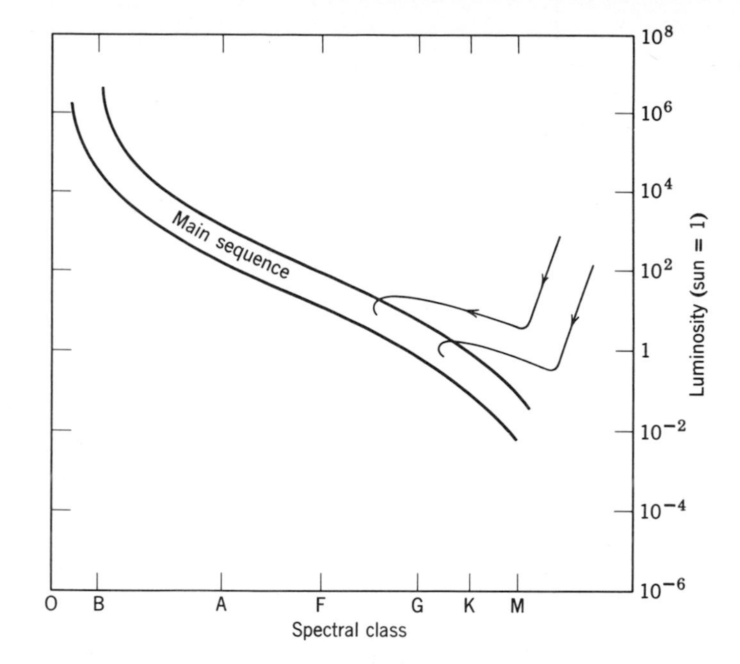

Figure 11-4
The evolutionary track of stars of moderate mass in the initial contraction stage.

reach the zero-age main sequence. One with 9 solar masses requires 21 million years before it becomes an early B-type star. A very massive star may require only ½ million years to contract from a parcel of gas in a nebula to become an O-type star.

Once on the main sequence a star remains there for a considerable length of time, but not indefinitely. It is, after all, converting the hydrogen of its core into helium. With time, therefore, the core becomes pure helium. Consequently, the conversion of hydrogen to helium can take place only on the surface of the helium core, for here is where the high core temperature meets the hydrogen surrounding the core.

Any changes that the star might undergo to remove it from the main sequence would have to result from the increased size of the helium core. The rate at which this core builds up depends on the rate of the hydrogen-helium conversion process. The rate at which this process takes place can be measured by the amount of energy the star radiates out into space. Since a main-sequence star is stable, all of the energy generated in the core must radiate from its surface. Stars with luminosities of 100 must convert hydrogen to helium 100 times faster than the sun; those with luminosities of 10,000 must use up their hydrogen fuel 10,000 times faster than the sun. An early B-type star has a luminosity of about 10,000, but it has a mass of only about 16 times that of the sun, so it has only 16 times the amount of fuel. The result is obvious: it runs out of fuel much sooner than the sun. The cost of its high living is a shorter life!

Table 11-1 gives some indication of how long stars of different mass maintain their equilibrium as main-sequence stars.

TABLE 11-1

Spectral Type	Mass (solar mass)	Time on Main Sequence (years)
O-type	20	10 million
B-type	16	17 million
G-type	1	10,000 million
M-type	⅓	30,000 million

A star's life on the main sequence ends when about 10% of its initial hydrogen has been converted to helium.

D. AFTER THE MAIN SEQUENCE When about 10% of the hydrogen fuel has been burned, the core begins to contract, releasing energy that forces the outer regions of the star to expand and hence cool. However, the increase in luminosity resulting from an increase in the size of the star exceeds the decrease in luminosity caused by the drop in surface temperature, so the star's net luminosity increases, moving it upward on the H-R diagram. The decrease in surface temperature simultaneously carries it to the right. At a certain point, depending on its mass, it ceases to expand, and in establishing an equilibrium it becomes a red giant (Figure 11-5). This equilibrium is, however, only a temporary one and sometimes not too stable. A number of red giants are long-period or irregular variables. The star is, after all, in a distorted situation.

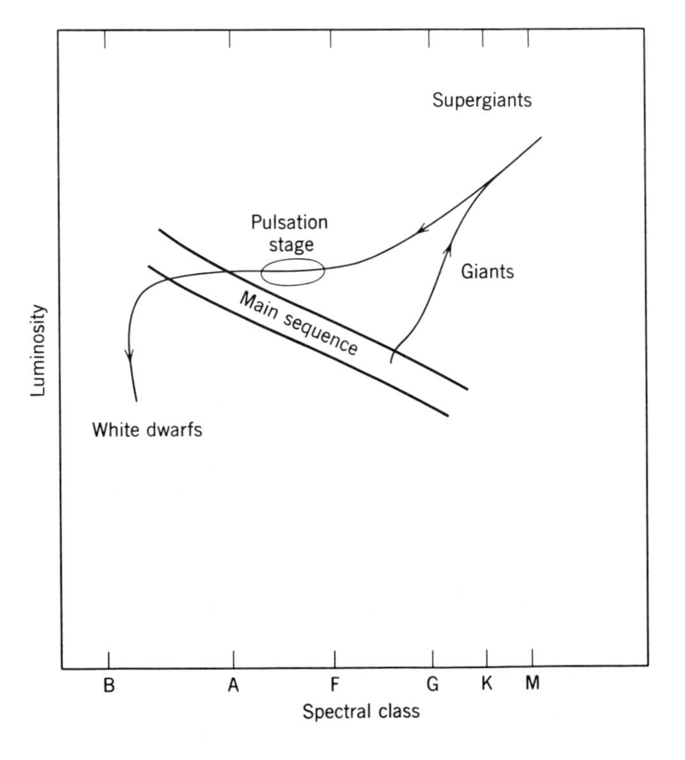

Figure 11-5
The evolutionary track of stars of moderate mass after leaving the main sequence.

The outer regions of the star are greatly expanded. This expansion is maintained by the core which is both hotter and smaller than it was when the star was on the main sequence. Because of its higher temperature and pressure, some of the helium of the core is converted to carbon and other elements. These new nuclear reactions supply enough energy to prevent the outer layers from collapsing, at least temporarily.

A star, however, cannot remain a red giant for as long a time as it remained on the main sequence. The argument to support this is the number of stars in each portion of the H-R diagram. There are many more main-sequence stars than red giants. Since we are saying that main-sequence stars become red giants, the number of each depends on the length of time a star spends as a main-sequence star and as a red giant. In fact, since there are many more stars on the main sequence than anywhere else on the H-R diagram, it follows that a star spends a longer part of its life as a main-sequence star than it spends in any other stage of its evolutionary track.

The reasons why a star leaves the main sequence and becomes a red giant are known to the extent that astronomers feel they are at least (and at last) on the right track. Calculations by R. Härm and M. Schwarzschild at Princeton University for a star with a mass 1.3 times that of the sun indicate that, by the time the star reaches the red-giant stage, its helium core has grown to nearly 40% the mass of the star. By this time the core material has been subjected to very high pressures and temperatures. Under these extreme conditions of temperature and pressure, the core material is compressed into an exceedingly compact form of matter.

Because of the exceedingly high temperature and pressure in the red-giant core, the atoms are stripped of their electrons. The core is composed not of atoms but of nuclei and electrons packed closely together. Material in this state is called *degenerate matter*.

Degenerate matter can form because the atomic nucleus is so much smaller than the atom. The nucleus has a diameter of the order of 10^{-13} centimeter, whereas an atom has a diameter of the order of 10^{-8} centimeters. Thus an atom's diameter is about 10^5 times that of the nucleus and its volume is 10^{15} times that of the nucleus; that is, an atom has sufficient volume to enclose 1,000,000,000,000,000 nuclei! In other words, atoms are composed mostly of empty space between very tiny atomic particles. If this space is eliminated, atomic particles can be packed much more closely together, and densities can reach prodigious values.

The density of degenerate matter can reach as high as 200,000 times that of water and more. A quart of material from a degenerate core of a red giant would weigh as much as 200,000 pounds if it were placed here on the surface of the Earth.

The hydrogen gas overlying the core of a red giant continues to be converted into helium, but the gases in the expanded outer layers do not transport all this energy to the surface. As a result, the core temperature increases. When it reaches 80,000,000°K (!), the helium starts to burn by converting into more massive elements. Because the core material is degenerate, it cannot yet expand; therefore, the temperature increases still more, and the helium-burning proceeds more rapidly. This rapid helium-burning is limited, however, for when the core reaches the fantastic temperature of 350,000,000°K it is forced to expand and the gases become nondegenerate.

E. THE HELIUM FLASH The very rapid burning of helium before the core expands is called the *helium flash*, since it lasts for only a few thousand years. The expansion of the core causes the temperature to drop and the helium flash ends.

Calculations indicate that the star's position on the H-R diagram does not remain constant during the helium flash. The star apparently moves from the red-giant stage slightly downward and to the left along the nearly horizontal giant branch of the H-R diagram (Figure 11-5). During part of this transition, instabilities arise and the star becomes a pulsating variable; which pulsating variable it becomes depends in part upon its mass. Presumably a more massive star becomes a long-period Cepheid, a less massive star either a short-period Cepheid or an RR Lyrae star. But the star's chemical composition may also be a factor in determining which type variable it becomes. There is theoretical evidence to suggest that the star may pass through the helium-flash stage more than once. Consequently, the star may move back and forth along the giant branch, becoming a variable more than once.

F. THE CORE SURVIVES Using the sun with its solar wind as an example, main-sequence stars eject gases back into space. Presumably the hotter stars eject more gases than the cooler ones. When each star uses up 10% of its hydrogen fuel, it leaves the main sequence and starts its final evolutionary changes. It first burns helium, but when that is used up it burns another fuel; when that is used up still another. As each fuel is consumed, the core contracts still more and becomes still hotter. The result is that more and more gases are ejected from the star's surface.

Eventually, the core becomes so hot and compressed that nearly all of the gases are blown off from around the core. If the star is very massive, this final display of energy is dramatic; the star becomes a supernova! All that remains is the core which by this time has become a neutron star (pulsar). If the star is only moderately massive, the final display of energy is less dramatic, perhaps it will become a planetary nebula. The remaining core is then a white dwarf.

The masses of white dwarfs are roughly equal to the mass of the sun or less; neutron stars may be three times more massive than the sun. Their sizes, however, differ considerably. A white dwarf may be the size of the Earth; a neutron star may have a diameter of only 15 miles. The density of degenerate matter in a white dwarf may reach 10^5 times that of water. The density of material in a neutron star, however, may reach 10^{15} times that of water. In neutron stars, the electrons have been forced inside the protons by the tremendous pressures.

Matter in a white dwarf and neutron stars is so compact that nuclear reactions are no longer possible without a massive explosion. The usual sources of energy that keep the surface of a stable star hot even as it loses energy to space no longer exist. This leaves one alternative: the white dwarf and neutron star are left to cool as they lose energy by radiation. The evolutionary path of a white dwarf is shown in Figure 11-6.

As a white dwarf cools, of course, it changes color. At least one white dwarf is a K-type star and its color is orange. It is called a white dwarf, however, for that term has come to mean a star whose mass, size, and composition classify it as a ball of degenerate gas.

But a K-type white dwarf still radiates energy and, therefore, it cools still more becoming first a red white-dwarf and then an infrared white-dwarf. Given enough time (billions of years) it will eventually become a black white-dwarf, that is, it will radiate as much energy out into space as it receives. It will then have cooled to the temperature of space. With all of its sources of energy locked up, it must remain forever as a cold blob of degenerate matter; it will have become a celestial tombstone.

G. STELLAR MASS AND THE EVOLUTIONARY TRACK The general shape of the evolutionary track described is the same for all stars of moderate mass, even though the location of the track differs for stars of different masses. As the

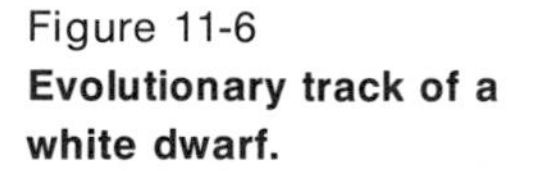
Figure 11-6
Evolutionary track of a white dwarf.

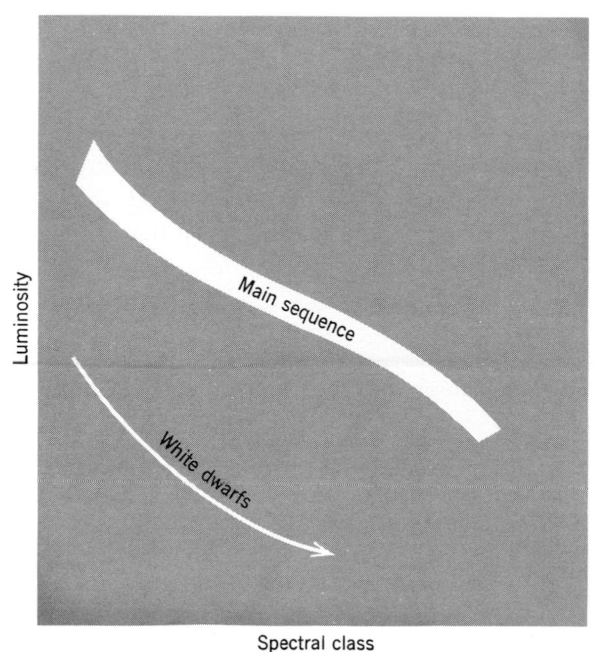

position of a star along the main sequence depends on its mass (the least massive star appears at the bottom), so does the location of the star's evolutionary track. With increasing mass, the evolutionary tracks appear ever higher on the H-R diagram. A more massive star will be more luminous at any point along its track than a less massive one at a corresponding point. Furthermore, the more massive a star is, the faster it will evolve along its evolutionary track. The stages through which a star passes during its evolution are represented in part by the different types of stars already discussed. If one stage of a particular evolutionary track is populated by more stars than any other, we can assume that stars of that particular mass must remain in the stage of evolution longer than in any other stage.

Very massive stars apparently evolve differently than stars of moderate mass. There is no evidence to indicate that they become super-super red giants, for none have been observed. The very massive stars become the supernovae, but the exact evolutionary path between the main sequence, the supernova, and the resulting neutron star is not clear.

Observational evidence indicates that novae (as against supernovae) are each members of a double star system. Apparently their eruption is a result of a complex interplay of gases between the two stars.

11.3 OBSERVATIONAL EVIDENCE OF STELLAR EVOLUTION

The observational evidence for such a hypothesis of stellar evolution is quite strong. If stars form in gaseous nebulae, then we ought to examine nebulae in search for evidence of recent or current star formation. If in that search we observe any number of different nebulae, for example, the Lagoon nebula (Figure 11-2), we find a large cloud of gas with many very hot luminous stars in the center. Surrounding those luminous stars are clouds of gas with many dark globules. It has been suggested that these dark globules may be parcels of gas in the process of contraction into protostars. But further evidence that stars are forming in clouds of gas and dust is difficult to obtain with optical telescopes. The dust simply blocks the light. Radiation of longer wavelengths, however, does pass through the dust clouds.

A. BEFORE THE MAIN SEQUENCE Observations in both radio and infrared have revealed the existence of strong sources of infrared energy in some of the dark clouds of dust, such as the ones near the star Rho Ophiuchi (Figure 11-3). Some of the sources of infrared coincide with localized clouds containing molecules of water (H_2O) and hydroxyl (OH). These hidden sources may well be blobs of gas in the very early stages of contraction.

Herbig-Haro objects are also found in clouds of gas and dust. Each such object has a stellar or semistellar nucleus surrounded by a small bright nebula. Figure 11-7 shows one group of Herbig-Haro objects. Figures *a* and *b* were taken with

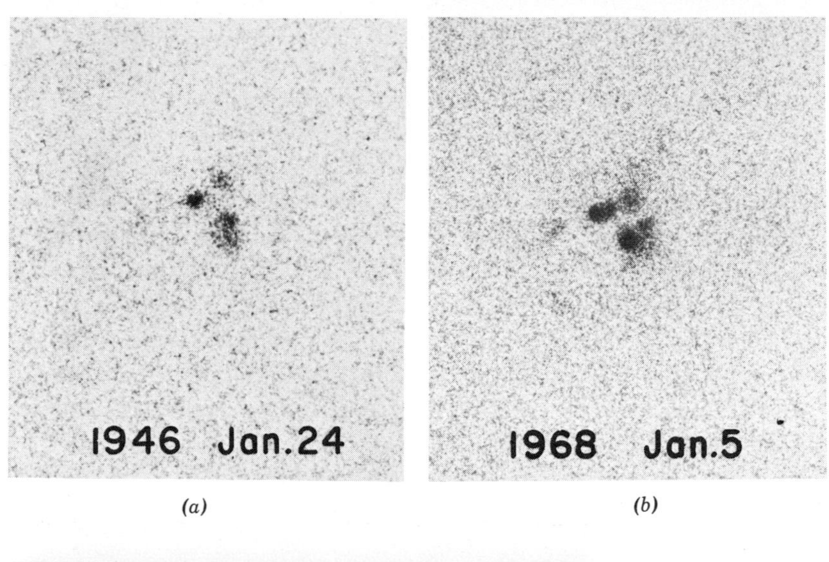

(a) (b)

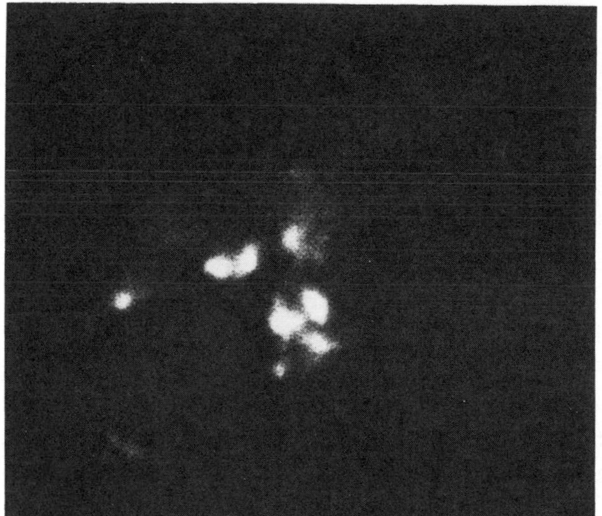

Figure 11-7
Three photographs of Herbig-Haro Object No. 2. Both photographs (a) and (b) were made with the 36-in. Crossley reflecting telescope at the Lick Observatory. These photographs are negatives; the bright portions appear dark. Changes in the objects are noticeable over the 12 years. (c) The same object photographed on Dec. 6, 1959 with the 120-in. telescope reveals more detail than those taken with the Crossley. (Courtesy G. Herbig, Lick Observatory.)

the same telescope (36-in. reflector) and show changes in this group. Figure 11-7*c* was taken with the 120-in. telescope and reveals detail. Presumably these are stars in the process of formation. Observations of their spectra indicate that these objects resemble in many ways another class of stars called the T Tauri stars.

In the constellations of Taurus and Orion there are other nebulae that contain *T Tauri stars* — a class of stars that are not very luminous, that have bright lines in their spectra, that are irregularly variable in light, and that are invariably associated with nebular material. All of the more than 500 T Tauri stars known are of a later spectral type than F8, and each is more luminous than a main-sequence star of the same spectral type. These two characteristics indicate that they may be to the right of, as well as above, the main sequence. Such a position on the H-R diagram is predicted for a newly contracting star following an evolutionary track that leads to a point on the main sequence to the left of, and slightly below, its present position as a T Tauri star.

T Tauri stars have bright lines in their spectra which indicates that they have either extensive atmospheres or nebulous material immediately surrounding them. The fact that they are invariably associated with nebular gases places them in what present theory considers the necessary environment for stars in the early stages of formation. However, they do move with respect to the clouds in which they are now immersed, which indicates that they will eventually leave these clouds. Since T Tauri stars are generally not found outside gaseous clouds, we must conclude that they lose their identifying characteristics before or when they leave the clouds. Thus, the stage of development represented by the T Tauri stars may be a short one, embracing perhaps only 10% of the time required for a star to contract onto the main sequence.

B. EVIDENCE FROM STAR CLUSTERS The small group of Herbig-Haro objects in Figure 11-7 are apparently forming a small group of stars. If these stars are close enough together and not moving rapidly amongst themselves, they may form a small cluster of stars. A *star cluster* is a group of stars held together by the mutual gravitational attraction of its member stars. There are two distinct kinds of star clusters: open clusters and globular clusters. *Open clusters* are found only in the plane of the Milky Way galaxy, their stars form a fairly loose assemblage and frequently contain some hot early A- and B-type stars such as the double cluster in Persius, H and Chi Persei (Figure 11-8). An open star cluster may have from a few dozen to several thousand stars.

Globular clusters, on the other hand, are a compact formation of as many as 1,000,000 stars (Figure 11-9). They are found mostly around the nucleus of our galaxy in the direction of the constellation Sagittarius. Many of the stars in globular clusters are red giants.

From the appearance of either kind of cluster it is generally concluded that the stars in any one cluster formed at the same time, or nearly so. Therefore, if the rate of evolutionary processes in stars depends on stellar mass, then clusters would be a good place to study those processes. H-R diagrams have been drawn for both globular and open clusters and, in fact, these diagrams led to some of the initial ideas of stellar evolution.

The differences between the H-R diagram for clusters (Figures 11-10 to 11-12) and for stars in the vicinity of the sun (Figure 9-8) are both striking and significant. The stars in the vicinity of the sun are of many different ages, those in a cluster are of a single age (or very nearly so).

The H-R diagram (actually a color-magnitude diagram, in which the color of the star replaces the spectral type) of a very young cluster (NGC 2264) shows that although the B-type stars have already reached the zero-age main sequence

Figure 11-8
The double cluster in Perseus. (Lick Observatory photograph.)

(Figure 11-10), the less massive stars that will finally form the lower part of the main sequence are still in their pre-main sequence stage of formation.

A color-magnitude array for several open clusters has been drawn on one diagram to show similarities (Figure 11-11). The most massive stars of these open clusters (for example H and Chi Persei) have already evolved to the right of the main sequence. It is assumed that each of the clusters plotted in Figure 11-11 is older than NGC 2264 in Figure 11-10. The less massive stars have not reached the main sequence of NGC 2264, the most massive have; whereas the most massive stars of NGC 2362 and H and Chi Persei have already left the main sequence.

Figure 11-9
The globular cluster M13 in Hercules which must contain more than 50,000 stars. (Photograph of the Hale Observatories.)

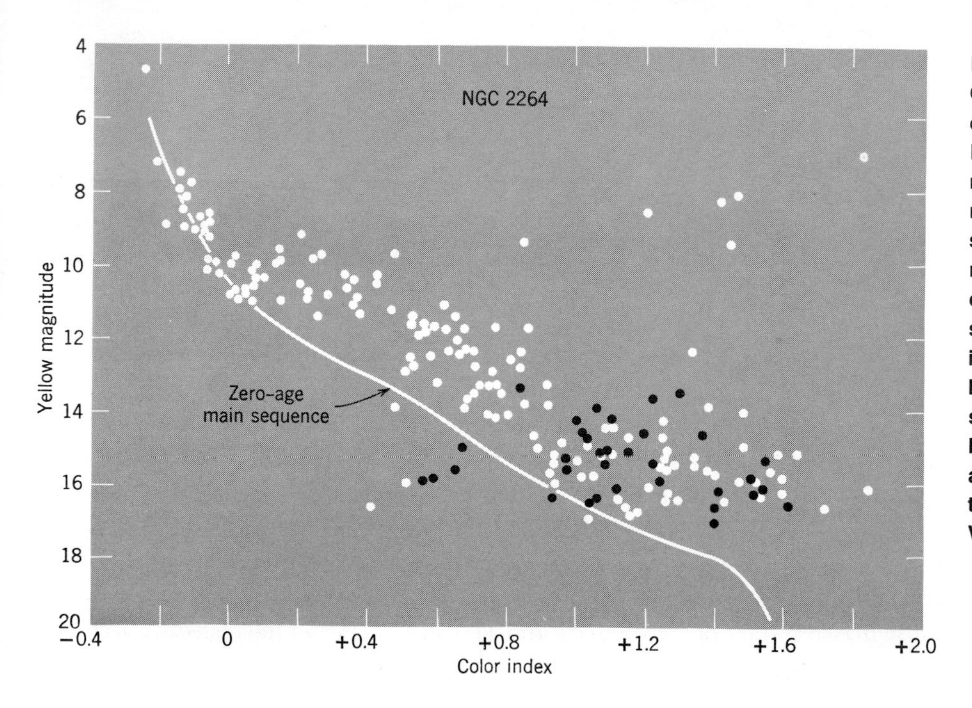

Figure 11-10
Color-magnitude diagram of stars in the open cluster NGC 2264. Only the most massive B-type stars have reached the main sequence. The stars represented by open circles are either T-Tauri stars or have bright lines in their spectra. The stars below the zero-age main sequence appear fainter because of clouds of gas and dust between us and those stars. (After Merle Walker.)

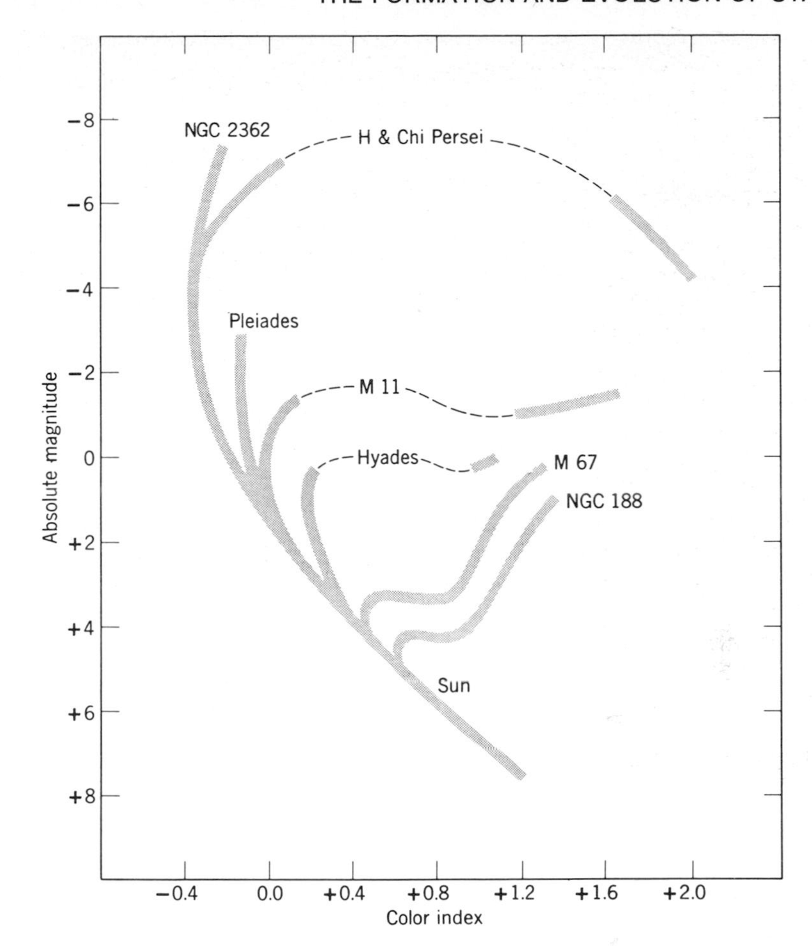

Figure 11-11
A composite color-magnitude diagram of several open clusters. (After Alan Sandage.)

The point on the main sequence where the stars branch off toward the right is called the *turnoff* point. Since less massive stars evolve more slowly, the location of the turnoff point should be indicative of the cluster's age: the further down the main sequence the turnoff point appears, the older the cluster. Hence, NGC 188 is older than M 67, which in turn is older than the Hyades, etc.

The H-R diagrams for globular clusters (Figure 11-12) are different from those of open clusters. Stars have progressed further along the evolutionary track and the turnoff point is still further down the main sequence, indicating a greater age for globular clusters. It should be noted that very few stars remain on the main sequence above the turnoff point; however, the fact that some stars are on the main sequence above the turnoff point leads us to suspect that these stars may have been formed later than the bulk of stars in the cluster. Therefore, not all the stars in the cluster are of precisely the same age. This diagram, then, does not represent the evolutionary track of any one star of given mass, but identifies points on evolutionary tracks of stars of differing masses whose ages are nearly the same. The stars that are still on the main sequence will

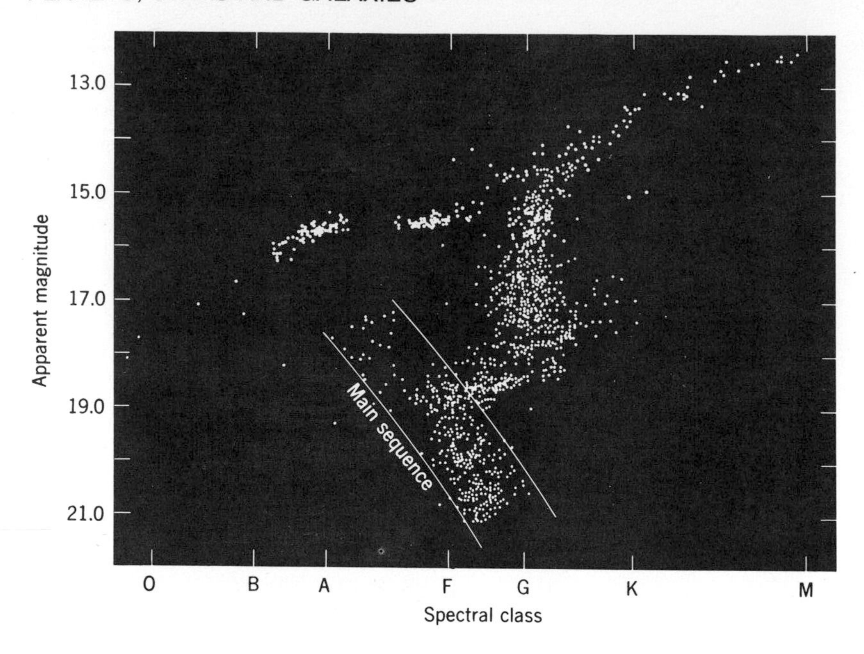

Figure 11-12
An H-R diagram of the globular cluster M3 according to H. C. Arp, W. Baum, and A. Sandage.

eventually evolve into red giants, but these less massive stars will presumably be less luminous than the red giants appearing in the cluster at the present time.

The gap in the upper branch of the diagram is of particular interest; it is here that the RR Lyrae stars in the cluster would be located, were they included on this diagram. According to current theory, the stars pass through the RR Lyrae gap (or Cepheid gap) during their helium flash when the stars move horizontally along the giant branch.

C. EVIDENCE OF NEW WHITE DWARFS A recent study of the central stars of 65 planetary nebulae by G. O. Abell of the University of California at Los Angeles indicates that these stars are just entering the white-dwarf stage. They have an average surface temperature of 50,000° K and an absolute magnitude of +5, placing them in the white-dwarf region of the H-R diagram. Their average radius is estimated to be about 0.1 the radius of the sun. It seems that eventually all of the gases overlying the degenerate core are blown away, and the core becomes the white dwarf.

11.4 THE ORIGIN OF THE SOLAR SYSTEM

It seems clear that the solar system was formed from the sun, thus, a consideration of how it formed should throw some light on the formation of stars in general.

A. FEATURES OF THE SOLAR SYSTEM A complete description of the origin of the solar system must be able to explain the main features of the system as it now exists, unless we assume (though it is very unlikely) that there has been some strong outside disturbing force since the origin. In general we could say that the solar system is fairly flat with almost everything spinning in the same direction. The following are the specific features which such a description must explain:

1. The motions (revolution and rotation) of nearly all the bodies that revolve about the sun are in the same direction, except those of Uranus and Venus, some of the satellites, and many of the comets.
2. The planes of revolution of these bodies are within a small angle of one another. The exceptions are the satellites of Uranus and most of the bright comets.
3. The planes of the equators of the planets as well as that of the sun are nearly parallel to the planes of revolution of the planets. Uranus is again an exception.
4. The densities of the four terrestrial planets are considerably greater than those of the four Jovian planets (Table 11-2).

TABLE 11-2
Densities of the planets in terms of water

Terrestrial Planets		Jovian Planets	
Mercury	5.4	Jupiter	1.3
Venus	5.1	Saturn	0.7
Earth	5.5	Uranus	1.3
Mars	4.0	Neptune	1.7

A. EARLY HYPOTHESES Before the sizes and distances of astronomical objects were understood, it was thought that the solar system may have formed as a result of a collision between the sun and a comet. But a comet would be lost without even a ripple on the sun. Then a collision or close approach between the sun and another star was considered. The probability of this happening, however, is extremely remote; it is estimated that during the lifetime of the Earth only 10 planetary systems could have formed by this process in our Milky Way galaxy of 200 billion stars.

The astronomer, faced by billions of stars, cannot fall heir to the belief that the Earth is something special, the planetary system unique. If the universe is to be investigated objectively at all, he cannot accept the belief that all the billions of stars exist solely for the benefit of mankind. He assumes that there are many other stars circled by planets that support life not too unlike our own. And for this reason he is very critical of hypotheses, like those of stellar collision or close approach, whose acceptance would imply the uniqueness or near-uniqueness of our solar system.

Furthermore it has been shown that even if two stars were to collide or suffer a close approach, the gases spewed out into space would not condense into planets; the density would be too low and the motions would only lead to further dispersal of the gas. Therefore, astronomers have sought to explain the formation of the solar system as a self-contained event with no outside influences.

B. NEBULAR HYPOTHESES The hypotheses that satisfy the restrictions of self-containment have one basic similarity: They all base their arguments on the supposition that the sun originated in a great nebulous cloud of gas and dust. Each theory assumes that the sun was once a dark globule which was both contracting under its own gravitational influence and rotating. The combination of both rotation and contraction leads to more rapid rotation. Recall how an ice skater spins rapidly. As she starts her spin her arms are outstretched; then as she brings her arms in close to her body, she spins more rapidly.

As the globule forming the *protosun* contracted, it rotated more rapidly, so rapidly, in fact, that it threw material off at the equator. A *nebular disk* was formed around the protosun's midriff. The flatness of this disk accounts for the observation that the planetary orbits lie in nearly the same plane. The rotation of the material in the disk accounts for the planets all revolving about the sun in the same direction. The protosun with its nebular disk may have looked like an out-of-focus picture of Saturn.

1. The Contraction Hypothesis The late G. P. Kuiper, of the Lunar and Planetary Laboratory at the University of Arizona, proposed that the planets formed in a manner not unlike the way in which the sun formed from a nebular cloud. This hypothesis, also called the *protoplanet hypothesis*, assumes that if portions of the nebular disk about the protosun reached a high enough density, then those portions ought to contract under their own gravitational influence.

Two opposing forces act on such a contracting parcel of gas in the nebular disk: (1) the force of the gas's own gravity, which tends to pull the eddy of gas together, and (2) the gravity of the sun, which exerts tidal forces in the eddy of gas that try to pull it apart (see Figure 11-13). If the density of the gas is high enough (higher than a critical value Kuiper called the Roche density), its gravitational field will gain control, and the material composing the eddy will contract into a *protoplanet*, a newly forming planet with a central nucleus surrounded by a disk of gas.

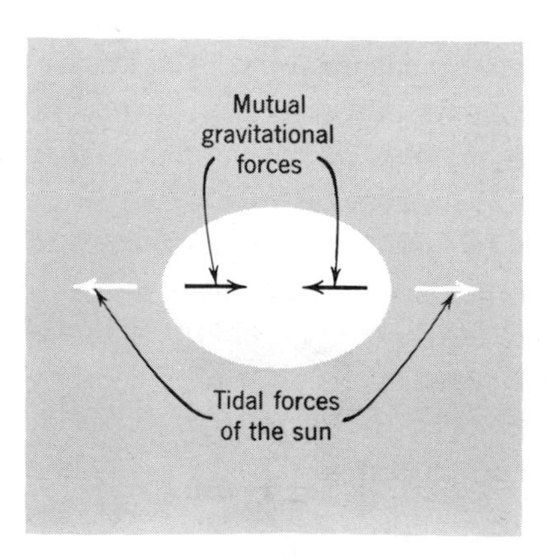

Figure 11-13
Tidal forces of the sun tend to pull a protoplanet apart; gravitational forces hold it together.

The tidal forces from the sun cause each protoplanet to become elongated with the long axis toward the sun, and consequently to rotate in the same length of time and in the same direction that it revolves about the sun (Figure 11-14). But as the protoplanet contracts under its own gravitational influence, it rotates more rapidly. Thus we see that the planets should rotate in the same direction in which they revolve. Similarly, the satellites form from the nebular disk about each protoplanet and thus revolve and rotate in the same direction that the planets rotate.

At first, this explanation of satellite formation does not appear to account for the rings of Saturn. However, E. Roche has shown that the tidal forces existing within a distance of 2.4 times the radius of a planet (the so-called Roche limit) are so great as to prevent the formation of a satellite. The concept of the Roche limit may then help explain the existence of Saturn's rings: the material originally composing the nebular disk surrounding Saturn which was within that limit has remained without coalescing into a satellite.

Kuiper suggests that the entire process of formation can be discussed in four stages.

The first stage is the collapse of the original nebula into a *protosun* with a nebular disk rotating about it. The protosun in this stage would still be dark.

The second stage includes the formation of the protoplanets in the nebular disk. The disk would break up into two parts: the inner part from which the four

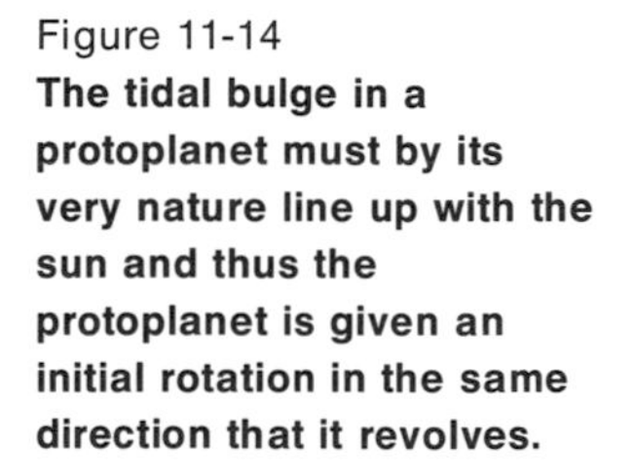

Figure 11-14
The tidal bulge in a protoplanet must by its very nature line up with the sun and thus the protoplanet is given an initial rotation in the same direction that it revolves.

terrestrial protoplanets form, and the outer part from which the four Jovian protoplanets form. In the region between these two parts the density of the nebular disk would not exceed the Roche density which is necessary for contraction to take place, and consequently the tidal forces imposed by the sun would prevent the formation of any one large body. It is conceivable that several bodies smaller than the planets would form, which upon colliding would break up and become the minor planets. The nebular disk beyond the Jovian protoplanets would also have a low density which would result in the formation of comets.

The third stage brings the sun from darkness to brightness. For the first time the sun emits intense radiation. This radiation would have ionized the gases in the inner part of the nebular disk, with consequences discussed in Chapter 12 (page 264).

The fourth stage considers the continued reaction of the solar radiation on the now ionized gases in the nebular disk. No doubt this radiation was composed in part by a solar wind. Just as comets' tails are forced away from the sun by the solar wind, so were most of the gases in the inner part of the nebular disk. Only those gases that are tightly held by the gravitational field of a protoplanet would remain in the vicinity of the sun; the rest of the material (actually the vast majority of it) would be ejected from the solar system back into interstellar space. Those planets closest to the sun and thus subject to more intense solar radiation and particle pressure would lose more material than would the four Jovian planets. Since the ejected material would be composed largely of lighter substances (mostly hydrogen) this would account for the greater density of the terrestrial planets.

The main difficulty with the contraction hypothesis is that it is hard to imagine how the terrestrial planets formed while the gas was being blown outward by the solar wind. In order for planets to form, the density of the gas in the nebular disk must exceed the Roche density, and this requires that hydrogen, which amounted to three quarters of the mass of the nebular disk, be present during the collapse process. On this basis, then, our planet should have a lot of hydrogen in it; but it doesn't, neither does Mercury, Venus, the moon, nor Mars. If hydrogen and helium were present and became trapped inside the terrestrial planets, how could the solar wind blow them away?

2. *The Accretion Hypothesis* The other general theory of the origin of the planets in the nebular disk is based on the idea that very small particles can collect together to form larger particles. These in turn combine to form eventually the planets and satellites of the solar system. Such a building up or growing by the collection of smaller particles is called *accretion*. It is clear from the impact craters on Mars, the moon, and the Earth that each of these bodies collected many smaller bodies during the first billion years or so of the solar system's existence. Some of those collected by the moon to form the maria, and the

Orientale Basin must have increased the moon's mass significantly. To this extent, then, the moon was formed by accretion.

The main difficulty with the accretion theory is that it is not easy to imagine how it all started. How did all those little tiny particles in the original nebular disk get together to form a few very large planets?

Apparently the first solid objects in the nebular disk were the chondrules found in the chondrite meteorites. These were somehow gathered into a matrix to form large bodies, which may in turn have collected to form perhaps minor planets, and eventually the planets. But the mechanism that permitted the first small particles to collect together is not at all clear.

The chondrules and the dust particles in the nebular disk were each very small particles; their mutual gravitational forces were not strong enough to effectively attract other particles.

Electrical forces, however, are much stronger than gravitational forces. Those particles were bombarded with high energy photons, with electrons and protons in the solar wind, and with cosmic rays. It is conceivable, therefore, that they could have become electrically charged. The Apollo astronauts have found many tiny particles clinging to their suits and equipment. These particles, it is presumed, where charged electrically.

Two objects with opposite electric charge attract each other, and an object with an electric charge (either positive or negative) will attract another particle with no charge. Try picking up bits of dust and paper with a plastic ruler rubbed with a paper towel. Certainly electrical forces might be strong enough to enable particles in the nebular disk to not only attract each other but to stick together once they collided.

In the accretion hypothesis, which has been developed to a large extent by H. C. Urey of the University of California and F. L. Whipple of Harvard University, many bodies with a diameter of several hundred miles could have formed which in turn collected to form the terrestrial planets and the moon. In fact, Urey suggests that the moon was originally in an orbit about the sun. It was then somehow captured by the Earth. The fact that the plane of the moon's orbit is more nearly parallel to the plane of the solar system than the plane of the Earth's equator is a strong argument in favor of the idea that the Earth captured the moon. The planes of the orbits of the Galilean satellites nearly coincide with the plane of Jupiter's equator, indicating that they were formed along with Jupiter. The outer satellites of Jupiter, however, have orbits that are considerably inclined from the plane of Jupiter's equator and in fact some of them revolve in retrograde motion. It is, therefore, generally accepted that Jupiter captured its outer satellites.

The Jovian planets evidently are composed primarily of hydrogen and helium. It can be surmised, then, that these gases, along with some rocky material perhaps, were collected together with the help of methane, ammonia, water, and carbon dioxide. Apparently, the nebular disk in this region had enough material to permit the formation of giant planets, whose gravitational fields soon became strong enough to gather in gases, and thus each planet grew bigger. The bigger each one grew, the faster it grew until the nebular disk became depleted with gases.

Beyond the region of the Jovian planets it is postulated that only smaller bodies could form, largely of the frozen compounds methane, ammonia, water, and carbon dioxide. These smaller bodies each revolve about the sun in its own orbit, which normally keeps it beyond the planets Neptune and Pluto. According to this hypothesis, however, one of these small bodies of frozen compounds is occasionally perturbed in its orbit so that it comes into the inner part of our solar system. It then becomes heated by the sun and forms one of the spectacular comets that surprise astronomers.

C. COMMENTS ON THE NEBULAR HYPOTHESES Neither the contraction theory nor the accretion theory is able to account for all of the features of the solar system. But this is the way science develops: observations lead to ideas, those ideas bring about further observations, which in turn breed new ideas. Certainly we need new observations before we can formulate a detailed theory on the origin of the solar system.

There are several aspects of the present theories, however, that need to be brought out. It is possible that the terrestrial planets formed by the accretion process, and the Jovian planets formed by the contraction process. This is supported to some extent by observation. Witness the extensive craters on Mercury, the moon, and Mars. And recall that the rings of Saturn and the inner satellites of both Jupiter and Saturn support the idea that these planets formed from a collapse of a dense portion of the nebular disk. It is perfectly possible that the terrestrial planets formed under conditions and by processes that were different from those that formed the Jovian planets. The gap between Mars and Jupiter could then be explained by the fact that because of the density, solar tidal forces, and temperature neither process operated in that region.

The next question to ask is why are the densities of the two major groups of planets so different?

The temperature of the nebular disk and the solar wind could have played a part in separating out the chemical elements to leave the rocky terrestrial planets in the center. Surely the inner part of the nebular disk was hotter than the outer regions, so it is reasonable to suppose that the rocky materials would have solidified there first. Certainly the gases methane, ammonia, water vapor, and carbon dioxide would not have solidified in the hotter parts of the nebular disk. The very lightest elements, hydrogen and helium, were probably blown away from the inner part of the solar system by the solar wind.

Just how the planets and other members of the solar system formed is clearly the subject of conjecture right now. But new studies have led to a new approach, an approach that carries with it the hope of achieving our goal of understanding. This new approach is the ability to make significant measurements of time, and of crystal structure and content. It is a result of cooperation between the astronomer, the physicist, and the crystallographer.

BASIC VOCABULARY FOR SUBSEQUENT READING

Accretion
Degenerate matter
Evolutionary track
Globular cluster
Herbig-Haro object
Nebula
Nebular disk
Open cluster
Protoplanet
Protostar
Stellar evolution
T-Tauri stars
Zero-age main sequence

QUESTIONS AND PROBLEMS

1. For each of the following situations, indicate how a star's central temperature, central pressure, and diameter would change:
(a) A decrease in the central pressure.
(b) An increase in the central temperature.
(c) An increase in the rate of energy generation.
(d) An increase in the rate of energy radiation from the surface.

2. What two pressures are in balance in a stable star?

3. Why must all of the energy generated in the center of a star be radiated out into space?

4. What would happen to a star if it radiated more energy into space than it generated in the center?

5. Explain why stars with larger mass form the upper part of the main sequence, and those with smaller mass form the lower part.

6. With the help of the H-R diagram explain the changes in temperature and luminosity of a star of moderate mass as it evolves from a parcel of gas and dust in a nebula.

7. What would be the differences in the brightest stars of a very young and a very old galaxy?

8. Cite observational evidence in support of the current theory of stellar evolution.

9. Give one way to estimate the age of a cluster, either open or globular.

10. What observational evidence is there in support of the accretion hypothesis of the origin of the solar system?

11. What observational evidence is there in support of the contraction hypothesis of the origin of the solar system?

FOR FURTHER READING

Cameron, A. G. W., "The Origin and Evolution of the Solar System," *Scientific American,* p. 33, September 1975.

Faul, H., *Ages of Rocks, Planets, and Stars,* McGraw-Hill paperback, 1966.

Gamow, G., *Biography of the Earth,* The Viking Press, New York, 1959.

Glasstone, S., *The Book of Mars,* National Aeronautics and Space Administration, Washington, D. C., 1968, Chapters 8 to 11.

Hurley, P. M., *How Old is the Earth?* Doubleday Anchor, New York, 1959.

Jastrow, R., *Red Giants and White Dwarfs,* Harper and Row Publishers, New York, 1967.

Page, T., and L. W. Page, eds., *The Origin of the Solar System,* The Macmillan Co., New York, 1966.

Shklovskii, I. S., and C. Sagan, *Intelligent Life in the Universe,* Holden-Day, Inc., San Francisco, Calif., 1966.

Struve, O., and V. Zebergs, *Astronomy of the 20th Century,* Crowell, Collier and Macmillan, New York, 1962, Chapter IX.

Whipple, F. L., *Earth, Moon, and Planets,* 3rd ed., Harvard University Press, Cambridge, Mass., 1968, Chapter 14.

Wood, J. A., *Meteorites and the Origin of Planets,* McGraw-Hill, paperback, 1968.

Bok, B. J., "The Birth of Stars," *Scientific American,* p. 49, August 1972.

Fox, S. W., and R. J. McCauley, "Could Life Originate Now?" *Natural History,* p. 26, August-September 1968.

Green, L. G., "Ordinary Stars, White Dwarfs, and Neutron Stars," *Sky and Telescope,* p. 18, January 1971.

Greenstein, J. L., "Dying Stars," *Scientific American,* p. 46, January 1959.

Herbig, G. H., "The Youngest Stars," *Scientific American,* p. 30, August 1967.

Neugebauer, G., and E. E. Becklin, "The Brightest Infrared Sources," *Scientific American,* p. 28, April 1973.

Sitterly, B. W., "Symposium on Abundances of Elements in Stars," *Sky and Telescope,* p. 11, January 1965.

"An Ex-variable Star?" *Sky and Telescope,* p. 323, June 1966.

"Dark Interstellar Clouds," *Sky and Telescope,* p. 227, April 1968.

"Infrared Glows Around Stars," *Sky and Telescope,* p. 158, March 1965.

"Lithium and Star Aging," *Sky and Telescope,* p. 82, August 1963.

"Lithium in Red Giant Stars," *Sky and Telescope,* p. 146, March 1968.

"Oldest Solids," *Scientific American,* p. 72, March 1963.

"Remarkable Variable Star V1057 Cygni," *Sky and Telescope,* p. 12, July 1971.

"Smallest White Dwarf?" *Sky and Telescope,* p. 17, January 1964.

"Some Early Results from Celescope," *Sky and Telescope,* p. 280, May 1969.

CHAPTER CONTENTS

LEARNING OBJECTIVES

BE ABLE TO:

1. GIVE EXAMPLES TO ILLUSTRATE HOW STELLAR MASS CAN BE DETERMINED FROM BINARY STARS.
2. GIVE OBSERVATIONS USED TO IDENTIFY SPECTROSCOPIC BINARY STARS.
3. GIVE OBSERVATIONS USED TO IDENTIFY ECLIPSING BINARIES.
4. GIVE THE RELATIONSHIP BETWEEN MASS AND THE DISTANCE FROM THE CENTER OF MASS FOR MEMBERS OF A BINARY SYSTEM.
5. COMPARE THE PROPER MOTION OF A SINGLE STAR WITH THAT OF A BINARY STAR.
6. DESCRIBE HOW PROPER MOTION OF A STAR MIGHT BE USED TO IDENTIFY THE EXISTENCE OF A PLANETARY SYSTEM.
7. EXTEND THE LUMINOSITY FUNCTION TO INCLUDE PLANETS.
8. EXPLAIN HOW STELLAR ROTATION SUPPORTS THE ARGUMENT FAVORING THE EXISTENCE OF MANY PLANETARY SYSTEMS.
9. GIVE THE CONDITIONS NECESSARY FOR THE EXISTENCE OF WATER IN A LIQUID STATE ON A PLANET.
10. GIVE OBSERVATIONS RELATED TO THE FORMATION OF LARGE MOLECULES AS A PRELUDE TO THE ORIGIN OF LIFE.
11. DISCUSS LIFE ON OTHER PLANETS, BOTH THE PROBABILITY OF EXISTENCE AND THE FORM THAT LIFE MAY TAKE.
12. BRIEFLY DESCRIBE A BLACK HOLE.
13. GIVE OBSERVATIONS SUPPORTING THE EXISTENCE OF BLACK HOLES.

an optical double, on the other hand, usually have different space motions and do not comprise a physical system.

A. STELLAR MASS The only direct way of determining the mass of a star is to study the influence of that mass on other objects. The influence of mass is transmitted through the star's gravitational field, and the only way to study that gravitational field is to observe how other objects move in that field. Double stars offer this opportunity (Figure 12-1), single stars do not.

Consider three examples of two objects each separated by 20 A. U., the average separation of visual binary stars and the distance of Uranus from the sun.

First, the sun and Uranus. The mass of Uranus is so small that the center of mass of the system is inside the sun. Uranus, therefore, is 20 A. U. from the center of mass which is the center of each orbit. Uranus's period of revolution is 84 years (Figure 12-2*a*).

Second, two stars each with a mass equal to that of the sun. The center of mass of this system is midway between the stars, since their masses are equal. Consequently, the radius of each orbit is only 10 A. U. (Figure 12-2*b*). Their period of revolution is 63 years.

Third, two stars each with a mass twice that of the sun. The center of mass is again midway between the two stars. Their period of revolution is 45 years (Figure 12-2*c*).

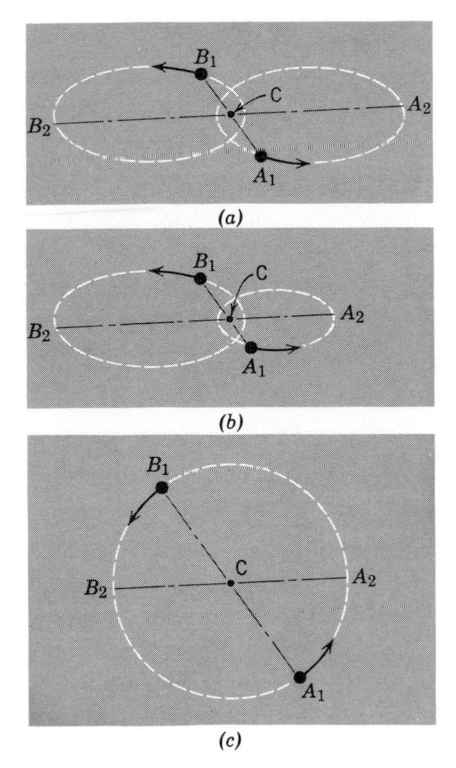

Figure 12-1
The motions of two stars *A* and *B*: (a) both stars are of equal mass and each moves in an elliptical orbit; (b) star *A* is more massive than star *B*; (c) both stars are of equal mass and each moves in a circular orbit.

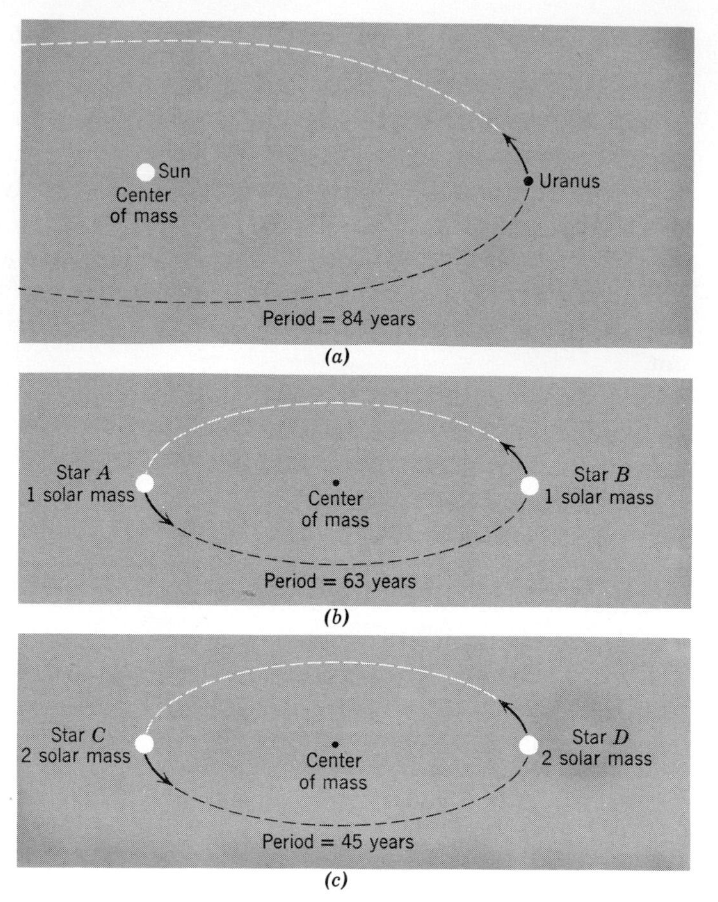

Figure 12-2
(a) The planet Uranus revolving about the sun at its distance of 20 A.U. (b) Two stars each of one solar mass separated by 20 A.U. (c) Two stars each of two solar masses separated by 20 A.U.

It becomes clear from these examples that the period of revolution depends on the masses of the two objects revolving about each other. The periods of revolution of the two binary systems in this example were calculated by using Newton's law of gravity and his second law of motion. The reverse is true when determining the mass of two stars in a binary system. From observations of both the period of revolution and the distance of each star from the center of mass of the system, the mass of each star can then be calculated.

The masses of the two stars in the Sirius binary system are: Sirius A, an A0 star, $m = 2.28$ solar masses; the mass of its F0 white dwarf companion, Sirius B, is 0.98 solar masses. The two stars are separated by a distance of 20.4 A. U. and their period of revolution is 50.1 years (Figure 12-3).

B. ORIENTATION OF ORBITS One complexity does enter in, however, and this is the orientation of the orbit in our sky. Are we looking at the orbits edge on, face on, or at some oblique angle? Radial velocities of each of the two stars permit us to determine this orientation.

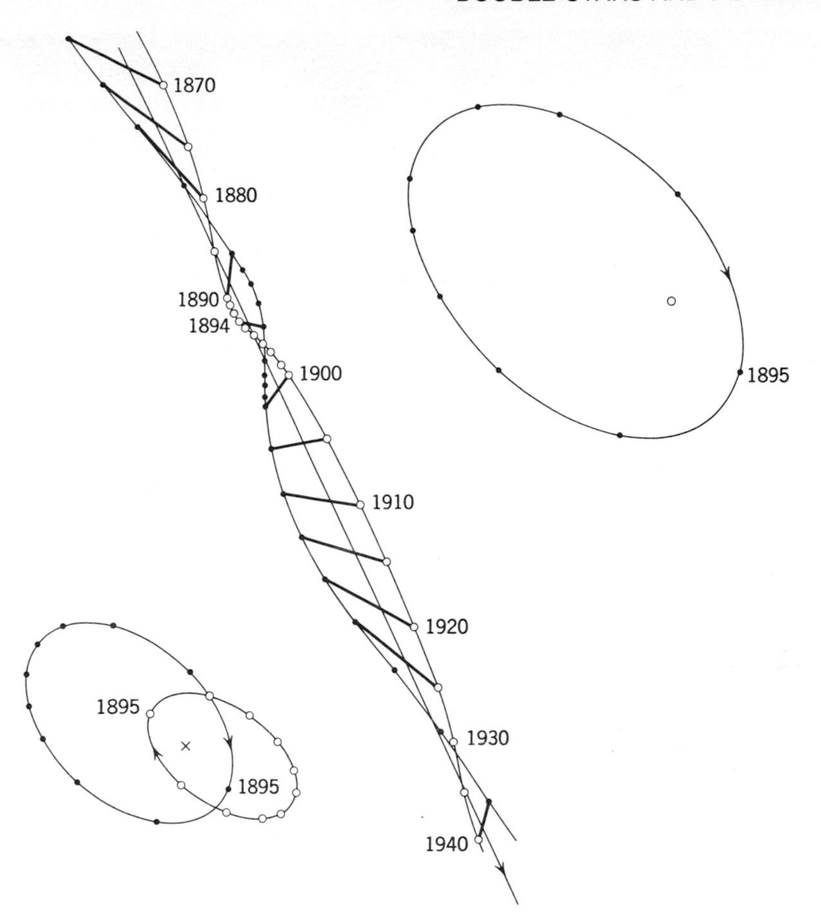

Figure 12-3
The motions of Sirius and its white-dwarf companion. Sirius is shown by a circle and its companion by a dot. (Upper right): The apparent relative orbit of the companion about Sirius. Dots are at five-year intervals. (Lower): Apparent orbits of Sirius and its companion about their common center of mass. Note that the center of mass (the focus of the elliptical orbits) is not at the focus of the apparent orbit. (Center): The motions of Sirius and its companion relative to the background stars. The center of mass moves on a straight line. The scale is not the same for the three drawings. (Cecilia Payne-Gaposchkin, *Introduction to Astronomy*, Copyright © 1954, Prentice-Hall, by permission.)

12.2 SPECTROSCOPIC BINARIES

If the inclination of the orbits of a binary star system is such that we see the orbits edge on, so to speak, the motion of the two stars will be revealed by a periodic Doppler shift of their spectral lines. Since a line joining the stars always passes through the center of mass of the system, one star will be approaching while the other is receding from us, and their spectral lines will be shifted in opposite directions. Consequently, since the Doppler shift of the spectral lines is independent of the apparent separation of the stars in the sky, a binary system will always be disclosed by the periodic Doppler shift in the spectra of two stars, although these may be so close together that they cannot be resolved even with the largest telescope. A binary system that can be recognized only by its spectrum is called a *spectroscopic binary*.

In Figure 12-4*a* we see two spectrograms of a binary system, each taken at a different time. In the lower stellar spectrogram the spectral lines are all shifted toward the violet, indicating that the brighter component of the pair is ap-

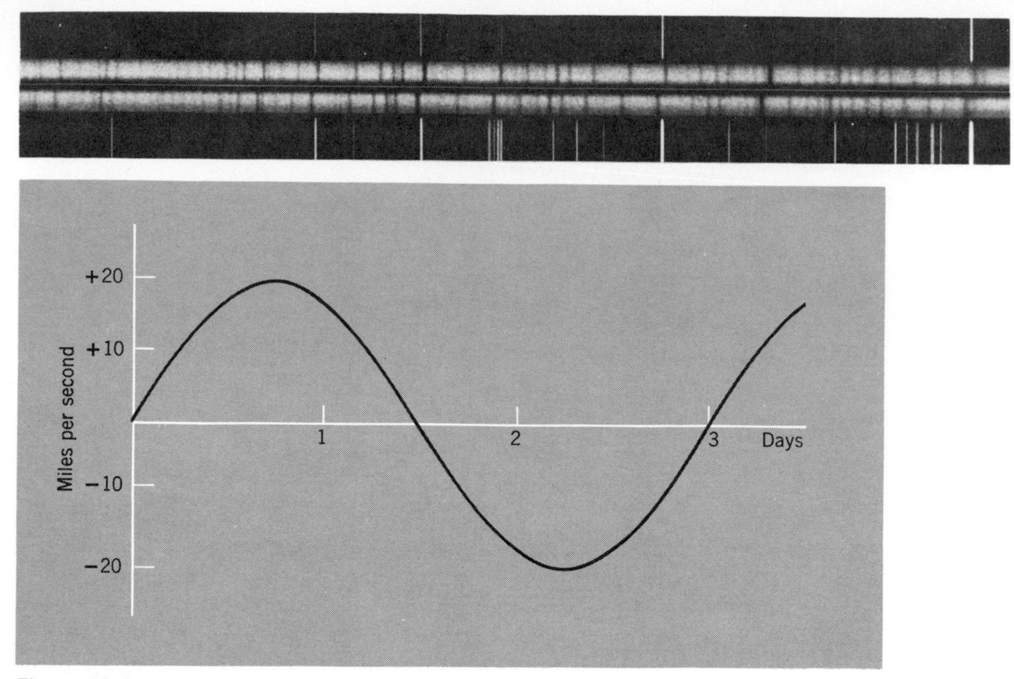

Figure 12-4

(a) Two spectra of the brighter component of the double star Castor, alpha-Geminorum, are included between two bright-line comparison spectra. This star is a spectroscopic binary. The shifting of the spectral lines from the longer wavelength (the upper of the two stellar spectra) to the shorter is evident. (Lick Observatory photograph.)

(b) A schematic drawing of the velocity curve for this spectroscopic binary.

proaching the Earth; the upper stellar spectrogram reveals that it is receding. The brighter component is so much brighter than the fainter component that the spectral lines of the fainter do not even appear. When the radial velocities are plotted on a graph against time, the velocity curve is obtained (Figure 12-4*b*).

The corresponding motions of the two stars are shown in Figure 12-5: *(a)* shows the brighter component receding from the Earth while the fainter is approaching; *(b)* shows the two stars traveling at right angles to the line of sight (at this time neither star shows a Doppler shift unless the entire system is moving relative to the sun); *(c)* shows the brighter component approaching while the fainter is receding.

If the two stars are of nearly equal magnitude, the spectral lines of the stars will be of nearly equal intensity, and both spectra will be visible. When the lines of one star are shifted to the violet the lines of the other star will be shifted to the red, but they will both be at their mean position at the same time. Therefore, as the stars continue to revolve about their common center of mass, the spectral

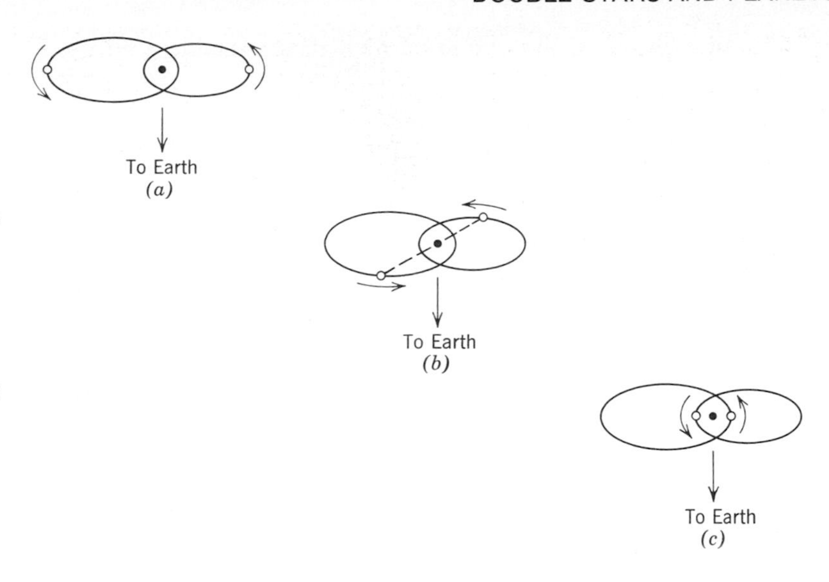

Figure 12-5
The motion of the two components in a spectroscopic binary. In (a) the primary star is receding and, therefore, its spectral lines are shifted to the red and those of the secondary (if visible) are shifted to the violet. The two stars in (b) are traveling in a direction perpendicular to the line of sight and, thus, their spectral lines are not shifted from their mean position. (c) shows the primary approaching and the secondary receding. The orbits are coplanar.

lines of each star will oscillate back and forth across their mean position so that we sometimes see double lines and sometimes only single lines.

The velocities of the two stars are proportional to their distances from the common center of mass. The more massive star is closer to the center of mass, so it travels more slowly in its orbit. When both spectra are visible, therefore, we can determine the velocity of each star and thus the relative distance of each from the center of mass. Without the inclination of the orbit, however, the mass of each star cannot be determined.

If one of the components of a spectroscopic binary system is much more luminous than the other, we see the spectral lines of only the more luminous star shifting back and forth. We can obtain its velocity curve but we can learn very little about the fainter component. We are nevertheless aware of the fainter star's presence, because the brighter component reveals a periodic shifting of its spectral lines without the same variation in brightness that occurs with the Cepheid. If the velocity curve of a Cepheid were mathematically treated as though it were a spectroscopic binary with only one velocity curve observable, the supposed fainter component would have to be inside the Cepheid. There are spectroscopic binaries, however, that have a Cepheid as one of their components.

There are spectroscopic binaries whose orbits we see very nearly edge on so that one star passes in front of the other, causing an eclipse. A great deal can be learned from these *eclipsing binaries.* If both spectra are visible, we can learn the masses of the individual stars. Eclipsing binaries are generally very close together, so may interact with each other. In some binaries, one star heats up that side of the other star facing it. In other pairs, gases stream from one star to the other. Some pairs are so close together that their gravitational forces produce permanent tides in each other, the two misshapen stars resemble a filled-out figure eight.

12.3
STARS, PLANETS, AND LIFE

It seems significant that the average separation of visual binary stars is about 20 A. U. Furthermore, it seems reasonable to presume that whatever forms in the nebular disk about a star depends upon the conditions of that disk. It is, consequently, reasonable to assume that either stars or planets may form in such a disk. Since double stars have formed, it is again reasonable to assume that planets have also formed about other stars.

There is, in fact, convincing evidence that planets are revolving about other stars. Work toward discovering such planets has been carried out at Sproul Observatory, Swarthmore College, under the directorship of Peter van de Kamp. Suppose, they reason, a single star moves through space. Astronomers make careful observations and plot its motion as successive positions on a chart. What sort of motion would those positions represent?

Recall Newton's concept of force and motion applied to moving objects: a moving object will travel in a straight line unless acted on by a force. The plot of the motion of a single star will, therefore, be a straight line.

Now, how about stars in a double-star system? Recall the motion of Sirius and its white-dwarf companion (see Figure 12-3). Consider the motion of these two stars against the background stars, that is, the two wavy lines and the one straight line going from the upper left to the lower right. The straight line represents the motion of the center of mass of the system. The wavy line with open circles represents the motion of Sirius; the wavy line with the dots, the motion of the white dwarf. From this drawing, which of the two stars is the more massive?

In a double-star system, the more massive star is always closer to the center of mass. For the same reason, suppose two weights are placed on pans suspended from the ends of a beam, one weighs 2 pounds and the other 6 pounds (Figure 12-6), and you are asked to lift the beam from one of the three points: A, B, or C. From which point would you lift the beam to keep it horizontal? The more massive object (the 6-pound weight) will be closer to the center of mass, and therefore lifting the beam at point C will keep the system in balance and the beam horizontal.

Now refer to Figure 12-3 again. Which star, Sirius or its white dwarf companion, is the more massive? Which of the two wavy lines is always closer to the center of mass? Sirius, of course, so it is the more massive.

Suppose now, a star is accompanied by one planet, and its motion through space is plotted on a chart. Would the star move along a straight line or along a wavy line as does Sirius? Its actual motion would be along a wavy line, but

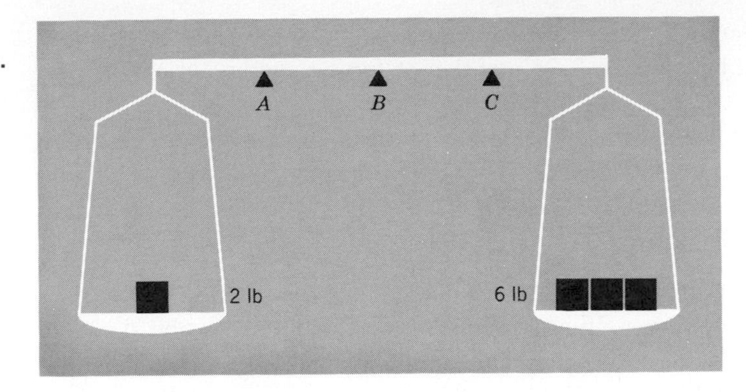

Figure 12-6
Beam balance.

would the wavy nature of the motion be detectable with our present equipment?

Whether the wavy motion of a star accompanied by one planet would be detectable depends on a number of factors. First the star-planet system must be relatively close to the Earth so that its proper motion is easily measured. There are a number of such stars about the sun.

Second, the difference in the mass between the star and the planet, although large, must not be too great. If the star is very massive, its motion about the center of mass will be too slight to be detectable. Therefore, a search for star-planet systems should concentrate on late-type main-sequence stars. For example, the sun is about 1000 times more massive than Jupiter, so its distance from the center of mass is only 0.001 times that of Jupiter; scarcely detectable. An M-type star, however, may have a mass of 0.1 times that of the sun, and if it were accompanied by a planet the mass of Jupiter, its distance from the center of mass would be 0.01 times that of its Jupiter-like planet. The wavy motion of this star might be detectable. It is possible, of course, that planets might be more massive than Jupiter, making the wavy motion of the central star more easily detectable. The astronomers at Swarthmore College have studied a number of likely stars.

A. BARNARD'S STAR Barnard's star is not only an M5 dwarf but it has the largest known proper motion of any star in the sky: 10.27 arc seconds per year (see Figure 9-3). At this rate it will move one apparent diameter of the moon (30 arc minutes) in 170 years, and it has already been observed for 60 years.

The proper motion of Barnard's star has been studied in detail to determine the nature of its invisible companion. Its proper motion is more subtle and complex than can be shown in Figure 9-3. So far the best solution to that motion is to assume that it has not one but two planets accompanying it and causing very slight wavy motions in its proper motion.

If the mass of Barnard's star is assumed to be 0.15 solar mass, then one of its planets, called B1, would have a mass of 1.1 times the mass of Jupiter and be revolving about Barnard's star in an orbit with a radius of 4.7 A. U. The second planet, B2, would have a mass 0.8 times that of Jupiter and be 2.8 A. U. from the central star.

There are other stars with large proper motions that also appear to have planets about them. One, known as BD+68°946, appears to have a companion with a mass of only 0.026 solar mass. (Jupiter's mass is about 0.001 solar mass.) Another star, 61 Cygni, may also have planets about it.

B. MORE PLANETS THAN STARS? The sun and apparently three of the stars nearest the sun have planets; how about the rest of the stars in the sky? We know from our stellar census (the luminosity function, see Figure 9-12) that the number of stars in the solar neighborhood increases as we consider stars of decreasing luminosity. This trend continues to about absolute magnitude +13, then the number begins to fall off for still less luminous stars. A star with a luminosity of +13 has a mass of about 0.07 solar mass.

A gaseous sphere with a mass greater than about 0.07 solar mass will, during the initial contraction stage, generate a high enough central temperature to ignite and sustain thermonuclear reactions. The energy so generated permits the object to radiate light into space and we call it a star. A gaseous sphere with a mass less than about 0.07 solar mass will not generate a high central temperature during the initial contraction; thermonuclear reactions will not be ignited and sustained. After contraction ceases the object cools down. Such a cold object we call a planet. Mass is the critical factor that differentiates a planet, such as Jupiter, from a star.

It would appear, therefore, that the number of faint stars in the solar neighborhood begins to decrease at about absolute magnitude +13 only because

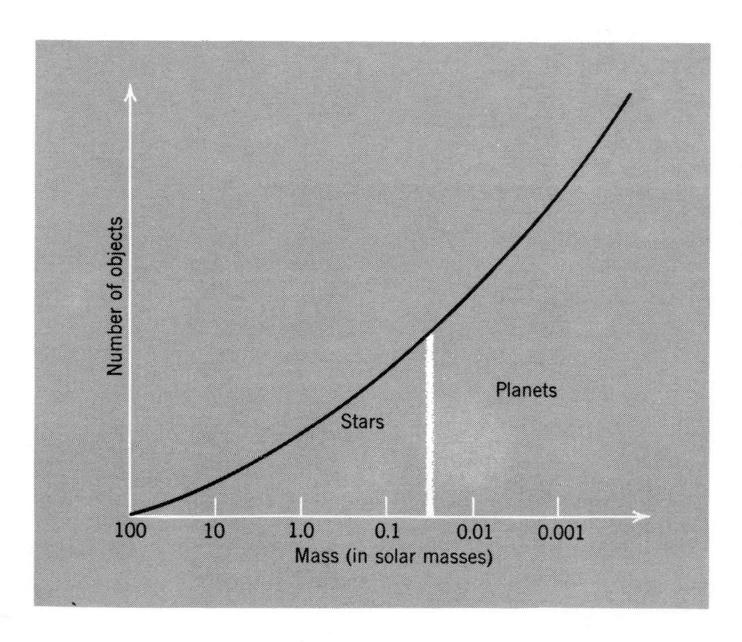

Figure 12-7
If the luminosity-function curve were redrawn as the number of objects against their mass, perhaps it would look like this. Are there more planets than stars in our galaxy?

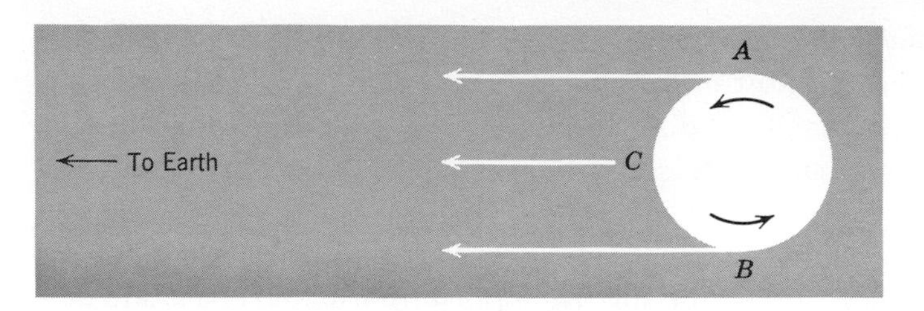

Figure 12-8
The spectral lines of a rapidly rotating star will be broadened because portion *A* of the star is approaching, *B* is receding, and *C* has the radial velocity of the star as a whole.

objects with mass less than about 0.07 solar mass become planets and are not visible from the Earth.

We are left to ask, then, what would the curve for the luminosity function look like if it were redrawn to represent the number of objects — stars and planets — against their mass. Would the number of objects continue to increase as shown in Figure 12-7? Are there more planets than stars in our galaxy? There certainly are more planets than stars in our own solar system! and Barnard's star? and the rest?

C. STELLAR ROTATION AND PLANETS Still another line of attack to this problem is brought forth from a study of stellar rotation. Stars reveal rotation by their spectral lines. A rapidly rotating star will have spectral lines that are broadened (Figure 12-8). Portions of the star are approaching the Earth while other portions are receding, and the resulting Doppler shifts yield broad spectral lines. This broadening can be distinguished from other forms of line broadening, such as pressure broadening. The average observed rotational speed at the equator of main-sequence stars of different spectral types is shown in Table 12-1.

It is clear from the table that the less luminous late-type main-sequence stars, let us say, F5 and later, rotate slowly. Early type main-sequence stars rotate rapidly. Why? Something must happen during the process of stellar formation that either increases the rotational speed of hot stars or decreases the rotation of cool stars.

Table 12-1
Stellar equatorial rotational velocities

Spectral Type	Main-Sequence Stars
B0	125 mi/sec
B5	130
A0	120
A5	100
F0	60
F5	16
G0	<10*
K, M	<10

*The symbol < means less than.

Stars form from whirling blobs of gas, so the rate of rotation of the final product must depend, at least in part, on how fast the gas was whirling in the first place. Measurements by radial velocity techniques indicate that the motions of gas within a nebula, such as the Orion nebula, are slow, 100 centimeters per second or less.

When a blob of rotating gas contracts, however, the speed of rotation increases. Studies indicate that if the sun had formed from a blob of gas and dust whirling as slowly as those in the Orion nebula, then during contraction its speed of rotation would now be less than ½ day rather than the observed 26 days.

The question then arises: What caused the sun to rotate more slowly? The best answer to this question is that the nebular disk about the sun supplied a magnetic brake. As the dark protosun became bright, the gases around the sun were first ionized and then blown away by the solar wind (see page 246). Rotating ionized gases set up a magnetic field, and the magnetic field of the inner nebular disk must have reacted with the sun's magnetic field to retard the rotation of the sun.

Extending this argument to all the main-sequence stars leads us to the obvious conclusion that all the stars from about F5 and later formed nebular disks about them in their early stages. In each, the ionized gases in the rotating nebular disk retarded the rotation of the central star. But if each of the main-sequence stars later than about F5 formed nebular disks which slowed the rotation of the central star, then would not each of these stars, like the sun, have formed planets in their nebular disks? It is, therefore, reasonable to speculate that planets form as a natural and inevitable part of stellar formation of those stars later than about F5, that is, the late F-type, the G-, K-, and M-type stars.

How many of the stars in the sky does this include? The luminosity function is the stellar census, it is a count of the number of stars in the sky of each absolute luminosity. A look at the luminosity function (Figure 9-12, p. 199) makes clear that the less luminous stars are by far more numerous than the very luminous ones. The absolute magnitude of an F5 star is +3.8; more than 95% of all the stars in the sky have a luminosity less than +3.8. This line of reasoning causes us to conclude that 95% of all the stars in the sky have planets revolving about them. Planets may indeed be very common in our galaxy.

12.4 ORIGIN OF LIFE

If 95% of the stars in the sky have planets, then what is the likelihood that at least one planet in each of those planetary systems has life on it; perhaps life not unlike our own? What are the essential factors that made life possible on Earth? Are these same factors apt to exist on those postulated planetary systems?

A. WATER ESSENTIAL FOR LIFE Water in the liquid state is the primary factor that not only permitted life to form on Earth but that has maintained it since then. For water to exist in the liquid state, a planet needs to have both a suitable temperature and an atmosphere.

1. Temperature The temperature on the surface of a planet depends on both the planet's distance from the central star and the temperature of that star. If the central star is a K- or M-type star, then presumably a planet would need to be closer than one astronomical unit for water to exist in the liquid state. But if planets in each of the postulated planetary systems are spaced as the planets in our solar system are, then perhaps at least one planet will be located such that puddles, ponds, and oceans will form.

2. Atmosphere The original atmosphere of the Earth was probably derived from the gases left over from the formation of the Earth to which were added the gases of early volcanic eruptions. To determine the composition of the prevolcanic atmosphere we must observe molecules, that is, chemical compounds, in the rest of the solar system and in the nebulae from which stars are born. Common chemical compounds in the other terrestrial planets, in the Jovian planets, and in comets are CH_4 (methane), NH_3 (ammonia), and CO_2 (carbon dioxide).

To detect molecules in distant nebulae we must make observations with suitable equipment. Molecules radiate and absorb energy in the radio and microwave region of the radiant energy spectrum, so we rely on the observations of radio telescopes. To the great surprise of astronomers, these observations have recently revealed a wealth — in both abundance and diversity — of molecules in nebulae and even about cool M-type stars. Common molecules found in the gas and dust clouds of our galaxy are: H_2O (water), NH_3 (ammonia), CO (carbon monoxide), HCN (hydrogen cyanide, H_2CO (formaldehyde), CH_3OH (methyl alcohol), and many others familiar to chemists here on Earth. Any of these gases could, therefore, have been present in the Earth's prevolcanic atmosphere.

The gases added to the atmosphere by volcanoes are familiar ones. Gases belching from the volcanoes on the Hawaiian Islands, for example, are composed of: water vapor, 71%; carbon dioxide, 14%; sulfur dioxide, 6%; nitrogen, 5%; and 4% other gases. This high percentage of water vapor is typical of volcanoes on the continents as well as those that form on the ocean floors. In fact, the high water content of the volcanic gases coupled with the abundance of volcanoes in past geological ages has led to the conclusion that the waters of the oceans came from volcanic eruptions in the first half-billion years or so of the Earth's existence.

B. THE FORMATION OF LIFE To speculate on the formation of life, a biologist can begin at either end. He can start with life as we know it today, search for the simplest organism, and then try to see how that organism might have been created. Or he could start with what astronomers believe were the most likely components in the early atmosphere of the Earth, and try to imagine how they might combine to form ever more complex molecules until, finally, the complexity permits life to be sustained. Both approaches have been tried but the latter has been more fruitful.

Which chemical compounds in the early atmosphere might have combined to produce more complex molecules? And what might those complex molecules have been?

The chemistry of living matter is totally dependent on the elements carbon, hydrogen, and oxygen. These three elements form the backbone, so to speak, of every giant molecule in every living organism. These three elements are present in three of the compounds that were very likely part of the Earth's early atmosphere. Those three compounds are: methane (CH_4), water (H_2O), and ammonia (NH_3). Molecules of these compounds can combine if they are given energy by some means or other. For example, energy might have been supplied by the sun's ultraviolet radiation that penetrated the early atmosphere but that is stopped by our present oxygen-laden atmosphere. Or energy might have been supplied by bolts of lightning flashing about cumulonimbus clouds and erupting volcanoes (Figure 12-9). Experiments have been performed to determine what happens when mixtures of these three substances are given energy.

S. L. Miller and Harold Urey have put methane, ammonia, and water (both liquid and vapor) in closed containers, supplied energy by means of a spark (to simulate lightning), and have cycled the gases through the liquid water. On analysis the water was found to contain some complex molecules of carbon, hydrogen, and oxygen. Among the findings were several amino acids. In other experiments, ultraviolet radiation was used to supply energy to the same three simple compounds and again amino acids were detected.

Amino acids are significant because they can be combined to form still more complex molecules: the proteins! Protein molecules are clearly one of the building blocks of living matter. For example, all enzymes are proteins.

However, the key molecule that had to form before life could exist is the DNA molecule. This very complex molecule carries the code that dictates the characteristics of an individual, that is, it guides the development of an individual from the fertilized egg. It determines whether the developing embryo will be a tree, a man, or a mouse. It also determines the color of the eyes, the shape of the leaf, or whatever.

The exact process by which a DNA molecule could have formed from rudimentary methane, ammonia, and water is not at all clear, but several of the steps have been worked out and performed in the biological laboratory. The process is not a direct one, nor is it likely to happen by the passing of simple

Figure 12-9
Volcanic eruptions, such as this one that formed the new island of Surtsey in Iceland, are often accompanied by lightning. (Courtesy Sigurgeir Jonasson, Vestmannaeyjar, Iceland.)

gases over a spark. However, if the oceans were formed within the first billion years of the Earth's existence, then life could have formed within the next billion. The solar system (and the Earth) is 4.6 billion years old, and it is estimated that life formed about 3½ billion years ago.

During the first billion years or so of the oceans' existence, chemical reactions occurred that converted the oceans into what has been called a broth: lifeless water teeming with complex molecules. Amino acids, proteins, sugars, and still more complex molecules must have formed. Given enough time, it is expected that DNA molecules could have formed in this broth. And the DNA molecule can replicate (duplicate) itself.

It is the vast amount of time that makes possible the development of DNA molecules from that primeval broth. Although the event is very unlikely to happen at any instant of time, or even in a week, month, year, or century; within a billion years, however, it is very likely to have happened.

By way of analogy, if you flip a coin once, there is a 50-50 chance that it will turn up heads. However, the chance that a head will appear at least once if it is flipped 10 times is 999 out of 1000 — almost a certainty. If an event is unlikely to occur in one reaction, it will be nearly certain to occur in 10,000 or 10,000,000 reactions. Given a billion years, the DNA molecule was certain to appear in that ancient broth that nourished the life it gave rise to. However, biologists don't expect to repeat the process in their laboratories.

Once life appeared in the seas, it multiplied and developed and, as it did, it began to alter the oceans and atmosphere that gave rise to it. After a billion years the forms of life began to diversify. A billion years ago plants appeared, and they began to add oxygen to the atmosphere. Our present atmosphere is quite different from the original one because of the influence of life.

C. LIFE ON OTHER WORLDS Since the molecules that existed in the early stages of the Earth are the same kinds of molecules that exist on comets and other planets in our solar system, and in the gaseous nebulae throughout the galaxy, they should combine to form the same kind of complex molecules. And, indeed, observations support this idea. Amino acids have been found that formed elsewhere in our solar system; they have been found in two meteorites. Both of these meteorites were seen to fall and soon afterwards were picked up, one in Kentucky in 1950, the other in Australia in 1969. Both are carbonaceous chondrites with a high percentage of carbon present (2.2%). Of the amino acids found in the meteorites, 6 are found in protein molecules normally found in living matter, the other 12 amino acids are not important to living matter. It has been shown that these 18 amino acids were in the meteorites before they fell; they were not contaminants from the Earth. Another complex molecule, formaldehyde, was found in another carbonaceous chondrite that fell in Mexico in 1969. The Mexican meteorite had only 0.25% carbon, apparently not enough to form amino acids, for none were found.

The amino acids here on Earth and in those two meteorites were formed from the same kind of molecules that exist in nebulae throughout the galaxy. There is, therefore, no reason why the same amino acids should not have formed on *any* planet in the galaxy. Furthermore, there is no reason why those same amino acids should not have combined to form the same kind of proteins we have here on Earth, so long as that planet has a temperature and atmosphere similar to the Earth's. If the same amino acids and proteins formed, then the same DNA molecules should also have formed. And the DNA molecule is the key to the development of the individual and the species.

The process by which life developed on the Earth was one of elimination. Of all the myriads of forms of life that have been tried by Mother Nature, only those that were successfully adapted to the environment have survived. Yet we have a diversity of life on Earth that is staggering — from the minute one-celled animals of the pond, the creatures on the ocean floor, to the plants, the insects, reptiles, birds, and mammals.

Given the same DNA molecules, therefore, and a planet with a similar temperature and atmosphere, the same factors should influence those DNA molecules. Thus the same development of life should take place on that planet as here on Earth. In other words, it makes sense to speculate that life on other planets is probably very much like the Earth's diverse forms of life.

With so many stars that may have planets, the formation of life elsewhere in our galaxy seems very likely. It did happen here on Earth, after all.

12.5 BLACK HOLES

Increased use of satellites and space probes have permitted astronomers to study objects in the universe that produce X rays. X rays are some of the very energetic forms of radiant energy that are blocked by our atmosphere. The satellite *Uhuru*, launched from the coast of Kenya in 1971, was designed specifically to observe the celestial sphere with X-ray detecting equipment. The results were startling. Among the objects detected are black holes!

Black holes had been predicted by the theory of general relativity. This theory predicts that light is attracted by a gravitational field; the stronger the field the stronger the attraction. At the extreme, it was predicted, the gravitational field could become so great that the escape velocity would become equal to the speed of light — that is, not even light could escape. Hence the name, black hole.

If light cannot escape, then how can astronomers ever hope to see a black hole? The observational prospects looked bleak, until "X-ray stars" were definitely established to exist by the Uhuru satellite.

This satellite could observe an X-ray source continuously and thus was able to identify a number of X-ray sources as stars that were members of spectroscopic and eclipsing binary systems. This gave the astronomers an observational method of learning a great deal about these strange objects.

The binary star is generally composed of a supergiant expanding and, therefore, leaving the main sequence and a companion that is an exceedingly compact object called the black hole. According to current theory the minimum mass of a black hole is about 3 times the mass of the sun. The diameter, however, hardly seems astronomical, somewhat less than 20 miles. The density of a black hole is of the order 10^{18} times the density of water.

The source of X-rays is apparently the gases from the supergiant that spill over and fall into the black hole. These gases are heated to tremendous temperatures during their fall and before reaching the "point of no return," X rays are emitted out into space.

The origin of a black hole may be very massive stars that evolve rapidly, their core collapsing not into a white dwarf or a neutron star, but into a black hole the existence of which can only be determined by observations of its influence on matter outside itself.

BASIC VOCABULARY FOR SUBSEQUENT READING

Amino acids
Black hole
Center of mass
DNA molecule
Eclipsing binary
Spectroscopic binary
Visual binary

QUESTIONS AND PROBLEMS

1. Which of the following two pairs of eclipsing binaries has the shortest period of revolution?

	Separation (solar radii)	Mass (solar masses) A	B
AR Aurigae	18.5	2.6	2.3
AH Cephei	18.7	16.5	14.2

2. Which of the following two pairs of eclipsing binaries has the shortest period of revolution?

	Separation (solar radii)	Mass (solar masses) A	B
YZ Cassiopeiae	19.4	3.3	1.6
Xi Phoenicis	10.0	3.0	2.1

3. A spectroscopic binary is composed of a main-sequence B-type star and a main-sequence K-type star. Would this be a single-line or a double-line spectroscopic binary?

4. Under what conditions do we observe: (a) a single-line spectroscopic binary, and (b) a double-line spectroscopic binary?

5. (a) Must a spectroscopic binary also be an eclipsing binary?
(b) Must an eclipsing binary also be a spectroscopic binary?

6. Under what conditions can a binary system cause:
(a) A total eclipse?
(b) A partial eclipse?
(c) An annular eclipse?

7. What observations have led to the belief that Barnard's star is accompanied by planets?

8. What observations indicate that stars of small mass are more common than stars of large mass?

9. What evidence indicates that late-type stars are more apt to have planets than early-type stars?

10. Discuss the likelihood that life forms on other planets would be very similar to life forms on earth.

FOR FURTHER READING

Kaufman, W. J., *Relativity and Cosmology,* Chapters 1, 2, 3, 4, and 6, Harper and Row, New York, 1973.

Sagan, C., *The Cosmic Connection,* Garden City, N. Y., 1973.

Struve, O., and V. Zebergs, *Astronomy of the 20th Century,* Crowell, Collier and Macmillan, New York, 1962, Chapter XIV.

Batten, A. H., and M. Plavec, "Two New Chapters in the Story of U Cephei," *Sky and Telescope,* part I, p. 147, September 1971, part II, p. 213, October 1971.

Brown, H., "Planetary Systems and Main Sequence Stars," *Science,* p. 1177, September 11, 1964.

Buhl, D., "Molecules and Evolution in the Galaxy," *Sky and Telescope,* p. 156, March 1973.

Eggen, O. J., "Stars in Contact," *Scientific American,* p. 34, June 1968.

Gursky, H. and E. P. J. vanden Heuvel, "X-Ray-Emitting Double Stars," *Scientific American,* p. 24, March 1975.

Hack, M., "Stellar Rotation and Atmospheric Motions," *Sky and Telescope,* p. 84, August 1970; p. 143, September 1970; p. 208, October 1970.

Huang, S.-S., "The Origin of Binary Stars," *Sky and Telescope,* p. 368, December 1967.

Johnston, K. J. et al, "Microwave Celestial Water-Vapor Sources," *Sky and Telescope,* p. 88, August 1972.

Lawless, J. G. et al, "Organic Matter in Meteorites," *Scientific American,* p. 38, June 1972.

Lippincott, S., and M. D. Worth, "The Double Star Sirius," *Sky and Telescope,* p. 4, January 1966.

Lovell, B., "Radio-emitting Flare Stars," *Scientific American,* p. 13, August 1964.

Meeus, J., "Some Bright Visual Binary Stars," *Sky and Telescope,* part I, p. 21, January 1971; part II, p. 88, February 1971.

Sagan, C., and F. Drake, "The Search for Extraterrestrial Intelligence," *Scientific American,* p. 80, May 1975.

Turner, B., "Interstellar Molecules," *Scientific American,* p. 51, March 1973.

van de Kamp, P., "Barnard's Star: The Search for Other Solar Systems," *Natural History,* p. 38, April 1970.

Walker, M. F., "New Observations of AE Aquarii," *Sky and Telescope,* p. 23, January 1965.

Warner, B., "Six Ultrashort-Period Binary Stars," *Sky and Telescope,* p. 358, December 1972.

"An Ex-Variable Star?" *Sky and Telescope,* p. 323, June 1966.

"Another Interstellar Molecule," *Sky and Telescope,* p. 157, September 1972.

"Dark Interstellar Clouds," *Sky and Telescope,* p. 227, April 1968.

"Formaldehyde in Allende," *Sky and Telescope,* p. 352, June 1972.

CHAPTER CONTENTS

LEARNING OBJECTIVES

BE ABLE TO:

1. GIVE AND DESCRIBE EACH OF THE VARIOUS KINDS OF GALAXIES.
2. EXPLAIN THE MEANING OF POPULATION I AND II OBJECTS.
3. GIVE OBSERVATIONS IDENTIFYING VERY ACTIVE GALAXIES.
4. GIVE OBSERVATIONS AND REASONING ESTABLISHING OUR MILKY WAY GALAXY AS A SPIRAL GALAXY.
5. DESCRIBE HOW ASTRONOMERS HAVE LOCATED THE NUCLEUS OF OUR GALAXY.
6. DESCRIBE GALACTIC COORDINATES AND COMPARE THEM WITH LATITUDE-LONGITUDE.
7. DESCRIBE HOW THE LOCATION OF SPIRAL ARMS IS DETERMINED.
8. DESCRIBE THE MANNER IN WHICH OUR GALAXY ROTATES.
9. DESCRIBE HOW THE LOCATION IN THE GALAXY OF POPULATION I AND II OBJECTS DIFFERS.
10. EXPLAIN THE SIGNIFICANCE OF THE OBSERVATION THAT OBJECTS OF DIFFERENT AGE GENERALLY OCCUPY DIFFERENT LOCATIONS IN OUR GALAXY.
11. EXPLAIN HOW DIFFERENCES IN ROTATION DURING THE EARLY STAGES OF GALAXIES MAY ACCOUNT FOR DIFFERENCES IN STRUCTURE.

Chapter Thirteen
GALAXIES

One of the spectacular sights in the summer sky is the Milky Way. This irregular band is our galaxy seen from the inside. Indeed, studies of our Milky Way galaxy are both helped and hindered by our intimate relation with it. What we do see, we see with great detail, certainly more detail than we can see in the other more distant galaxies. But that dark band running down the middle of the Milky Way is composed of clouds of gas and dust that restrict our visual observations to our part of the galaxy, a part quaintly described by the late Walter Baade of the Hale Observatories as the "local swimming hole."

The problem is not unlike that of placing your head between the spokes of a large wagon wheel and then trying to figure out what the rest of the wheel looks like. Without previous knowledge of wagon wheels in general, the problem would be a difficult one. Correspondingly, a knowledge of galaxies in general is useful to the astronomer. Astronomers therefore study other galaxies in great detail, for not only do these studies help us interpret observations of our own galaxy, but they also add to our knowledge of how the universe is put together. Let us investigate the structure of galaxies in general before we consider our own in detail.

13.1 HUBBLE'S CLASSIFICATION OF GALAXIES

Galaxies appear in every portion of the sky except along the Milky Way where the dust and gas of our own galaxy conceal the galaxies beyond. Hundreds of millions of galaxies are seen in the universe; the most distant ones visible in our large telescopes are estimated to be some 12 billion light years away. But galactic structure can be studied in only those that are much closer to us. The late Edwin P. Hubble, of the Mount Wilson Observatory, made an intensive study of galaxies and recognized three basic structures for the nearer and brighter galaxies: *elliptical*, *spiral*, and *irregular*. Each basic structure is further subdivided according to variations of shape within that group.

The elliptical galaxies (chapter opening photo of this chapter) seem to be similar in appearance to the globular clusters although they are much larger and contain many more stars. They vary in shape from spheres to ellipsoids whose major axis is about five times longer than the minor axis. Since rotation causes flattening, the faster a galaxy rotates the more ellipsoidal it is.

Hubble's classification of elliptical galaxies is shown in Figure 13-1; the letter E (standing for elliptical) is followed by a number that indicates the degree of ellipticity. An E0 galaxy represents a circular disk. With increasing ellipticity

Chapter Opening Photo
Elliptical galaxy — a satellite of the Andromeda galaxy. (Photograph from the Hale Observatories.)

Figure 13-1
The shapes of elliptical galaxies vary from spherical to ellipsoidal. (Photographs from the Hale Observatories.)

the number following the E increases to a maximum of 7. Since it is very difficult to determine the orientation in space of an elliptical galaxy, this classification applies to its shape only as it is seen from the Earth.

The spiral galaxies present a dynamic picture (Figure 13-2) that stands in contrast to the rather uniform elliptical galaxies. The first impression received upon looking at a spiral galaxy is that it is rotating about a central nucleus from which *spiral arms* curve out into space. Spiral galaxies were classified by Hubble into three subgroups: Sa, Sb, and Sc, according to the amount of material in the arms relative to that of the nucleus, as well as to the degree of openness of the arms.

In the same group with the normal spirals are the *barred spirals* (Figure 13-3), so called because they have a "bar" of stars running through the nucleus. From the ends of this bar spiral arms may form, sometimes into a ring about the nucleus. Why some spiral galaxies assume the shape of normal and others of barred spirals is not known. The normal spirals, however, are more numerous: of the 600 galaxies studied by Hubble, 17% are elliptical; 50% are normal spirals; 30% are barred spirals; 3% are irregular.

The third type, the irregular galaxy, has no definite form. The two Magellanic Clouds are often considered to be members of this group, although some astronomers believe that the Large Magellanic Cloud should be classified as a barred spiral (Figure 10-2, p. 208).

Figure 13-2
Four different normal spiral galaxies. (a) Type Sa, (b) Type Sab, (c) Type Sb, (d) Type Sc. (Photographs from the Hale Observatories.)

A. ELLIPTICAL GALAXIES It was not until 1944 that new photographic techniques made it possible to resolve one of the nearest elliptical galaxies, a companion of the Andromeda galaxy, into separate stars (opening photograph, Chapter Thirteen). The distribution of stars in elliptical galaxies is such that the overall shape of the galaxy does not change on photographs taken with different time exposures. A shorter exposure records only the more densely populated central regions, whereas a longer exposure records the outer regions of the galaxy. Thus it appears that the stars are placed rather symmetrically about the center but decrease in number from the center out.

The similarities of an elliptical galaxy, the nucleus of a spiral galaxy, and a globular cluster are striking. The brightest stars in each are red supergiants; the

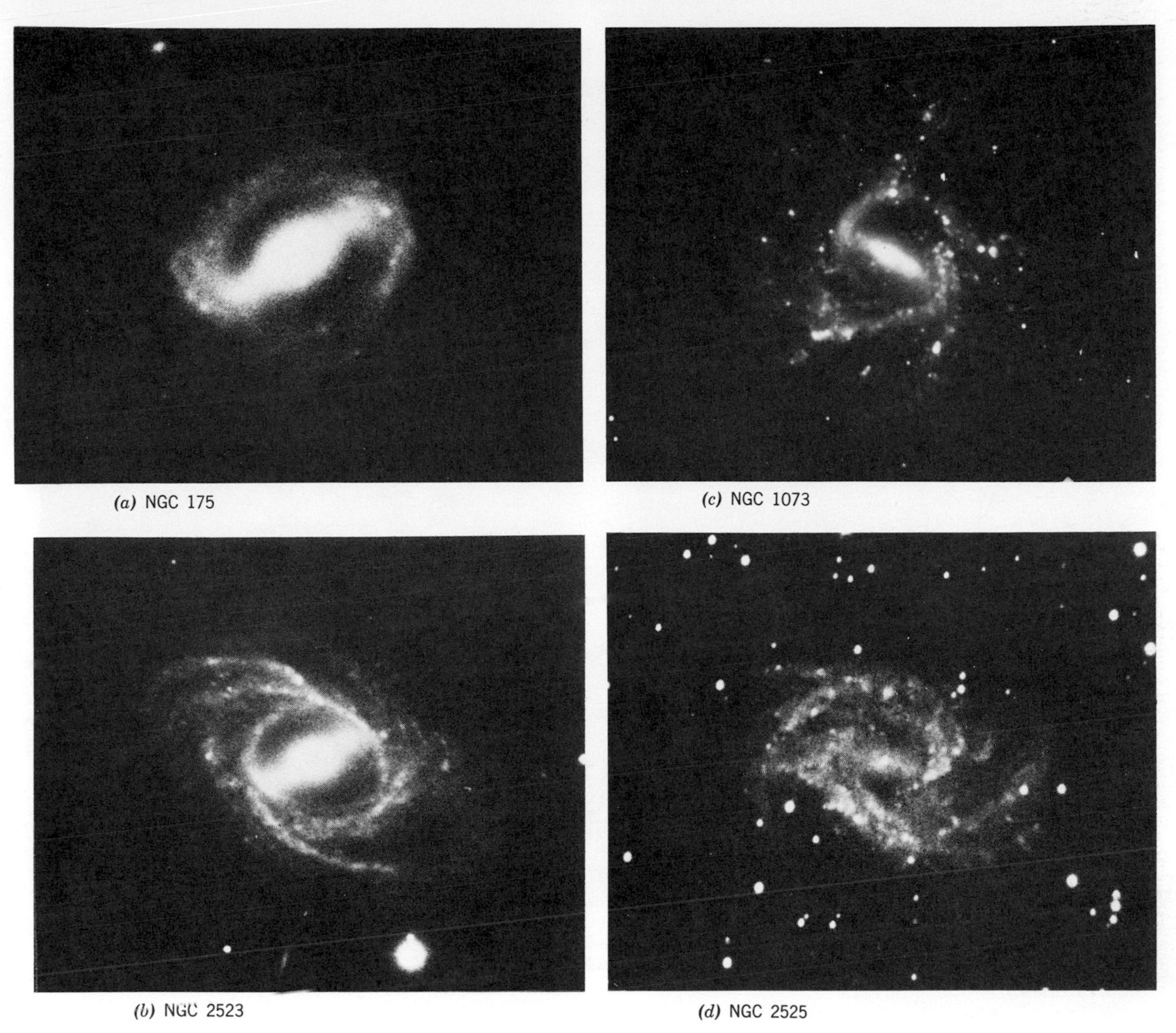
(a) NGC 175

(c) NGC 1073

(b) NGC 2523

(d) NGC 2525

Figure 13-3

Four different barred spiral galaxies. (Photograph from the Hale Observatories.)

blue giants are conspicuous by their absence. It is not surprising, therefore, to learn that elliptical galaxies also contain Cepheids and are practically devoid of interstellar material. Since the lack of interstellar material makes them nearly transparent, more distant galaxies can sometimes be seen through their outer regions. It has become clear that the largest and most massive galaxies observed are a few exceptionally large elliptical galaxies.

B. NORMAL SPIRAL GALAXIES Spiral galaxies differ considerably from elliptical galaxies both in structure and in stellar content. The nucleus, of course, resembles an elliptical galaxy, but here the similarity ends. The Andromeda Galaxy, M31, (Star Plate 5) is typical of the normal spirals and is close enough to our own galaxy (2.2 million light years from the sun) to be faintly visible to

Figure 13-4
Detail of the nucleus of the Andromeda galaxy. (Photograph from the Hale Observatories.)

the naked eye as a hazy patch in the sky. Moreover, in a photograph the details of the Andromeda galaxy can be clearly seen (Figure 13-4). The central portion is the nucleus which is composed of red giants and other stars that are generally considered old stars. The spiral arms, on the other hand, can be traced by following the lanes of bright nebulae, dark clouds of gas and dust, and open clusters with their blue-stars. The spiral arms are the breeding grounds of new stars.

Despite the brightness of the spiral arms, most of the light from the region outside the nucleus comes from a substratum of stars whose luminosity is equal to or less than that of giants. The arms are superimposed upon this substratum, which is nearly transparent, since more distant galaxies are visible through the parts of it that lie between the spiral arms. Observations indicate that the shape of the substratum is somewhat ellipsoidal. It extends above and below the plane of the galaxy, includes many globular clusters, and RR Lyrae stars both indicative of an older age than the spiral arms.

Since we see the Andromeda galaxy at an oblique angle (the plane of the galaxy makes an angle of about 13° with the line of sight), we are able to determine by the Doppler shift the rate of rotation of those parts that are either approaching or receding from us. N. U. Mayall, retired director of the Kitt Peak National Observatory, and more recently, Vera C. Rubin of the Carnegie Institution of Washington have studied the rotation by observing nebulae in the spiral arms and the recently discovered gas diffused throughout the nucleus. The spectra of the bright nebulae have spectral lines that are sharper than the dark lines in stellar spectra. Radial velocity measurements from these sharper bright spectral lines are more accurate than those made from the dark fuzzy stellar lines.

The results of these studies indicate that the outer regions of the galaxy rotate in a manner not unlike the motion of the planets about the sun; that is, the speed of rotation decreases with increasing distance from the center. The rotation of the inner regions of the galaxy, however, is more complex.

Since we know the distance of the Andromeda galaxy, we can determine its size from its angular diameter. The portion that appears in Plate 5 has a diameter of about 100,000 light years. Measurements made with a photocell, however, indicate that there is a fringe of stars extending beyond the obvious photographic boundaries. The diameter including this fringe is about 180,000 light years. From the size and rotation of the Andromeda galaxy, the mass of this system has been estimated to be about 3×10^{11} solar masses. Since the sun is roughly an average star in mass, we can assume that there are some 300 billion stars in the Andromeda galaxy!

C. BARRED SPIRAL GALAXIES Barred spirals are similar to normal spirals in that their nuclei contain red-giant stars. Their spiral arms are composed of bright nebulosities as well as open clusters, lanes of dark obscuring matter, and blue-giant stars. The significant difference between normal and barred spirals can be seen by comparing Figures 13-5 with 13-6. The bar and the nucleus seem to be rotating as a unit, like two spokes extending from the hub of a wheel. The spiral arms seem to lag behind the bar and the nucleus; thus they appear to rotate in a manner similar to that in which our planets revolve. At present there is no explanation for the structure of barred spirals.

D. IRREGULAR GALAXIES The irregular galaxies are quite different from both the elliptical and the spiral galaxies. They have no central nucleus, no spiral arms, and apparently no plane of symmetry. Common examples of this type of galaxy are the Magellanic Clouds (Figure 10-2).

The two *Magellanic Clouds* are located fairly close to our own galaxy. The Large Cloud is about 150,000 light years from us, and the Small Cloud about 200,000 light years away. The diameter of the main body of the Large Cloud is roughly 32,000 light years; that for the small cloud is 25,000 light years. The

Figure 13-5
The spiral galaxy in Triangulum, M33. (Photograph from the Hale Observatories.)

mass of the Large Cloud is about 0.1 times that of our Milky Way Galaxy, that of the Small Cloud is about 0.02. The luminosities of the two clouds seem to be average for galaxies in general, our galaxy is much more luminous than most.

The Large Magellanic Cloud is a most interesting object for astronomers concerned with the evolution of stellar systems. It has an abundance of Population I objects: blue-giant stars, open clusters, obscuring gas and dust, and emission nebulae. In fact, the largest emission nebula known, the Tarantula nebula, lies in the Large Cloud. It is so big that were it to replace the Orion nebula, it would cover the entire constellation of Orion and be an extremely obvious feature in our night skies. But the most curious objects in the Large Cloud are the globular clusters, for many of these globular clusters have blue-giant stars! These globular clusters must be young and some may still be in the process of formation. These clusters are very regular and globular in shape but their color-illuminosity diagrams look like those of the open clusters in our galaxy. These color–luminosity diagrams have a thin line of stars running right up the main sequence to the blue giants. Other globular clusters in the Large Cloud appear to be a billion years old, but still younger than any in our galaxy. It would appear as if star formation in the Magellanic Clouds has either started late or for some reason has been retarded.

The Clouds have radial velocities of recession of 170 and 100 miles per second respectively when referred to the sun, which moves within our own galaxy. When their motions are referred to the center of our galaxy, however, we find that the Clouds have essentially zero radial velocity. This fact, coupled with their nearness to our galaxy, has led most astronomers to believe that they are revolving about our galaxy as satellites. Radio observations of neutral hydrogen indicate that the two Clouds are enclosed in a common envelope of hydrogen gas. But unlike some systems (Figure 13-7), there does not appear to be a bridge of stars connecting the Magellanic Clouds to our galaxy. The Andromeda galaxy also has two satellite galaxies, but both of these are elliptical galaxies.

E. DWARF GALAXIES Surprisingly enough the most common galaxy of all is the *dwarf galaxy* (Figure 13-8). These galaxies are difficult to detect, for they are not as obvious nor as luminous as the bigger galaxies, yet they exceed the bigger ones in number. There are more dwarf galaxies in the immediate neighborhood of our galaxy than all other kinds put together. Large numbers of dwarf galaxies have been detected in nearby clusters of galaxies. Some of these dwarf galaxies are regular in shape, and these tend to have older stars; the irregularly shaped dwarfs tend to have very luminous blue stars.

F. POPULATION I AND II OBJECTS Studies made in the mid-1940s by the late Walter Baade, then studying the Andromeda galaxy with the 100-inch telescope at the Mount Wilson Observatory, led to some generalizations about the stars

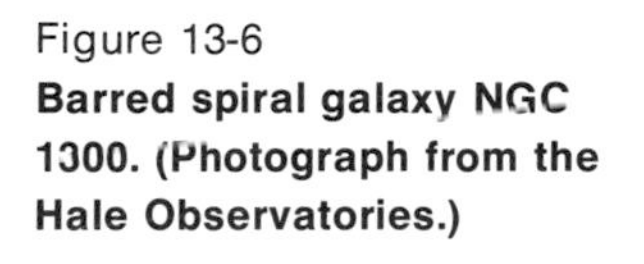

Figure 13-6
Barred spiral galaxy NGC 1300. (Photograph from the Hale Observatories.)

and other objects that make up all galaxies. He noted that the most luminous stars in the nucleus of a spiral galaxy are red giants; the most luminous stars in the spiral arms are blue giants. The spiral arms also contain bright nebulae, dark lanes of gas and dust, and open clusters.

In recognition of this difference a terminology was coined by Baade. The blue giants, the bright and dark nebulae, and the open clusters he called Population I objects. The red giants of the nucleus, the globular clusters, and elliptical galaxies he called Population II objects.

One of the significant differences between Population I and Population II objects is age. Population I objects are younger than Population II. But not all Population I objects are of the same age. Nor, in fact, are all Population II objects equally old. It seems that there is a range of age within each population.

There are other significant differences between the two populations. One of these is location within a spiral galaxy. In addition, the spectra of some of the

Figure 13-7
Spiral galaxy NGC 5194 with a bridge of stars connecting it to its satellite galaxy NGC 5195. (Photograph from the Hale Observatories.)

Figure 13-8
The dwarf galaxy in Sextans. (Photograph from the Hale Observatories.)

Population I stars indicate that they have a higher abundance of heavier elements such as calcium, iron, magnesium, etc., than do the rest of Population I stars. The reasons for all of these differences and their interrelationships is the subject of a great deal of current research.

13.2 VERY ACTIVE GALAXIES

The discussion of galaxies so far has not introduced any new element into our concept of stellar systems. Population I objects appear in spiral arms, Population II objects in nuclei and elliptical galaxies. Galaxies rotate, but they move rather majestically. Even with an occasional supernova, galaxies seem to be rather placid; but this is not a safe conclusion. The galaxy M 82 (Figure 13-9*a*) was once classified as a peculiar galaxy; it certainly appears to be neither an elliptical nor a spiral galaxy. As observations with radio telescopes became more and more precise, it became clear that M 82 is a very strong radio source. Such discrete radio sources are of particular interest to astronomers because they are invariably a result of electrons moving at very high speeds. Subsequent observations of M 82 have revealed that it is far from placid. A photograph taken with only the light of the red H-alpha line of hydrogen reveals a fine structure of filaments extending outward from the nucleus in a direction perpendicular to what appears to be the plane of the galaxy (Figure 13-9*b*). Since the plane of the galaxy is inclined by about 8° from the line of sight, observations of radial velocity of these gases can be made. Their velocity indicates that the gases have been ejected by the nucleus. Furthermore, the velocity of the ejected gases increases in proportion to their distance from the center of the nucleus. The gases near the ends of the filaments are traveling outward at a velocity of 6000 miles per second, and judging from their present distance — 14,000 light years from the nucleus — they must have left the nucleus about 2 million years ago. The gases only half as far out have just one-half that velocity, so they, too, must have left the nucleus about 2 million years ago. In fact, it seems that all of the gas left at the same time — the galaxy suffered an explosion!

These ejected gases emit not only a bright-line spectrum but a continuous spectrum as well. This observation coupled with the fact that M 82 is a strong nonthermal radio source leaves little doubt that this galaxy emits synchrotron radiation. It would appear as if the magnetic field passes through the nucleus parallel to the axis of rotation of the galaxy, and the high-speed electrons are spiraling about the magnetic lines of force as they proceed outward from the nucleus.

Figure 13-9
Exploding galaxy M 82 photographed in (a) blue light, and (b) red light of hydrogen. (Photograph from the Hale Observatories.)

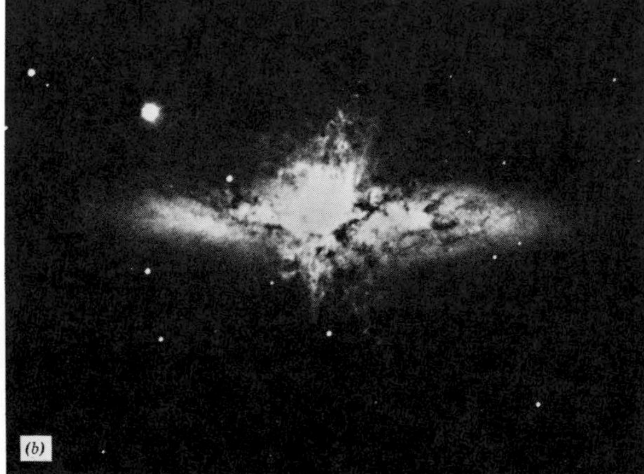

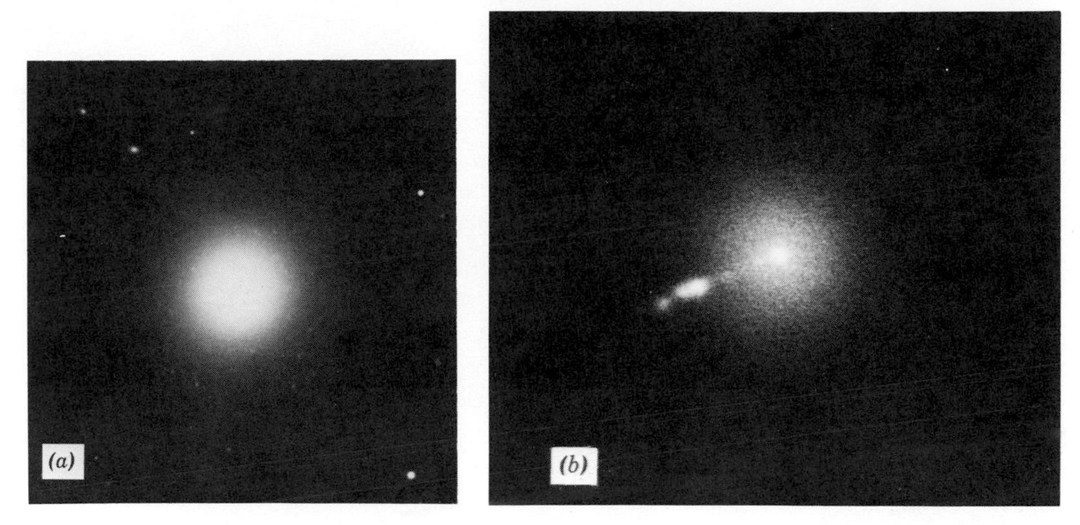

Figure 13-10
An elliptical galaxy, M 87, with (a) a long exposure, and (b) a short exposure revealing the jetlike source of radio energy. (Lick Observatory photograph.)

Another galaxy, M 87, when photographed on a long-time exposure, appears as a normal elliptical galaxy (Figure 13-10*a*). But it, too, proves to be a source of strong radio energy, so it was investigated more thoroughly. A short-time exposure revealed a jet of gas extending outward from the nucleus (Figure 13-10*b*). This galaxy is also a source of X-ray radiation, indicating that the jet must contain very high-energy electrons, indeed. The galaxy NGC 5128 (Figure 13-11*a*) is also a radio galaxy, but the radio energy does not all originate in the optical portion of the galaxy. A great deal of the radio energy comes from invisible gases on either side of and several diameters from the center of the galaxy (Figure 13-11*b*).

Details of these unusual galaxies are still being sought; the reasons why these galaxies emit such vast quantities of energy will become clear only after more details have been acquired, and discussed among the astronomers and then digested. Then perhaps a fresh idea will appear which may serve to describe these exploding (if they are) galaxies.

13.3 THE MILKY WAY GALAXY

Which, if any, of these many and varied galaxies does our own Milky Way galaxy resemble?

Since our galaxy has a great deal of interstellar gas and dust and bright nebulae, it cannot be an elliptical galaxy. Since it has a very definite plane of

Figure 13-11
A strong radio galaxy NGC 5128 in Centauris. (a) Photograph. (b) Regions marked *R* indicate where the radio energy originates. (Photograph from the Hale Observatories.)

symmetry, namely the galactic plane as defined by the Milky Way, it does not appear to be an irregular galaxy. We are left, therefore, with the tentative conclusion that our galaxy has a spiral structure. This conclusion is in harmony with a cursory glance at our galaxy, the shape of which can best be imagined by looking at the Milky Way, which forms a fairly narrow band around the celestial sphere. It must therefore be a fairly flat system, perhaps similar in cross section to the spiral galaxy in Figure 13-12 which is seen "edge on." The similarity is made more apparent by comparing this galaxy (NGC 4565) with a wide-angle photograph of the Milky Way (Figure 13-13). The difference between the two photographs, of course, is that one is viewed from the outside and the other is viewed from the inside.

Ours, then, is a spiral galaxy. We want to determine its size and structure — that is, locate the galactic nucleus (locate the position of the sun relative to the nucleus), observe evidence of spiral arms, and determine its mode of rotation.

A. THE GALACTIC NUCLEUS In finding the nucleus of our galaxy we automatically locate the relative position of the sun. This has been done by several methods, one of which is based on the location of globular clusters (Figure 13-14). Of the 121 known globular clusters about 30 are concentrated in the constellations of Sagittarius. The remaining 90 are spread over a much larger part of the sky, the vast majority of them being in that half of the celestial sphere that is centered on Sagittarius. The distances of these globular clusters can be determined with a fair degree of accuracy, for not only do they contain RR Lyrae stars but many also inhabit regions of the sky that are relatively free from interstellar material. Since the globular clusters form a spherical system about the nucleus of our galaxy, it is presumed that the center of the system of globular clusters is the same as the center of the nucleus of our galaxy.

Other studies substantiate the idea that the system of globular clusters is concentric with the nucleus of our galaxy. The RR Lyrae stars not in the globular clusters form a similarly shaped system, although it is flattened slightly in the direction of the galactic plane, but again centered in Sagittarius. From these studies, the nucleus is estimated to be 31,000 light years from the sun.

Studies using Cepheids, O- and B-type stars, and RR Lyrae stars led to the conclusion that the diameter of the galaxy is about 100,000 light years. This places the sun about two-thirds the way out from the center (the radius is 50,000 light years).

B. THE GALACTIC PLANE Having located the nucleus of our galaxy, we are left to find the central plane. A rough estimate of its location can be made by bisecting the Milky Way. It can be more precisely located by studying the distribution of certain objects over the celestial sphere. The objects that give the best results are the ones that can be seen for the greatest distance and, therefore, include the globular clusters, the O and B stars, and the Cepheids. The distribution of each of these is such that their number increases toward the galactic equator. Radio studies of the distribution of hydrogen gas, both ionized and neutral, have been invaluable in locating the central plane of our galaxy.

From observations that locate the central plane of our galaxy, it has become evident that the galaxy in the region of our sun is about 5000 light years thick. The sun lies about 40 light years from the central plane and is oriented such that the south pole of the Earth is closer to the galactic plane.

C. GALACTIC COORDINATES Having located both the nucleus of our galaxy and its central plane, it now becomes convenient to establish a system of galactic

Figure 13-12
Spiral galaxy NGC 4565 seen "edge on." (Photograph from the Hale Observatories.)

coordinates that can be used by astronomers when they discuss the location of objects in our sky but want to consider those objects relative to our galaxy, rather than the solar system.

The *galactic equator* has been established as the central plane of our galaxy. *Galactic latitude* is measured north and south from the galactic equator through 90° to the north galactic pole and the south galactic pole. By definition, the north galactic pole lies in that hemisphere which includes the north celestial pole, although the two poles are 62° apart. The north galactic pole is between the constellations Bootes and Leo, in a region of very few stars.

Galactic longitude must be measured from some chosen point on the galactic equator. At a meeting of the International Astronomical Union in Moscow in 1958, the point from which galactic longitude is measured was selected as that point on the celestial sphere which is defined as the galactic center.* The location of this point was determined by observations of the 21-cm line and lies in the direction of Sagittarius. Galactic longitude is measured from this point eastward along the galactic equator through 360°.

*Previous to this the origin of galactic longitude was in the constellation of Aquila at one of the two intersections of the galactic equator with the celestial equator.

Figure 13-13
Wide-angle photograph of the Milky Way. (Washburn Observatory.)

Figure 13-14
Our galaxy is surrounded by a halo of globular clusters.

There are, therefore, three coordinate systems on the celestial sphere. These are based on the three principal planes: the plane of the Earth's equator, the plane of the Earth's orbit about the sun, and the plane of the galaxy. The angle formed by the intersection of the galactic equator with the celestial equator is about 62°. Thus the plane of the Earth's equator makes an angle of 62° with the plane of the galaxy. The two planes intersect at two points in diametrically opposite parts of the sky. One of the intersections is in the constellation of Aquila, just south and west of the star Altair. The other intersection is just east of Orion, between Orion and the star Procyon (see the star charts).

D. THE LOCATION OF SPIRAL ARMS To locate the spiral arms, we can consider first the O- and B-type stars, since they are very luminous. The intrinsic luminosity of an O or B star can be determined by placing it on the H-R diagram, and its apparent brightness can be determined by means of photoelectric measurements. Its distance, however, cannot be determined directly by the inverse square law (as discussed on p. 194), for astronomers have accepted the fact that they can see these distant stars only through the scattering and absorbing interstellar material. This scattering and absorbing makes the stars fainter and thus appear more distant than they would be were there no interstellar material.

In order to determine the true distance as accurately as possible, some estimate must be made of the amount of light that has been lost in transit between the star and the Earth. One method is to determine the *color excess* of the star. The color excess is a measure of the amount of light scattered by interstellar material. The greater the color excess the more light lost in transit, and the farther away the star appears. The amount of light absorbed can be estimated from the amount scattered. By considering both of these factors, astronomers are able to make reasonable estimates of stellar distances by applying corrections to the inverse square law.

When the distances and directions of a large number of distant O and B stars have been determined we can see that they are not distributed randomly about the sun but fall into patterns that strongly suggest the existence of spiral arms. This indication of spiral structure has been verified by studies involving the space distribution of emission nebulae, which when coupled with the work on O and B stars and clouds of ionized hydrogen gas disclose the spiral features shown in Figure 13-15. The sun, it appears, is on the inner edge of the Orion arm.

The constellation Orion, with all of its activity of stellar formation has a galactic longitude of about 210°, so it would appear between the upper and the left-hand sides of Figure 13-15. The constellation Auriga has a galactic longitude of about 170°, Cassiopeia about 120°, and Cygnus about 60°. Sagittarius, of course, has a galactic longitude of 0°. It is in this direction that the great star clouds shown in Figure 13-16 appear.

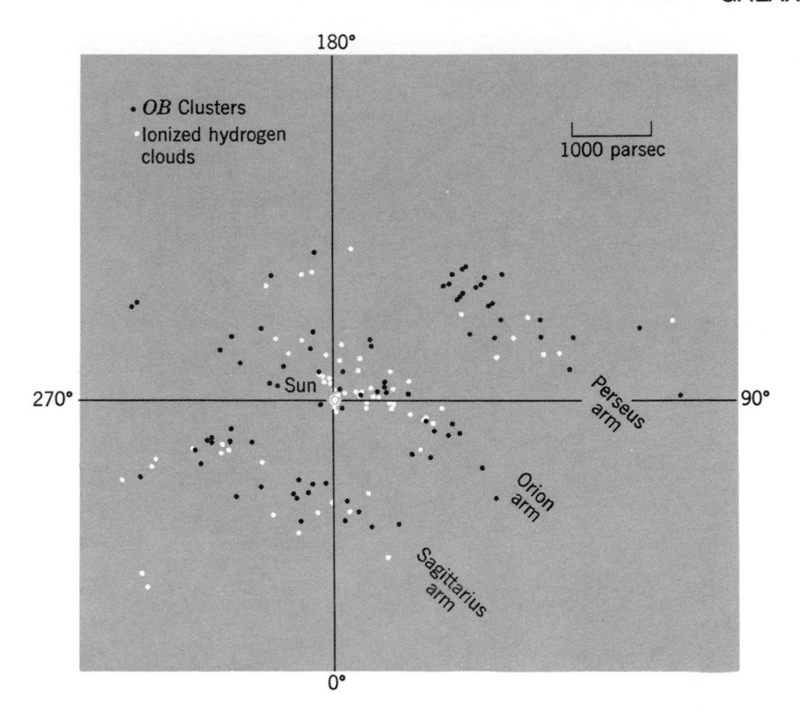

Figure 13-15
Spiral structure within about 3000 parsecs (roughly 9000 light years) of the sun as obtained from observations of early type stars and clouds of ionized hydrogen. The center of the galaxy is toward 0°, 180° is diametrically away from the center. (After W. Becker and R. Fenhart.)

E. GALACTIC STRUCTURE AND ROTATION Studies of the rotation of our galaxy cannot be separated from studies of its structure. Observations have been carried on with both optical and radio telescopes. Observations of the O and B stars along the galactic equator permit us to see a distance of about 15,000 light years. The radius of our galaxy, however, is 50,000 light years, so the O and B stars, as bright as they are, do not permit much of a view; there is too much intervening gas and dust that absorbs their light.

Studies with radio telescopes, however, have permitted us to see nearly the whole galaxy. Radio waves travel right through the gas and dust in the plane of the galaxy. In a like fashion, radio waves travel right through the fog, smog, and clouds here on Earth.

Fortunately for astronomers, the hydrogen atom both emits and absorbs a spectral line in the radio region of the spectrum. This spectral line has a wavelength of 21 centimeters, and has been extremely useful in observing clouds of neutral hydrogen. By way of analogy, each radio station emits energy as a bright spectral line in the radio region of the electromagnetic spectrum. The intensity or frequency of that spectral line changes slightly so that it can carry the messages we receive on our radio and TV receivers.

Observations of the 21-cm line indicate that clouds of neutral hydrogen are arranged in our galaxy as shown in Figure 13-17. This information combined with studies of O and B stars indicates that our galaxy is a fairly open spiral, perhaps not unlike the one shown in Figure 13-18.

Figure 13-16
Star clouds of the Milky Way are best seen in the constellation of Sagittarius. The dark lanes are intervening clouds of gas and dust. (Photograph of the Hale Observatories.)

The rotation and structure of any galaxy depends to a large extent on the mass of the entire complex assemblage of stars, gas, and dust. From all of the information acquired thus far and from theoretical studies of the structure and rotation of our galaxy, it appears that our Milky Way galaxy has a mass of 2×10^{11} suns! There is enough material for roughly 200 billion stars in our galaxy!

Each star revolves in an orbit about the center of the galaxy. These orbits are not all circular, most are slightly elliptical, and they tend to carry the star on both sides of the central plane of the galaxy, first one side then crossing over to the other.

Figure 13-17
A drawing of the Milky Way galaxy as observed by the 21-cm line. The circle marks the center of the galactic nucleus, and the dot in the small circle marks the position of the sun. This drawing was prepared by Westerhout of the Leiden Observatory, but the portion of the Milky Way visible only in the southern hemisphere was observed by Kerr in Australia. The portion omitted lies behind the galactic nucleus. (Leiden Observatory.)

Stars in the sun's neighborhood travel with a speed of about 150 miles per second, some a little faster, others a little slower. It is significant that all of the Population II stars travel slower than the Population I stars. Population II stars have a speed of about 100 miles per second in their orbits about the galactic center.

We know the sun's speed as it revolves about the galactic center, and we know its distance from that center. Assuming that the orbit is circular, we can easily calculate the length of time it takes for the sun to make one complete revolution. The radius of the sun's orbit is 31,000 light years (its distance from the galactic center), so the circumference (length of the orbit about the galaxy) must be about 200,000 light years ($C = 2\pi R$). Traveling at a speed of 150 miles per second, it should take the sun about 230 million years to make that trip. The solar system is 4.6 billion years old, so it must have made about 20 revolutions already.

Figure 13-18
NGC 5457, an open spiral galaxy in the constellation of Ursa Major. (Photograph from the Hale Observatories.)

13.4
SPIRAL ARMS

One of the many unsolved problems of the structure of spiral galaxies is the very existence of spiral arms. If galaxies rotate as they are observed to do, then each galaxy must have made many revolutions since its formation. If the sun has made 20 revolutions about the galactic center why aren't the spiral arms twisted about the galaxy so tightly as to be unrecognizable? How do open galaxies maintain their open shape? What mechanism produces the stability which maintains the barred spiral arms? A different and yet related question concerns the existence of vast amounts of gas and very young stars in the spiral arms. Why hasn't all of this gas condensed into stars long ago? Why are stars still forming in a galaxy that is perhaps 10 billion years old?

Observations by radio telescopes of the 21-cm line indicate rather conclusively that hydrogen gas is being ejected by the nucleus of our galaxy (and the nuclei of other spirals as well). The amount of gas ejected is estimated to be about 1 solar mass per year, and it leaves the nucleus with a velocity of about 30 miles per second. Perhaps this is enough to replenish the gas which has formed the stars now visible, but by itself it does not account for the structure and stability of the spiral arms.

One of the suggestions to account for the existence of the spiral arms considers the speed of revolution of the material about the galactic center. It has been proposed that because of irregularities in the gravitational field of the galaxy the speed of revolution of the stars and clouds changes slightly. Presumably, their speed decreases in certain regions. This decrease in speed results in an accumulation of material.

By way of analogy, a reduction in the speed of cars on a freeway will cause an accumulation of automobiles called a traffic jam. The cars don't have to stop. An accident alongside the freeway may cause the traffic to slow from an average speed of 55 to 40 mph. The resulting accumulation has been experienced by many exasperated commuters. Even after the accident has been cleared up, the accumulation of cars persists. So a decrease in the speed of the material revolving about the center of the galaxy may cause an accumulation that we see as the spiral arms. That is, the spiral arms may be celestial traffic jams!

Measurements of motions of stars and gases within a spiral arm indicate that the two classes of objects move differently. It would seem as if the stars form in the gases only to acquire a velocity different from the gas which nurtured them; the stars must leave the gases and populate the regions between the spiral arms. And between the spiral arms there has indeed been observed a substratum of red-dwarf stars. These stars must have left the spiral arms billions of years ago. The gas which forms these stars is then replenished by the ejection of gases from the nucleus. This leaves the question of how long can a nucleus supply gas to support spiral arms? Do galactic nuclei shrink with age? Do spiral arms then lose whatever strength they have?

13.5 GALACTIC FOSSILS

It is very striking indeed that there seems to be a significant relationship between the orbital motions and locations of stars, and their ages (Figure 13-19):

1. The globular clusters are the oldest members of our galaxy and they form a spherical system concentric with the galactic nucleus. Their orbital velocities are low, and their orbits are eccentric and inclined at large angles from the galactic plane.
2. The O and B stars are the youngest members and they form the flattest system, being nearly coincident with the galactic plane. Their orbits are very nearly circular and their orbital velocities are higher than those of any other stars.
3. The other stars seem to form a continuum between these two extremes.

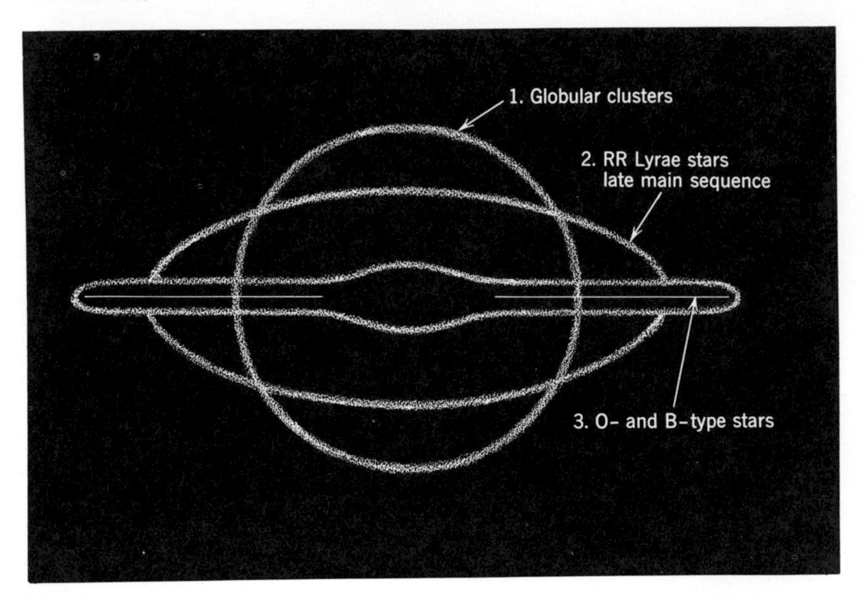

Figure 13-19
Speculation leads to these three stages of galactic evolution, because a star's location in our galaxy as well as its orbital inclination from the galactic plane and its speed of revolutuion about the galactic center all depend to a certain degree on the age of the star.

At one extreme are the O and B stars, which are very young Population I stars. The late main-sequence stars (K and M) are usually older than the O and B stars. The later main-sequence stars form a system that is fairly flat, but their orbits are more likely to carry them farther from the galactic plane than do the orbits of the O and B stars. Their orbits are also less circular than those of the youngest stars, and their orbital velocities a little lower. The RR Lyrae stars are Population II stars but are younger than the globular clusters. They form a system that, although it extends far from the galactic plane, is slightly flattened and thus does not extend as far as the globular clusters. Their orbits are also eccentric and inclined from the galactic plane but not as much as the orbits of the globular clusters. Their orbital velocities are low but not as low as those of globular clusters.

The ages of stars are also related to their location in the galaxy in that the older stars are more likely to congregate near the galactic nucleus. Thus their orbits are more likely to be smaller than those of younger stars.

All these facts assembled are wonderful material for speculative thought; it appears that our galaxy, like a tree, has left rings that tell of its past. When our galaxy was at the age during which its globular clusters were forming from gaseous clouds of hydrogen, it must have had a shape quite similar to that now defined by the system of globular clusters — a nearly spherical shape. Furthermore, it must have rotated with a velocity nearly equal to that of the globular clusters — a low velocity. Astronomers speculate further that apparently the galaxy then began to contract, but the globular clusters, having formed, maintained their orbits. As the galaxy contracted it rotated more rapidly. In rotating more rapidly it began to flatten out and after roughly 12 billion years it has become the flat system that we now observe. The youngest stars and the remaining clouds of hydrogen are all part of this flat system; the older stars are not. With increasing age, the older stars form ever more spherical systems and on the average have smaller orbits and lower orbital velocities. Their orbits, however, can be more eccentric and more inclined from the galactic plane than those of younger stars.

This is not meant to imply that there are no Population II objects in the galactic plane, for there are. Some may even have orbits that lie in the galactic plane. Those Population II objects in orbits inclined from the galactic plane must still cross it twice in one revolution. During this crossing they are temporarily (thousands of years) close to the galactic plane.

That so many globular clusters are concentrated near the galactic nucleus indicates that during the time when the globular clusters were forming, the shape of our galaxy may have resembled that of a small elliptical galaxy of type E0, with a high concentration of stars near the center, that gradually thins out with increasing distance from the center.

This comparison of our galaxy at this stage of development with elliptical galaxies is interesting; unfortunately, it is misleading. Although our galaxy in the early stages of its development may have had a shape similar to that of elliptical galaxies, it must have been different because elliptical galaxies contain little apparent interstellar gas and dust. Thus star formation must be nearly at an end in these Population II objects.

13.6 EVOLUTION OF GALAXIES

Galaxies have a very strong tendency to form in clumps and clusters. Very often galaxies will form together in a very tight assemblage (see Figure 13-20). It is presumed that galaxies so close together are under the gravitational influence of one another. If this is true, then the relative masses of each galaxy can be estimated from their motions; the more massive galaxies will move the slowest, the least massive will move the fastest. So reasoned Thornton Page then of Van Vleck Observatory, Wesleyan University, who set about to determine the relative masses of many galaxies from a number of such close assemblages. Surprisingly enough he found that the E0 galaxies, the very spherical ones, are consistently nearly 30 times as massive as the open Sc spirals. Furthermore he found a gradation in mass from the E0 galaxies to the Sc galaxies. The dwarf galaxies, however, are not part of this gradation, their masses are even less than those of the Sc galaxies.

It is well known that elliptical galaxies do not rotate as fast as the spiral galaxies. However, the measure of rotation should include the amount of matter rotating, so we employ angular momentum as this measure. Angular momentum relates not only the speed of rotation of an object to the amount of mass of that object, but also to the distribution of mass. For example, consider two

Figure 13-20
A group of galaxies called Stephan's quintet. (Lick Observatory photograph.)

wheels of the same mass and radius. Let one of them be a flat disk and the other one a spoked wheel with nearly all of the mass in the rim. If these two wheels rotate with the same speed, the spoked wheel will have more angular momentum than the disk wheel. If the speed of rotation of each wheel were increased, the angular momentum would also increase.

H. C. Arp, of the Hale Observatories, plotted the mass of the galaxies against their angular momentum per unit mass and found a very interesting relationship (Figure 13-21). It is significant that the most massive galaxies have the least amount of angular momentum per unit mass; and that there should be a relatively uniform gradation between the E0 and the Sc galaxies. Arp suggests that this gradation results from the conditions present during the formation of the galaxies.

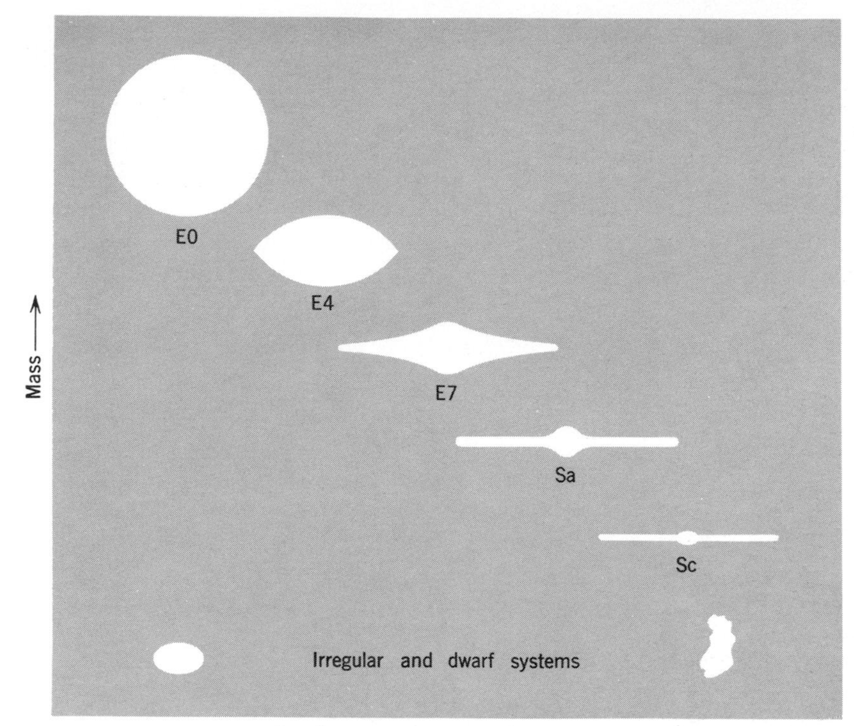

Figure 13-21
As the angular momentum per unit mass of galaxies increases, their total mass decreases.

If the galaxies form from blobs of gas of nearly equal mass, then all galaxies should have equal mass; but if the gas from which each galaxy forms has a different amount of angular momentum, then the resulting galaxies might be different. Arp suggests that during the contraction of a large nonrotating blob of gas the density of the gas remained fairly uniform so that star formation proceeded at a fairly uniform rate and for only a relatively short period of time. All of the gas would have condensed into stars and all of the stars would have nearly the same age. In other words, an elliptical galaxy would form from a nonrotating blob of gas. Those blue-giant stars formed during this early age would long since have evolved into white dwarfs, neutron stars, or black holes. There being no more gas to form new stars, no successive generations of blue giants would form.

On the other hand, if the initial blob of gas were rotating — if it had an appreciable amount of angular momentum — then as the gas contracted it would have rotated still faster. As it rotated faster, gases would have been thrown out to form an equatorial bulge, and later a nebular disk reminiscent of the nebular disk about the protosun. The formation of the nebular disk would have decreased the density of the gas and consequently decreased the rate of star production. With the increased rotation much of the gas may have been thrown free of the forming galaxy resulting in a decrease in mass. Evidence for this idea is supported in the stages of evolution of our own galaxy.

Consequently, it is suggested that the amount of angular momentum of the initial blob of gas is the main factor determining whether the gas will become an E0, E1, E2, . . . , Sa, Sb, or Sc galaxy.

It is generally accepted that during the lifetime of a spiral galaxy the production of supernovae and other gas-ejecting stars replenishes some of the interstellar material used up in star formation. If that initial interstellar material is nearly pure hydrogen, however, these gas-ejecting stars "contaminate" the interstellar gases. During the lifetime of a star, hydrogen is converted into helium, which is in turn converted into carbon, oxygen, and the heavier elements. By ejecting gases back into space stars spew forth some of these heavier elements as well. Consequently, the first stars to form in a spiral galaxy should have a low percentage of metals; stars formed later in the life of a galaxy should have a higher percentage of metals. The most recently formed stars should have the highest percentage of metals of all the stars. It is well established that Population I stars have a higher percentage of metals than do the Population II stars, so the idea of an increasing metallic content of the interstellar gases is well supported. Presumably, all of the stars of the elliptical galaxies have the same, or nearly the same, chemical composition.

This suggestion of galactic evolution is one of the first to really bring some of the various shapes and forms of galaxies into a unified principle of formation. There are certainly many galaxies which do not fall into the straight sequence E0 to Sc, such as the barred spirals, the dwarf systems, the irregular systems, the exploding galaxies, etc. But more progress is made when astronomers have a framework — a hypothesis — upon which to work. If that hypothesis is later shown to be incorrect, astronomers have still learned and benefited from it. This suggestion of galactic evolution will serve astronomers very well, even if it cannot answer all of their questions now. The details will be forthcoming with additional studies, both observational and theoretical. This is the manner in which ideas develop into good descriptions of our universe.

BASIC VOCABULARY FOR SUBSEQUENT READING

Barred spiral galaxy
Dwarf galaxy
Elliptical galaxy
Galactic equator
Galactic latitude
Galactic longitude
Irregular galaxy
Normal spiral galaxy
Population I objects
Population II objects

QUESTIONS AND PROBLEMS

1. What are the criteria used to classify spiral galaxies into Hubble's Sa, Sb, and Sc types?

2. In what ways does an elliptical galaxy resemble a globular cluster?

3. Why, when studying the Milky Way galaxy, is it convenient to devise still another coordinate system?

4. What observations led to the discovery of exploding galaxies?

5. Describe three basic differences between Population I and Population II objects.

6. What observations have led to our understanding of the general shape of the Milky Way galaxy and to a knowledge of the location of the sun with respect to the center?

7. Which objects can best be used to trace the spiral arms of our galaxy? How can each of these objects best be observed?

8. Describe the general shape of the spiral arms of our galaxy near the sun.

9. Describe the "galactic fossils" of our Milky Way galaxy.

10. How has it been possible to estimate the relative mass of galaxies when they form in a group?

FOR FURTHER READING

Baade, W., *Evolution of Stars and Galaxies,* Harvard University Press, Cambridge, Mass., 1963.

Bok, B. J., and P. Bok, *The Milky Way,* Harvard University Press, Cambridge, Mass., 1957.

Hodge, P. W., *Galaxies and Cosmology,* McGraw-Hill paperback, New York, 1966.

Page, T., and L. W. Page, ed., *Beyond the Milky Way,* The Macmillan Co., New York, 1969.

Page, T., and L. W. Page, ed., *Stars and Clouds of the Milky Way,* The Macmillan Co., New York, 1968.

Page, T., ed., *Stars and Galaxies,* Prentice-Hall paperback, Englewood Cliffs, N.J., 1962.

Struve, O., and V. Zebergs, *Astronomy of the 20th Century,* Crowell, Collier and Macmillan, New York, 1962, Chapters XIX and XX.

Arp, H., "On the Origin of Arms in Spiral Galaxies," *Sky and Telescope,* p. 385, December 1969.

Bok, B. J., "The Large Cloud of Magellan," *Scientific American,* p. 32, January 1964.

Bok, B. J., "The Spiral Structure of Our Galaxy," *Sky and Telescope.* p. 392, December 1969; p. 21, January 1970.

Hodge, P. W., "Dwarf Galaxies," *Scientific American,* p. 78, May 1964.

Hodge, P. W., "The Sculptor and Fornax Dwarf Galaxies," *Sky and Telescope,* p. 336, December 1964.

Iben, I., "Globular-Cluster Stars," *Scientific American,* p. 26, July 1970.

Irwin, John B., "Some Current Problems Concerning Galaxies," *Sky and Telescope,* p. 287, November 1973.

Page, T., "The Evolution of Galaxies," *Sky and Telescope,* part I, p. 4, January 1965; part II, p. 81, February 1965.

Sanders, R. H., and G. T. Wrixon, "The Center of the Galaxy," *Scientific American,* p. 67, April 1974.

Strom, R. G. et al, "Giant Radio Galaxies," *Scientific American,* p. 26, August 1975.

Toomre, Alar and Juri, "Violent Tides between Galaxies," *Scientific American,* p. 39, December 1973.

Westerlund, B. E., "Report on the Magellanic Clouds," *Sky and Telescope,* p. 23, July 1969.

Weymann, R. J., "Seyfert Galaxies," *Scientific American,* p. 28, January 1969.

"Distribution and Motions of Supergiant Stars," *Sky and Telescope,* p. 162, March 1970.

"Galaxies with Bright Infrared Cores," *Sky and Telescope,* p. 357, June 1970.

"Gould's Belt: An Expanding Group of Stars," *Sky and Telescope,* p. 93, February 1968.

"The Heart of the Milky Way," *Sky and Telescope,* p. 282, May 1965.

"Helium and the Galaxy's Age," *Sky and Telescope,* p. 219, April 1965.

"Intergalactic Matter and the Milky Way," *Sky and Telescope,* p. 283, May 1972.

"Movie of the Milky Way's Hydrogen Clouds," *Sky and Telescope,* p. 92, February 1970.

"Our Galaxy as Mapped with RR Lyrae Stars," *Sky and Telescope,* p. 95, February 1968.

"Very Young Galaxies," *Sky and Telescope,* p. 91, February 1973.

"X-ray Galaxies," *Scientific American,* p. 50, April 1966.

CHAPTER CONTENTS

LEARNING OBJECTIVES

BE ABLE TO:

1. DESCRIBE, IN GENERAL, HOW GALAXIES ARE DISTRIBUTED IN THE OBSERVABLE UNIVERSE.
2. EXPLAIN THE SIX MAIN METHODS OF DETERMINING DISTANCES OF GALAXIES.
3. GIVE THE MOST WIDELY ACCEPTED INTERPRETATION OF THE RED-SHIFT.
4. GIVE THE TWO BASIC POSTULATES OF THE PRINCIPLE OF SPECIAL RELATIVITY.
5. EXPLAIN WHY THE WORD "NOW" HAS NO MEANING IN THE OBSERVABLE UNIVERSE.
6. GIVE THE MAIN ARGUMENTS OF THE BIG-BANG THEORY.
7. GIVE THE MAIN ARGUMENTS OF THE STEADY-STATE THEORY.
8. GIVE OBSERVATIONS IN SUPPORT OF THE FAVORED THEORY, THE BIG-BANG OR THE STEADY-STATE.
9. EXPLAIN THE SIGNIFICANCE OF THE RATE OF EXPANSION OF THE OBSERVABLE UNIVERSE.
10. EXPLAIN HOW QUASARS SUPPORT THE BIG-BANG THEORY.

Chapter Fourteen
THE UNIVERSE

We define the universe as that which includes all matter in existence. Since the amount of matter and the size of the universe are not yet known, we must start with what we know about the *observable* universe. From their analyses of the observable universe astronomers are able to formulate hypotheses about the structure of the entire universe (the study of *cosmology*) and its origin, evolution, and even its future (the study of *cosmogony*).

14.1 THE DISTRIBUTION OF GALAXIES

The basic building block in the universe is the galaxy, just as the basic building block in a galaxy is the star. Astronomers observe all the visible galaxies projected onto the celestial sphere, and from this projection conclude that galaxies are not distributed at random throughout the universe.

C. D. Shane, of the Lick Observatory, has presented conclusive evidence that the galaxies are clumped together into clusters, Figure 14-1*a*. Using plates that had been taken for a large-scale study of the proper motions of stars, Shane and his co-workers counted the number of galaxies in each square degree of the sky and plotted these counts on a large "celestial" map. They then drew contour lines joining those regions of the sky that had the same number of galaxies. A small portion of their "galactic contour map" is shown in Figure 14-1*b*. Each of the areas of closed contour lines represents a cluster of galaxies, and this is only a small portion of the entire celestial sphere! Although there also appear to be stray galaxies that are not contained in any cluster, they are the exception and not the rule.

Fritz Zwicky, of the Mount Wilson and Palomar Observatories, has concluded that many of these clusters have spherical symmetry; that the space density of the galaxies increases toward the center of the cluster; and that the very bright elliptical galaxies are concentrated toward the center, with the spiral and irregular galaxies more prevalent in the outer regions. Some of the clusters may contain as many as 100,000 galaxies but these are exceptionally rich.

Each cluster of galaxies is a group bound within itself, held together by the mutual gravitational attraction of all the members. But this gravitational field does not restrict the motion of the member galaxies within the cluster; galaxies will have intrinsic velocities within the cluster of as much as 600 miles per second. Zwicky has commented that each cluster appears to be an isolated unit, that no clusters form in pairs, groups, or clusters of clusters.

Within the cluster, there is overwhelming evidence that galaxies tend to

Chapter Opening Photo

A group of five galaxies with clouds of material connecting them. (Photograph from the Hale Observatories.)

Figure 14-1
(a) A cluster of galaxies in the constellation Hercules. (Photograph from the Hale Observatories.)

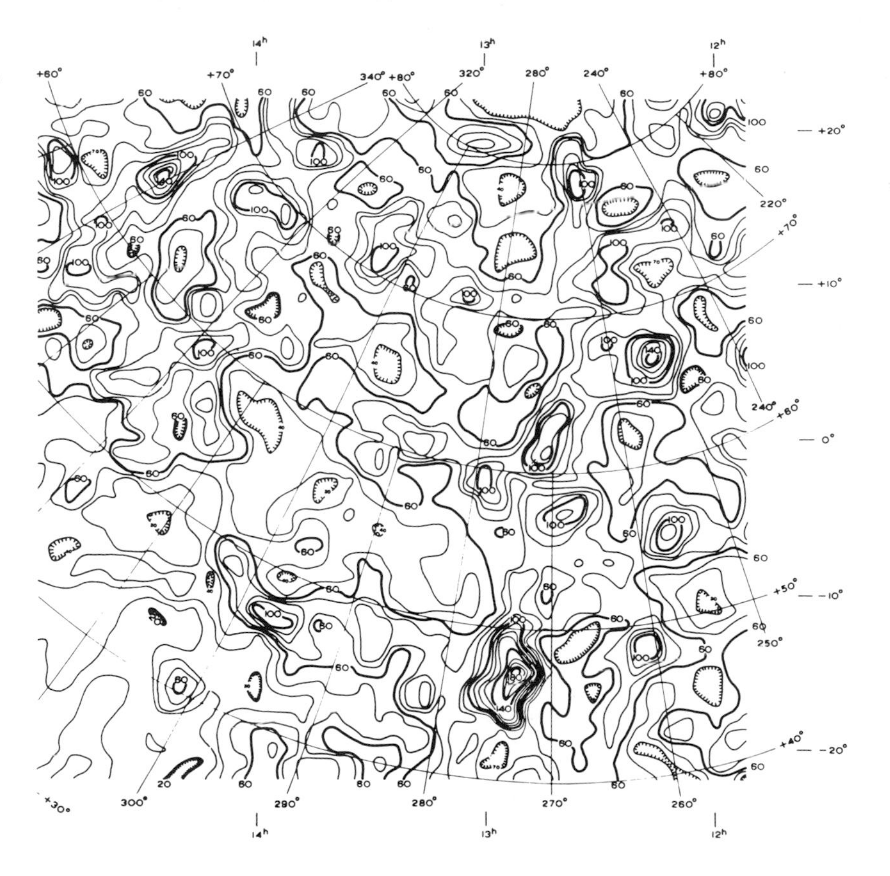

(b) The distribution of galaxies in a portion of the celestial sphere as drawn by C. D. Shane indicates that galaxies tend to form in clusters. (Lick Observatory photograph.)

associate closely with one another. There are many galaxies that exist as pairs, others as three and more (opening Figure, Chapter 14). The distance between galaxies in a cluster is not as great in comparison to their size as is the distance between the stars in comparison to their size. On the average, galaxies have a diameter of about 5×10^4 light years and in a cluster they have an average separation of roughly 30×10^4 light years. Thus they are separated, on the average, by a distance roughly six times their diameters, whereas stars in the region of the sun are separated by a distance of about 30 million solar diameters.

Zwicky amply shows that intergalactic gas and dust does exist. He points out that the number of faint clusters of galaxies is much reduced in regions of the sky covered by nearby rich clusters of galaxies. For example, there is an average of only 8 faint clusters seen through the central region of the Coma Berenices cluster in an area of the celestial sphere of 36 square degrees, but in regions where no nearby rich clusters exist there is an average of 70 clusters for the same area.

14.2 DISTANCES OF THE GALAXIES

Although we can learn a great deal from the study of the distribution of galaxies on the two-dimensional celestial sphere, we must consider the universe as being three-dimensional and study the distances of the galaxies.

Let us review the ever-present problem of distance determination. For only the nearest stars does the method of heliocentric parallax do us any good. Beyond about 50 light years, heliocentric parallax begins to fail and other less direct methods must be used. An astronomer can measure the apparent brightness of a star whose intrinsic luminosity is known, such as a Cepheid or RR Lyrae star. From the brightness and luminosity the distance can be determined by the inverse square law. The astronomer may have to account for the loss of light from scattering by interstellar matter, in which case his results are likely to be less accurate.

A. RR LYRAE STARS AND CEPHEIDS The distances of the nearest galaxies, the Magellanic Clouds, have been determined by using RR Lyrae stars, for their period–luminosity relation is more reliable than that of the Cepheids. The Cepheids must be used for more distant galaxies such as the Andromeda galaxy and M33, since RR Lyrae stars cannot be seen in them. For still more distant galaxies not even the Cepheids can be seen and the astronomer must resort to other methods.

B. TEN BRIGHTEST STARS One method that has been used in the past is to estimate the luminosity of the brightest stars in a given galaxy. This method is based on the assumption that the brightest stars in each galaxy are of nearly the

Figure 14-2
A supernova appeared in the galaxy NGC 7331 in 1959. The upper photograph was taken before the supernova erupted. (Lick Observatory photograph.)

same luminosity. This luminosity can be estimated from studies of the stars in our galaxy as well as of the brightest stars in nearby galaxies of known distance. From this luminosity and the measured brightness of the stars the distance of the galaxy can then be estimated. But even this method fails with galaxies at still larger distances, so astronomers must resort to other procedures.

C. BRIGHTEST GLOBULAR CLUSTERS Many galaxies are surrounded by a halo of globular clusters similar to ours which, because each cluster contains thousands of stars, can be seen at greater distances than can the brightest stars in any galaxy. The brightest globular clusters can be used to estimate distances by the same method used for the brightest stars. Still more distant galaxies, however, are too far away for even the globular clusters to be seen, much less individual stars, except for an occasional supernova.

D. SUPERNOVAE But an occasional supernova is of great help to the astronomer. As indicated on p. 215 supernovae can be used as a tool to estimate distance, since supernovae of either type I or II have fairly consistent luminosities within these types. Supernovae of type II reach luminosities of about 200 million suns and may thus be used to determine the distances of galaxies too far away for globular clusters to appear (Figure 14-2). Although

Figure 14-3
Photographs of clusters of galaxies all enlarged the same amount to show the decrease in apparent size and brightness of the galaxies with increasing distance. (a) Part of the Coma Berencies cluster estimated to be 220 million light years away. (b) Corona Borealis cluster estimated to be 740 million light years away. (c) Hydra cluster 2100 million light years away. (Photographs of the Hale Observatories.)

supernovae, when observed, prove very valuable for the measurement of distances, they are sporadic and thus not readily available. Therefore other distance indicators are more frequently used.

E. TEN BIGGEST GALAXIES The realization that galaxies form in clusters had led astronomers to try to use the galaxies themselves as distance indicators. The biggest galaxies in each cluster are of about the same size and luminosity. This luminosity and size have been determined and used to estimate the distances of clusters of galaxies. Figure 14-3 indicates how this method works. The clusters shown in each plate are at different distances from us; thus the galaxies in the Coma Berenices cluster, which is the closest of the three, appear larger than those in the other two clusters. The galaxies in the Hydra cluster appear the smallest, since it is the most distant.

F. THE RED-SHIFT There is yet another method of estimating distances of galaxies. In 1929 Hubble announced the results of studies which clearly indicated that with the exception of our Local Group *all* the galaxies have positive radial velocities, that is, a Doppler shift to the red. Furthermore, the farther the galaxy is from us the larger the red-shift. The spectral lines normally used to measure the red-shift are the H and K lines of calcium, and the extent of their shift can be seen in Figure 14-4. The galaxy in the top spectrogram has a velocity of *only* 750 miles per second, and the lines have been shifted the distance shown by the short arrow. In each succeeding spectrogram the H and K lines are shifted ever farther to the red until in the last plate they are shifted almost to the limit of sensitivity of the photographic emulsion used.

It will also be noticed that the apparent sizes of the galaxies in Figure 14-4 yielding each spectrogram decrease as the red-shift increases. This decrease in size is caused by an increase in distance. Hubble discovered that the red-shift of the spectrum increased for galaxies of increasing distance from us. The interpretation he placed on this is the one adopted today: ever more distant galaxies are receding from us at ever greater radial velocities.

Hubble carried his study further. He determined that there is a direct relationship between the distance of a galaxy and the red-shift of its spectrum. By 1936 his observations permitted him to estimate that for an increase of every 100 miles per second in the radial velocity, the distance of the galaxies would increase by one million light years. In other words, a galaxy with a radial velocity of 500 miles per second should be 5 million light years away; one with a radial velocity of 3000 miles per second would be 30 million light years distant. The relationship between the distance of galaxies and their red-shift is called the *Hubble constant*.

In the intervening years, larger telescopes and better auxiliary equipment have increased not only the number of observations, but these new observations are more precise. Consequently, the value of the Hubble constant has changed several times.

By the early 1960s, the Hubble constant had been reduced to 14 miles per second for every one million light years. In 1970, it was estimated to be 18 miles

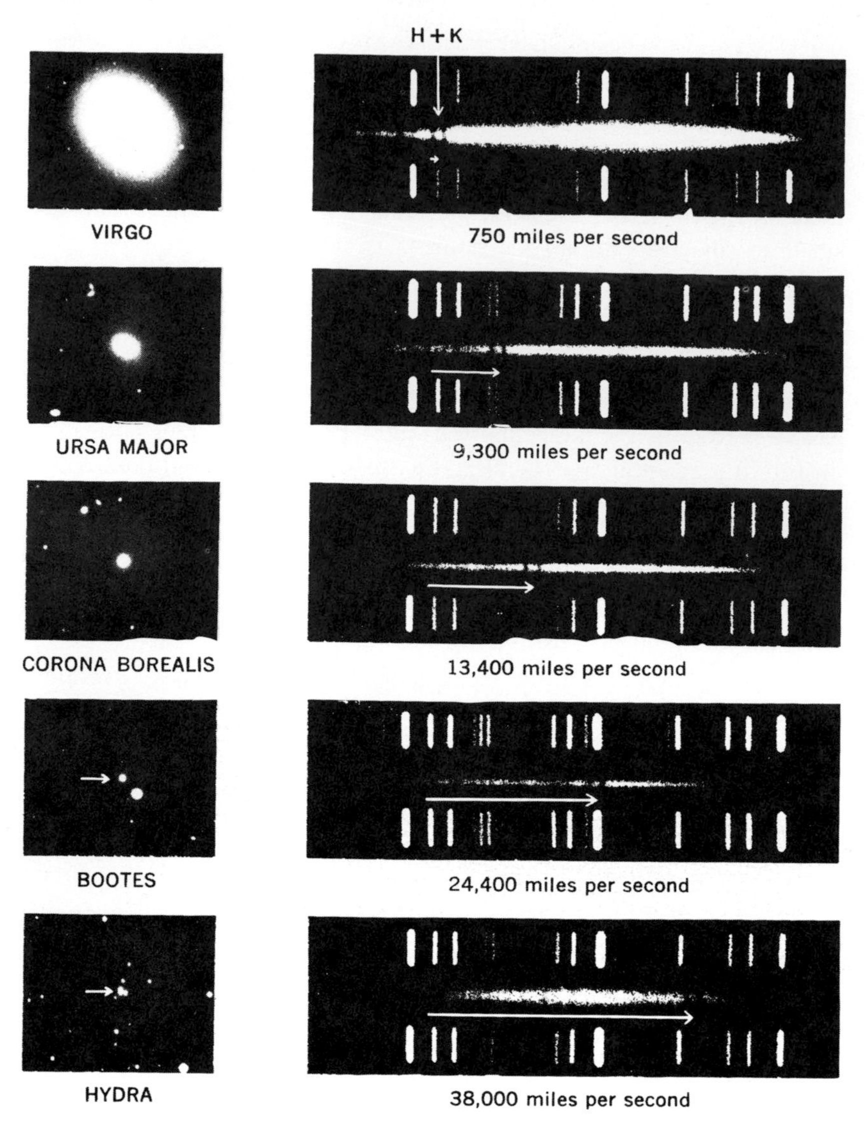

Figure 14-4
The actual spectrograms of five galaxies showing the extent of the shift of the H and K lines of calcium to the longer wavelength. The apparent diameter of the galaxies decreases with increasing distance and they appear ever fainter. (Photographs from the Hale Observatories.)

per second for every one million light years. The most recent value is an increase in radial velocity of only 11 miles per second for an increase of distance of one million light years. This is shown graphically in Figure 14-5.

A decrease in the Hubble constant means that our estimation of the size of the universe is increasing. For example, the galaxy in Hydra yields a radial velocity of 38,000 miles per second; using Hubble's original value that galaxy would be (38,000/100 = 380) 380 million light years away. Using the most recent value, that same galaxy appears 3500 (38,000/11) million light years away. In short, the increased number and quality of observations, along with further studies, indicate that the universe is perhaps 10 times bigger than Hubble thought it was in 1936.

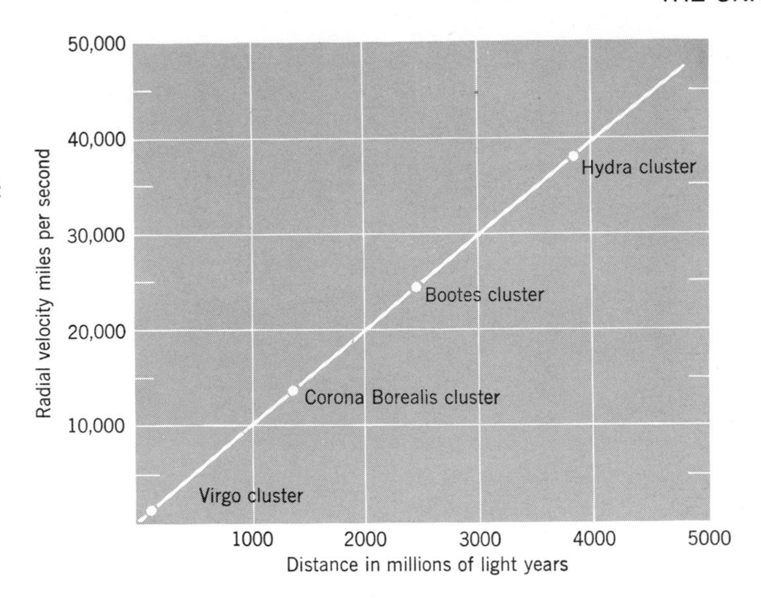

Figure 14-5
The velocity of recession of galaxies increases with increasing distance from us. It must be remembered, however, that these distances represent the location of these clusters hundreds of millions of years ago.

The obvious question arises: Is the most recent value of the Hubble constant reliable? Certainly it is expected to be much more reliable than Hubble's original value, but it would not be surprising if there were further modifications, although they are expected to be small. According to Sandage and Tammann of the Hale Observatories who derived this value, it is accurate to within 10%.

According to the new Hubble constant, the distances of three of the galaxies shown in Figure 14-4 are: Virgo, 75 million light years; Corona Borealis, 1340 million light years (1.34 billion light years); and Hydra, 3800 million light years. And that is a long ways off!

14.3 THE MEANING OF THE RED-SHIFT

It appears that the red-shift, discovered by Hubble, is a fundamental characteristic that may help man in his endeavor to understand the universe: (1) With the exception of our Local Group, *all* the galaxies have their spectra shifted to the red. If this is interpreted as a Doppler shift we are left with the conclusion that all these galaxies are receding from us! (2) The farther the galaxy is from us the larger the red-shift. Thus the velocity of recession increases with increasing distance. (3) These velocities of recession reach values interestingly close to the velocity of light.

It is possible to interpret the red-shift as a consequence of light losing energy as it travels through the universe. It may, for example, lose energy by interaction with matter. Such an interpretation would not demand an expanding universe. But, then, the problem of stability of the universe arises. With the gravitational forces acting mutually on galaxies, why don't they all attract each other and the universe collapse? In other words, it is easier to understand an expanding universe or a collapsing universe than it is one whose size does not change. For these reasons, astronomers generally agree that the radial velocity interpretation is the most plausible. The natural consequence of this interpretation is that the universe is expanding. Thus the term red-shift has become synonymous with expansion of the universe.

The fact that the red-shift increases with increasing distance from our galaxy places a maximum limit to the size of the observable universe. If galaxies exist at such distances that their radial velocities equal or exceed that of light, we shall never be able to observe them because the light will never reach us.

Since all the galaxies are receding from us, we could infer that our galaxy is at the center of the universe, were it not that such an inference would place us only a step away from Plato, who put the Earth at the center of the universe. Astronomers are unwilling to accept any interpretation that places us in such a unique position. In general they prefer to base their thinking on the *cosmological principle: the appearance of the universe does not depend on the observer's position.* That is, with the exception of local and small differences, the universe should appear the same to all observers no matter where they may be located — in our galaxy or in a very distant galaxy.

To explain the observed expansion on these grounds, the structure of the universe must be considered. Discussions of the structure of the universe, however, involve the theory of relativity expounded by Albert Einstein. Therefore it becomes necessary to consider certain aspects of that theory before proceeding with the structure of the universe.

14.4
THE PRINCIPLE OF SPECIAL RELATIVITY

At the turn of this century there were many observations conflicting with the physics derived since the time of Galileo and Newton. Yet that physics — now called *classical physics* — has proved extremely successful. For example, astronomers used Newtonian physics to predict the existence and location of an unseen planet, Neptune. Scientists still use it to place satellites in the orbit of their choosing, and to aim lunar probes at a particular location on the moon. It works. But it has its limitations.

In 1888 the famous Michelson-Morley experiment was performed which showed that the measured velocity of light is independent of the motion of the observer. For example, if we were to measure the velocities of light emitted by star *a* (Figure 14-6) and by star *b*, we would find that those two velocities are equal! The measurement of each of those two velocities is independent of the motion of the Earth. This is not true of cars moving along a freeway; two cars

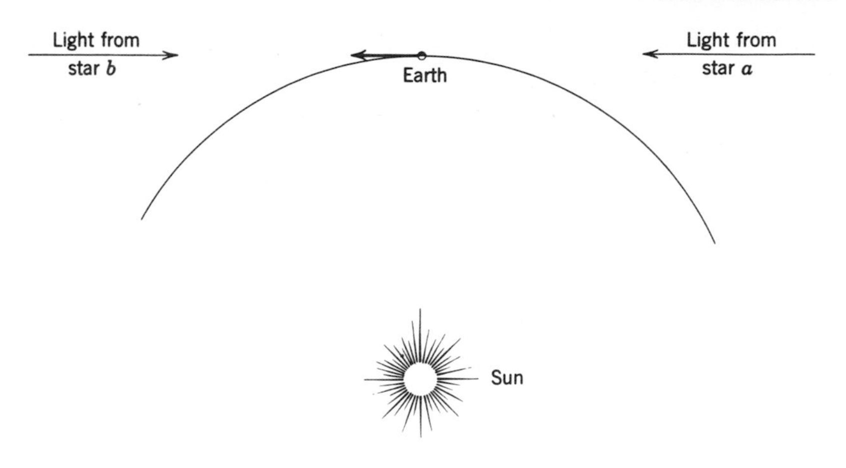

Figure 14-6
The Michelson-Morley experiment showed that the velocity of light as measured from the Earth of either star *a* or *b* is the same despite the motion of the Earth.

moving at a velocity of 55 miles per hour but in opposite directions have a velocity with respect to each other of 110 miles per hour. But automobiles do not travel at speeds approaching that of light, and that is what makes the difference.

Einstein derived his *principle of special relativity* to describe the actions of systems moving relative to each other at *constant* velocities. The consequences of this principle are many, the concepts are profound and in conflict with our common sense, that is, the everyday garden variety of common sense.

The two basic postulates of special relativity are: (1) the velocity of light in free space is finite and constant for all observers, and (2) the laws of physics should be valid for all observers even if those observers move relative to one another. By adopting these two postulates Einstein showed that the measured values of mass, length, and time change when transferred from one system to another if those two systems are moving at relativistic speeds (speeds close to that of light) with respect to each other.

If people outfit two spacecraft with essentially identical items and embark on a very rapid space flight such that the two spacecraft *A* and *B* travel at a relativistic speed with respect to each other, they will observe that objects on the *other* spacecraft appear both more massive and yet smaller — but smaller in only one direction, the direction of relative motion — than those once identical objects on their own spacecraft. They will find that the clocks on the *other* spacecraft will run slowly compared to the clocks on their craft. Passengers in both craft *A* and *B* will notice the same effects; the measurements of mass, length, and time depend upon the motion of the observer relative to that which he observes.

Ridiculous! says the skeptic. A kilogram is a kilogram, and a second of time is a second of time! But it has been proved beyond the shadow of a doubt, claims the physicist. Electrons have been accelerated to speeds close to that of light and their mass does indeed increase; that is, a greater force is required to obtain the same acceleration to still greater speeds. Atomic particles called pions which serve as good clocks, because they decay into other particles in a measurable duration of time, have been accelerated to relativistic speeds and they decay more slowly!

In proposing these fundamental ideas Einstein did not give any reason for the apparent change in mass, length, and time when the observer moves relative to that which he observes. He simply showed conclusively that *these observations are the natural consequence of our living in a universe in which the velocity of light is finite and constant for all observers regardless of their motion.* The principle of special relativity has been proved valid and is the foundation for much of modern physics and astrophysics.

One of the many consequences of these fundamental concepts is the equivalence of energy and mass: $E = mc^2$. Recall that it is the conversion of mass to energy which permits astronomers to account successfully for the tremendous outpouring of energy from stars for billions of years. Another consequence of the postulates of special relativity is the fact that in free space no material object can travel with a velocity equal to or greater than that of light. This applies to electrons — this has been verified — and to galaxies.

One of the nonrelativistic implications of the finite velocity of light is that astronomers really are cosmic historians; today they *observe* the past. Astronomers can only say what happened on the sun 8 minutes ago; they give the date of the supernova that resulted in the Crab nebula as A. D. 1054, even though the light from that explosion had already been traveling in space for about 6000 years before it reached the Earth. When they observe galaxies estimated to be 2 billion light years from the Earth, they see those galaxies as they were 2 billion years ago, not as they are "now." Therefore what use is it to say a certain galaxy is 2 billion light years away from us? None whatsoever, for all that an astronomer can say is that the galaxy's position 2 billion years ago is 2 billion light years away from our position at the present time. He cannot and never will be able to measure how far apart we are at any one given instant of time. There is no "now" in the universe because light has a finite velocity, and measurements of such vast distances become misleading.

This condition results from the fact that we cannot really separate time and distance as two separate and independent entities. If we wish to identify a certain event we must not only say where it occurred but when. If two people arrange to meet in the northwest corner of the 50th floor of the Empire State building, they have specified the three space coordinates: one along the north-south line, the other along the east-west line and the third along the up-down line. But they must also state at what time they will meet. Thus it takes four dimensions to identify a particular event, three of space and one of time, and these four dimensions are inextricably bound together.

14.5 THE HISTORY OF THE UNIVERSE

Even if we do see the universe stretching out in both space and time, we still ask where the expansion is taking us. If the universe is now expanding, can we trace it backward in time to figure out whether it was all one big blob that exploded some eons ago. Or was its beginning more prosaic? Or did it ever being? Has it always existed? If it had a beginning, how did it begin? And when?

We must be careful, however, not to ask questions we know we cannot answer. One reason for the successes of science is that scientists have restricted their studies to questions that bore fruit; questions that were not only pertinent to their field of study, but which they knew they had a reasonable chance of answering.

In the past two decades there have been two main theories attempting to explain the structure and history of the universe. One is called the *big-bang theory*, since it assumes a definite beginning of rather dramatic proportions. The other, the *steady-state theory*, maintains that the universe has an infinite age and thus no beginning or ending.

A. THE BIG-BANG THEORY According to the big-bang theory all the matter in the universe was once contained in a *primeval nucleus*, the density of which was perhaps equal to the density of the nucleus of an atom. Our present laws of physics are not adequate to describe details of the primeval nucleus, but clearly at some time and for some reason, this theory demands that the primeval nucleus exploded. From an initial temperature in the billions of degrees, the expanding universe cooled to its present chilly state.

By determining the rate of expansion and the size of the observable universe, that is, by determining the value of the Hubble constant, it is possible to retrace the process of expansion to find the date of the explosion. The most recent value of the Hubble constant, 11 miles per second per million light years, places the explosion about 18 billion years ago. However, this does not take into account the possibility that the rate of expansion might change with time. The rate of expansion might decrease with time, or increase, or even remain constant.

1. Decreasing Rate of Expansion As the speed of a rock thrown into the air decreases until it reaches its greatest height, so the universe might continue to expand at a decreasing rate until it stops expanding. And then what?

The rate of expansion can decrease only if a mutually attractive force acts between each of the galaxies. The only such attractive force known is the gravitational force. It is the gravitational force between the rock and the Earth, after all, that causes the speed of the upward thrown rock to decrease until it stops. The rock then falls back down to the Earth. Therefore, should the expansion of the universe be brought to a halt by gravitational forces, the universe would then "fall back down," that is, it would collapse back into a primeval nucleus.

The primeval nucleus, however, is where it all began; hence, presumably it would explode again and expand into another universe. New galaxies would form, but all traces of the present universe would be erased in the intervening primeval nucleus. Such a universe is called a *pulsating universe*, for it could conceivably continue pulsating forever, each explosion creating a new universe only to be erased by the collapse into a primeval nucleus.

Since the pulsations could go on and on, the pulsating universe is attractive to the scientist, since it dismisses the questions "How and when did it all begin?" and "How and when will it all end?" These are questions that the scientist asks only in rhetoric, for he does not seriously expect to find the answers.

To pulsate, however, is not the only possibility for a universe that is expanding at a decreasing rate. Rockets, after all, are something like rocks, each is held to the Earth by the force of gravity. But a rocket has fuel, which when ignited can send it off with a speed so great that even though the speed decreases with increasing height, the rocket escapes from the Earth completely, that is, the rocket can exceed the escape velocity of the Earth.

It is possible, therefore, that the rate of expansion of the universe is so great that it exceeds the escape velocity of the universe. The universe would simply continue expanding forever. The space density of the clusters of galaxies would continue to decrease until in some future epoch the only galaxies visible from the Earth (should it still exist) would be those in our own cluster, the Local Group. The rest of the universe would have vanished off into space, leaving scarcely a trace.

A universe that expands infinitely from an explosion demands a single beginning yet an infinite lifetime. This rather lopsided existence does not seem to have much appeal to the scientist who would prefer an ending of similar nature to the beginning. But the scientist must accept the reliable observations of his colleagues and interpret them as best he can.

2. Increasing Rate of Expansion The fate of a universe that expands at an increasing rate is the same as that of the universe which is expanding faster than the escape velocity of the universe. They both proceed on to infinity, only the one expands faster than the other.

Instead of the force of gravity attempting to collapse the universe, a universe that expands at an increasing rate requires that a force of repulsion push the clusters of galaxies apart.

3. Constant Rate of Expansion A universe that expands at a constant rate has best been described by the steady-state theory.

B. THE STEADY-STATE THEORY The idea of the steady-state universe results from an extension of the cosmological principle. Proponents of this theory make the point that Einstein showed that the universe is a four-dimensional space-time continuum. Therefore, the cosmological principle should be restated: *the appearance of the universe does not depend on the observer's position in either space or time.* That is, with the exception of local and small differences, the universe should appear the same to all observers no matter where they may be located or when — at what astronomical epoch — they should observe it. The universe should appear the same, in its overall aspects, if viewed 15 billion years ago or 5 billion years from now. Surely some stars will have been born and died out; individual galaxies will have changed, but the overall aspects of the universe should remain constant throughout all space and time.

A basic assumption in the steady-state theory is that the universe never began, or to be more precise, that it is in a process of continual creation. As the universe expands, its overall density decreases, but there is a universal lower limit to this density, below which it cannot exist. As the density approaches this lower limit, more matter is created to increase the density once more. Thus as the universe continues to expand, new matter is created to fill the void. The newly formed matter is hydrogen that eventually forms clusters of galaxies. Each new cluster of galaxies lives out its life as the universe continues to expand, while new clusters of galaxies form. New galaxies form, and old ones die, but the universe is always at the same density and there are always galaxies of different ages. Thus the universe will be the same if examined at any epoch. Although individual galaxies and clusters will have changed, the overall picture will not have changed. This is what is meant by a "steady-state" universe that is not only infinite in age but infinite in extent.

14.6 OBSERVATIONS OF THE UNIVERSE

We have presented several models or theories of the universe. The question now is: Which one does our observable universe most nearly resemble?

To settle any scientific question, we have learned to rely on observations. A statement should be questioned unless it can be supported with observations, and the more observations the better.

A. QUASARS The correlation of discrete radio sources with images on photographic plates has led astronomers down many new paths. Exploding galaxies were one such path. Another path is the very puzzling objects, discovered in 1960, which have been dubbed *quasars,* short for quasi-stellar sources. (When an object is discovered it must be named for purposes of classification and conversation, but its name may have nothing to do with its actual structure — since its structure may not be known when the name is applied. For example, quasi-stellar source means essentially the same thing as asteroid! But there the similarity ends.)

Quasars were observed first as discrete radio sources and so individuals are named according to their number in the *Third Cambridge* catalog: 3C 48, 3C 273, etc. They were later identified as bluish "stars" (Figure 14-7), with unrecognizable spectral lines. Maarten Schmidt of the Hale Observatories, however, showed that these strange spectral lines are really lines of the far ultraviolet

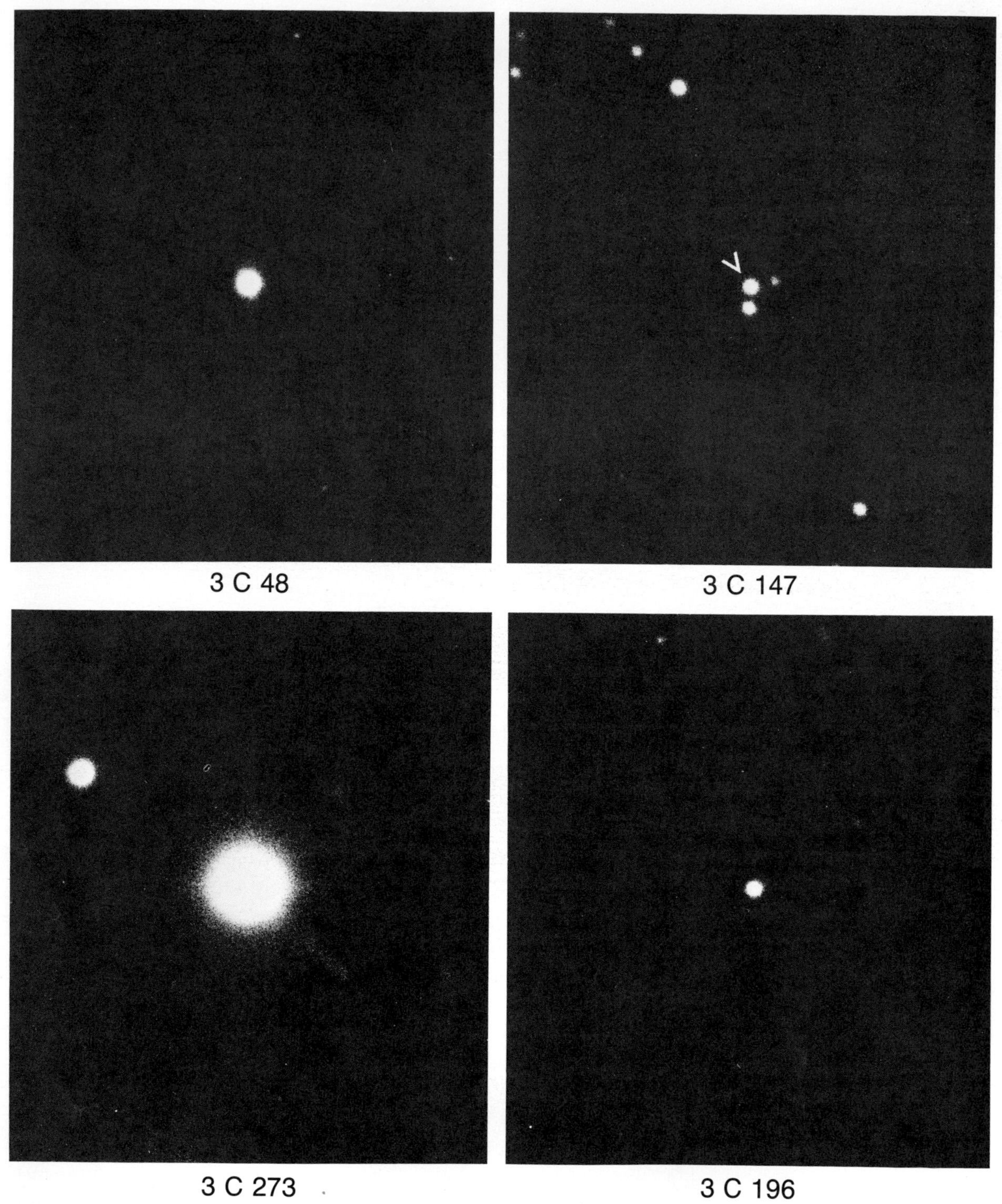

Figure 14-7

Four different quasars. The jet can be seen extending from 3C 273. (Photographs from the Hale Observatories.)

displaced into the near ultraviolet and visible part of the spectrum. Quasars suffer a severe case of red-shift.

The red-shift of 3C 9 is so large that the change in wavelength exceeds the unshifted wavelength by a factor of 2! This means that the Lyman-alpha line of hydrogen with an unshifted wavelength of 1216 Å has an apparent wavelength of 3648 Å (1216 + 2 × 1216). The Balmer H-alpha line with an unshifted wavelength of 6563 Å appears in the infrared region at 19,689 Å. To translate this into a radial velocity by using the classical relationship for the Doppler shift

$$\frac{\Delta\lambda}{\lambda o} = \frac{v}{c}$$

would yield a velocity of recession twice that of light! (See Problem number 5, page 32). How can we see it, if the Earth and that quasar are receding from each other at twice the speed of light?

Clearly there is a discrepancy! This discrepancy, however, is easily resolved. All we must do is to realize that the speeds involved are too great to use Newtonian physics; we must employ the principle of special relativity. It, too, produces an algebraic expression for the Doppler shift:

$$\frac{\Delta\lambda}{\lambda o} = \sqrt{\frac{c + v}{c - v}} - 1$$

where v is the relative velocity of the source and c is the velocity of light. By using this equation from special relativity, we determine the radial velocity of 3C 9 to be $0.8c$, or 80% the speed of light!

It is easier, however, to discuss the Doppler shift of quasars in terms of the fraction $\Delta\lambda/\lambda o$, called simply the *red-shift*. If the red-shift is 1, the spectral lines are shifted an amount equal to their laboratory wavelength λo. The Lyman-alpha line, for example, would have an observed wavelength of 1216 Å + 1216 Å = 2432 Å. The observed wavelength of the Lyman-alpha line of a quasar with a red-shift of 2 would be 1216 Å + 2432 Å = 3648 Å.

Continued observations of quasars indicate that their density in space increases with increasing red-shift until a red-shift of 2.5 is reached. Then their numbers seem to decline even though their brightnesses have not yet reached the limit of our telescopes. Apparently the decline in numbers is real.

If the red-shift results from the expansion of the universe, then as we consider ever greater red-shifts, we are considering ever more distant quasars. But as we observe ever more distant quasars, we are also looking back in time. If their numbers begin to decrease after a red-shift of 2.5, then their density in space — and thus in time — must actually decrease. But why?

Recognizing uncertainties in estimating their distances and thus their age, but using the Hubble constant as 11 miles per second per million light years, and assuming a constant rate of expansion, quasars with a red-shift of 2.5 are roughly 14 billion light years away from us. We see them as they were 14 billion years ago. If the age of the universe is 18 billion years, then most of the quasars formed about 4 billion years after the big bang. As the universe aged, their numbers declined, so their activity must not be long lasting. We see no quasars nearby, so they must not be forming this late in the age of the universe.

The observations of quasars clearly tip the scales toward the big-bang theory. Quasars seem to have been a passing phase, early in the life of the universe, which because of the expansion and great distances are only now becoming visible to us. Astronomers are cosmic historians.

If quasars are so distant from us, however, why are they so easily visible? They simply must radiate fantastic amounts of energy, perhaps 100 or 1000 times the amount of energy radiated by the Milky Way galaxy! Just how they manage this vast expenditure of energy is as yet an unsolved mystery.

The mystery is further complicated by the observations that their brightnesses and thus their luminosities vary significantly in a matter of days. How such an apparently large, massive, and luminous object can change its luminosity so suddenly is not at all understood.

The mystery of the universe is still very much with us, even if the nature of that mystery is changing.

B. THE PRIMEVAL FIREBALL The quasars, even though they remain unexplained, support the big-bang theory of the universe. Another unexpected discovery, this one in 1965, also gives definite evidence in favor of the big-bang theory.

Two radio astronomers at Bell Telephone Laboratories, A. A. Penzias and R. W. Wilson, detected what was for them a disturbing background noise in their radio telescope. Since the radio signals come uniformly from all directions in the sky, it was proved not to be of terrestrial origin. Thorough investigation showed that it was radiation from a source whose temperature is about 3° K.

At the same time that this investigation was proceeding, Robert Dicke and some of his friends at Princeton University (only 30 miles away), were independently investigating theoretically what would happen to the energy of the *primeval fireball*, the big bang, that started our universe expanding. They had come to the conclusion that although the initial temperature of the explosion may have been 10^{10} degrees Kelvin, and the wavelength of the radiation correspondingly very short, the temperature would have cooled because of expansion. The temperature they predicted from their theoretical studies was about 5° K, nearly the same as that causing the radiation found by Penzias and Wilson at Bell Telephone Laboratories.

Continued studies have indicated that the observed radiation may, indeed, be left over from the initial explosion of the universe. The radiation has remained in the universe and has continually lost energy, until it now represents not a body at 10^{10} degrees Kelvin, but one of only 3° K. Radiation of this nature is strong evidence in favor the big-bang theory, for it plays no role whatsoever in the steady-state theory.

C. A DISTURBING NOTE During the past decade or so, Allan Sandage of the Hale Observatories has carefully reexamined the problems of distance measurements in the universe. He has carefully arrived at a more precise value of the Hubble constant.

In the late 1930s and 1940s Hubble believed that the observable universe was about 1.8 billion years old. In 1958 and again in 1968 Sandage published his more precise value of the Hubble constant. He had, after all, the use of the then recently completed 200-inch Hale telescope on Mount Palomar, as well as contributions of the research of other astronomers. His revised value of the Hubble constant yielded a new age of the universe: about 13 billion years. This much greater age, of course, only increased the size of the observable universe. All those galaxies appeared to be spread out over a much larger volume of space. The space density of galaxies appeared much less than before.

In 1974, Sandage published a still more precise value of the Hubble constant and, again, our horizons were stretched. It now appears that the universe is 18 billion years old and proportionately larger. The universe appears to be 10 times older than Hubble imagined it to be, and the radius of the observable universe is 10 times larger.

It is this increase in the estimate of the size of the observable universe that is disturbing. The rate of expansion of the universe depends on the density of matter in the universe. Although the determination of the radius of the observable universe has increased tenfold, the amount of matter has not increased in proportion. This leads to a new estimate of the density of matter, and this new estimate is so low that astronomers wonder if the expansion of the universe can be stopped. It now appears that the universe will expand forever, because there is not enough mass to produce a strong enough gravitational field to stop it.

On the other hand, studies of clusters of galaxies indicate another conflict. Clusters of galaxies are supposedly held together by the mutual gravitational forces of the member galaxies. The individual galaxies move about, and if they were to move fast enough they could break away. Here is the conflict: there does not appear to be enough mass in some of the clusters to hold its members together. Estimates are that some clusters appear to need 7 to 10 times the amount of mass than is actually observed. Where is the missing mass? If it were detected, would it be enough to put the brakes on the expansion of the universe?

Having learned something of the changes that observations have forced astronomers to make in their thinking of the universe, we should be ready to expect an explanation of quasars, of the expanding universe, or of the clusters of galaxies, or of black holes that stems not from the present ideas of physics and astronomy, but perhaps from some fundamental mechanism not yet imagined. Startling observations demand bold explanations; new ideas stimulate new observations. This is the way science progresses.

BASIC VOCABULARY FOR SUBSEQUENT READING

Big-bang theory
Cosmogony
Cosmology
Hubble constant
Primeval fireball
Primeval nucleus
Pulsating universe
Quasars
The red-shift
Steady-state theory

QUESTIONS AND PROBLEMS

1. Give a general description of a cluster of galaxies, indicating something of its size, shape, types of member galaxies, and motion of galaxies within the cluster.

2. Of those galaxies in a cluster, which move with the greatest velocity? Which move the slowest?

3. List in order the methods of distance determination used by astronomers. Begin with the method that can be used only for the nearest stars and end with the method used for the most distant galaxies.

4. Give some indication of the accuracy of each of the methods of distance determination listed in Question 3.

5. The cluster of galaxies in Ursa Major has an average radial velocity of 26,000 miles per second. Using the Hubble constant (or the graph in Figure 14-5) estimate the distance of this cluster.

6. Outline the steady-state theory and the big-bang theory of the universe. Compare the predictions made by each that lend themselves to observations, and that might be used to support one theory or the other.

7. If the number of quasars increases with increasing distance from our Milky Way galaxy, which of the two main theories of the universe would this observation support and why?

8. If a quasar has a red-shift of 1.5, what is the observed wavelength of the Lyman-alpha line?

9. A quasar is observed to have a red-shift $\Delta\lambda/\lambda o = 1.5$. What would be its velocity of recession using the classical equation for the Doppler shift? Now calculate its velocity of recessing using the Doppler shift from special relativity. Use $c = 3 \times 10^{10}$ cm/sec.

FOR FURTHER READING

Barnett, L., *The Universe and Dr. Einstein,* William Morrow and Co., New York, 1957.

Hodge, P. W., *Concepts of the Universe,* McGraw-Hill paperback, New York, 1969.

Hodge, P. W., *Galaxies and Cosmology,* McGraw-Hill paperback, New York, 1966.

Munitz, M. K., *Theories of the Universe,* The Free Press, New York, 1957.

Page, T., and L. W. Page, ed., *Beyond the Milky Way,* The Macmillan Co., New York, 1969.

Aller, L. H., "The Chemical Composition of Cosmic Rays," *Sky and Telescope,* part I, p. 285, May 1972; part II, p. 362, June 1972.

Bunner, A. N., "High-Energy Cosmic Rays," *Sky and Telescope,* p. 204, October 1967.

Burbidge, E. M., and C. R. Lynds, "The Absorption Lines of Quasi-stellar Objects," *Scientific American,* p. 22, December 1970.

DeWitt, B. S. et al., "A Relativity Eclipse Experiment Refurbished," *Sky and Telescope,* p. 301, May 1974.

Ginzburg, V. L., "The Astrophysics of Cosmic Rays," *Scientific American,* p. 51, February 1969.

Green, L. C., "The Fourth 'Texas' Symposium" (The Primeval Fireball), *Sky and Telescope,* p. 153, September 1967.

Green, L. C., "Observational Aspects of Cosmology," *Sky and Telescope,* p. 199, April 1966.

Green, L. C., "Quasars Six Years Later," *Sky and Telescope,* p. 290, May 1969.

Green, L. C., "Relativisitic Astrophysics," *Sky and Telescope,* part I, p. 145, March 1965; part II, p. 226, April 1965.

Hack, M., "The Abundance of Helium in the Cosmos," *Sky and Telescope,* part I, p. 154, September 1972; part II, p. 233, October 1972.

Pacini, F., and M. Rees, "Rotation in High-Energy Astrophysics," *Scientific American,* p. 98, February 1973.

Pasachoff, J. M., and W. A. Fowler, "Deuterium in the Universe," *Scientific American,* p. 108, May 1974.

Penrose, R., "Black Holes," *Scientific American,* p. 28, May 1972.

Schmidt, M., and F. Bello, "The Evolution of Quasars," *Scientific American,* p. 54, May 1971.

Webster, Adrian, "The Cosmic Background Radiation," *Scientific American,* p. 26, August 1974.

"Black Holes," *Scientific American,* p. 54, October 1970.

"Dead Galaxies," *Scientific American,* p. 50, April 1969.

"The Extragalactic Distance Scale," *Sky and Telescope,* p. 93, February 1970.

"Hubble Constant Revised," *Sky and Telescope,* p. 229, April 1972.

"Light Variations in Quasars," *Sky and Telescope,* p. 18, July 1969.

"The Local Group of Galaxies," *Sky and Telescope,* p. 21, January 1964.

"New Measurements of the Cosmic Microwave Background," *Sky and Telescope,* p. 10, January 1968.

"Pulsating Stars and Helium Abundance," *Sky and Telescope,* p. 145, September 1973.

"A Radio Test of Relativity," *Sky and Telescope,* p. 138, September 1970.

"Through a Black Hole Darkly," *Scientific American,* p. 46, March 1971.

INDEX

INSTRUCTIONS FOR USE OF STAR CHARTS

To use these star charts, step outside at 9 P.M. (standard time) and face south. Each column of stars will extend up into the south from the south point on the horizon at 9 P.M. during the month at the bottom of each column. For example, Cetus will be fairly high in the southern sky at 9 P.M. in December; Orion at 9 P.M. in February. In the charts other than the polar star charts, south is to the bottom, north to the top, west is to the right, and east is to the left.

THE NORTHERN CONSTELLATIONS

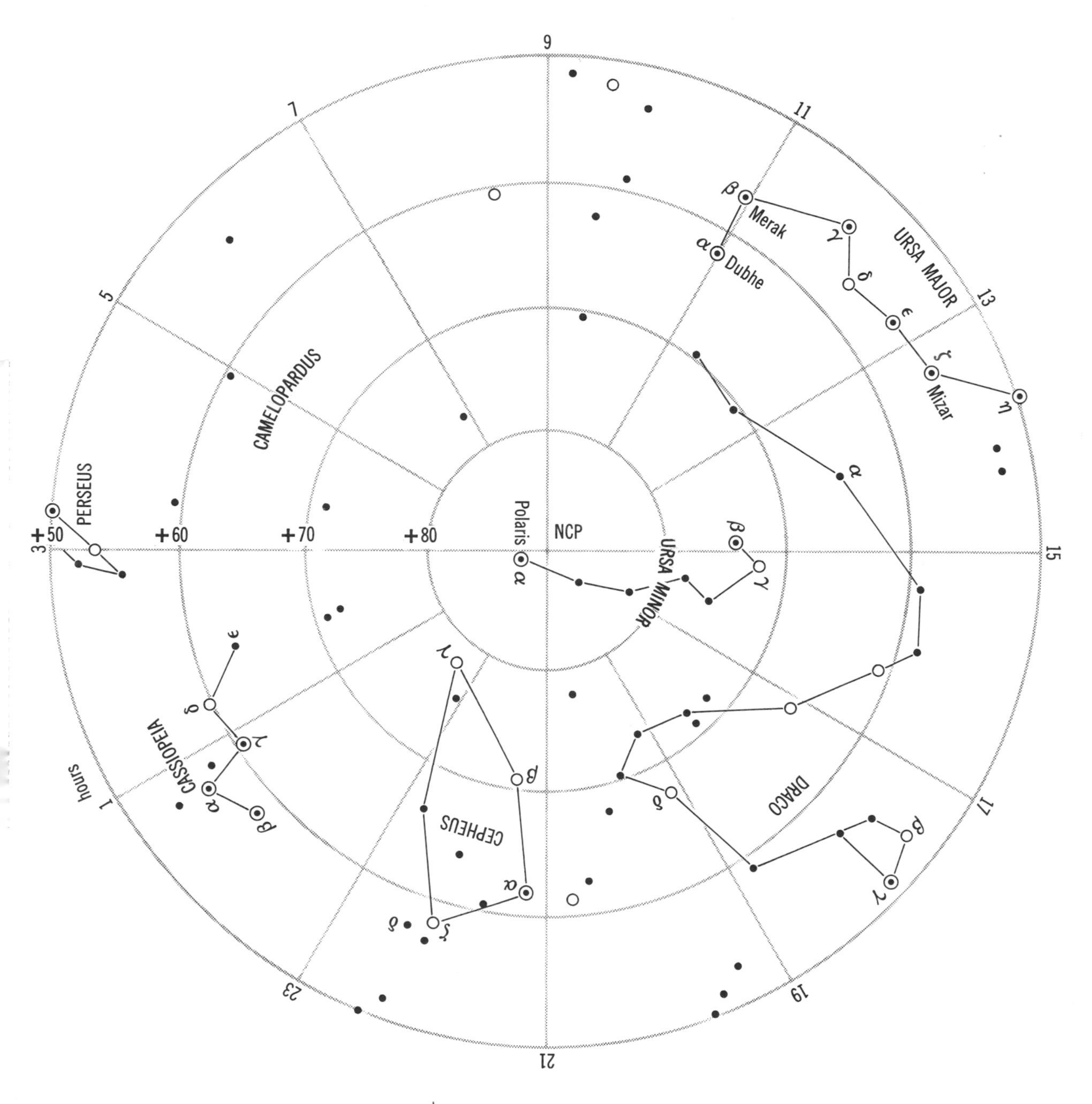

brighter than magnitude 1.5

between magnitudes 1.5 and 2.5

between magnitudes 2.5 and 3.5

fainter than magnitude 3.5

THE AUTUMN CONSTELLATIONS

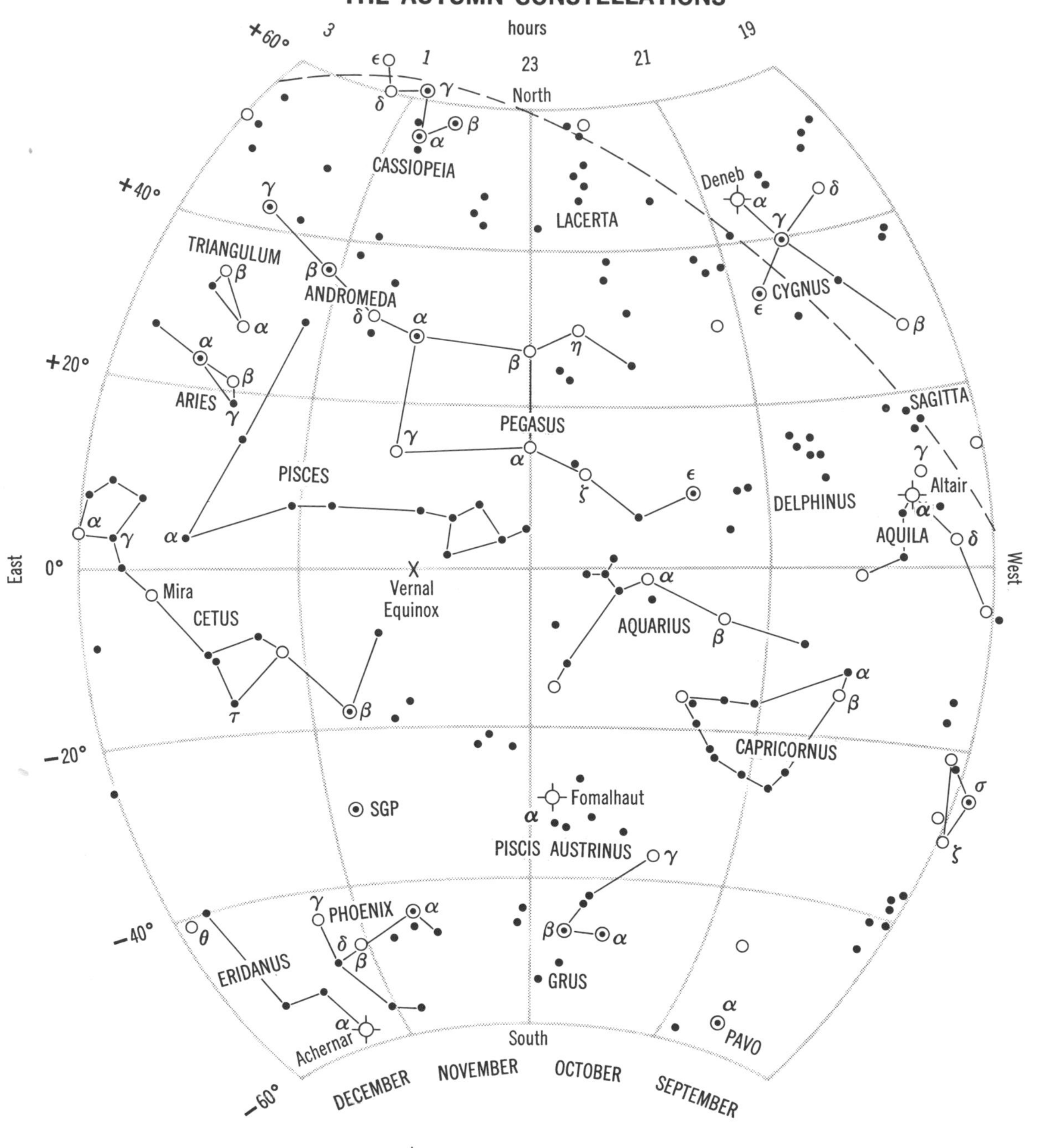

brighter than magnitude 1.5

between magnitudes 1.5 and 2.5

between magnitudes 2.5 and 3.5

fainter than magnitude 3.5

plane of Milky Way Galaxy

THE WINTER CONSTELLATIONS

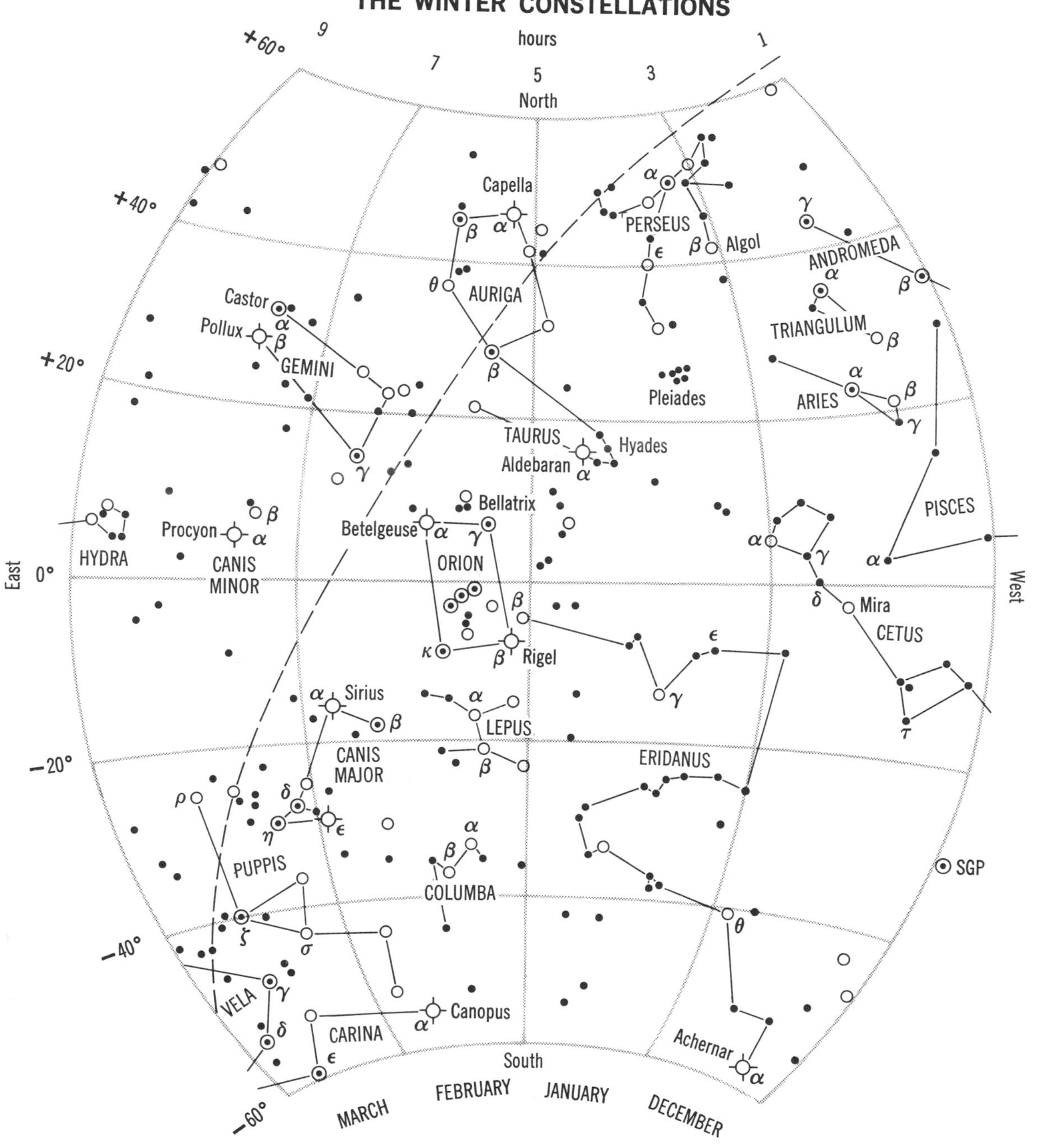

- brighter than magnitude 1.5
- between magnitudes 1.5 and 2.5
- between magnitudes 2.5 and 3.5
- fainter than magnitude 3.5
- – – – plane of Milky Way Galaxy

THE SPRING CONSTELLATIONS

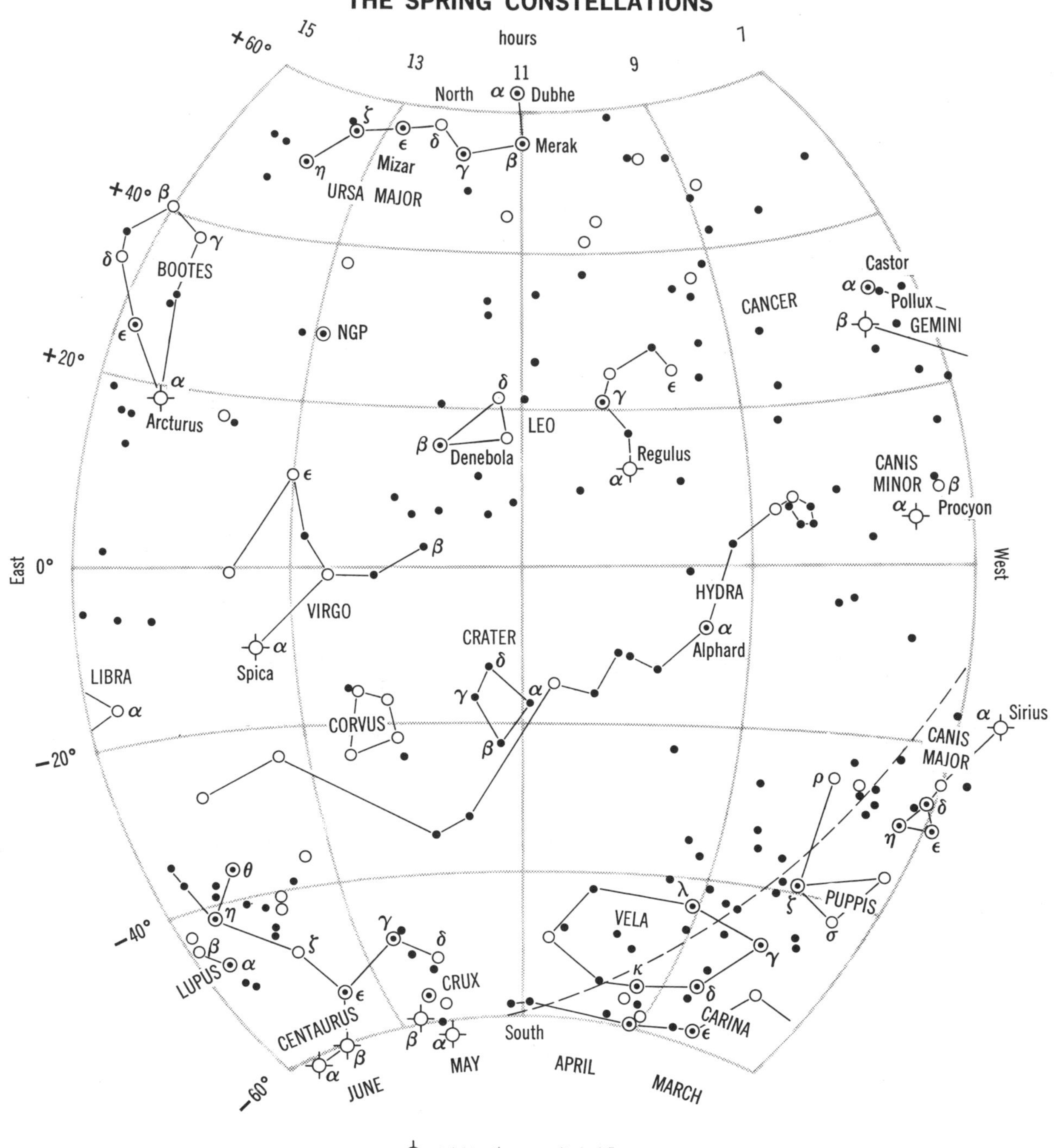

brighter than magnitude 1.5

between magnitudes 1.5 and 2.5

between magnitudes 2.5 and 3.5

fainter than magnitude 3.5

--- plane of Milky Way Galaxy

THE SUMMER CONSTELLATIONS

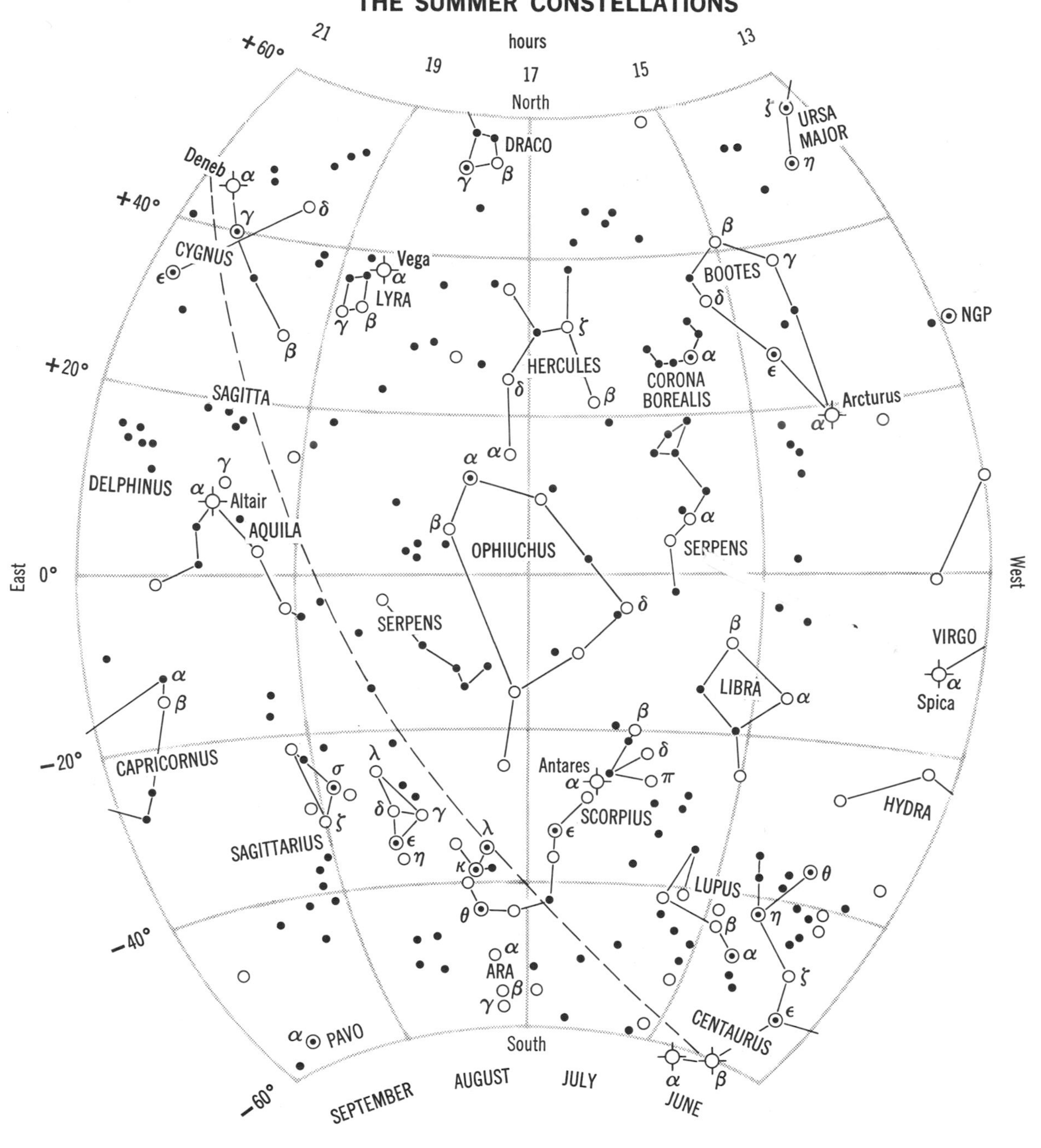

- brighter than magnitude 1.5
- between magnitudes 1.5 and 2.5
- between magnitudes 2.5 and 3.5
- fainter than magnitude 3.5
- ——— plane of Milky Way Galaxy

THE SOUTHERN CONSTELLATIONS

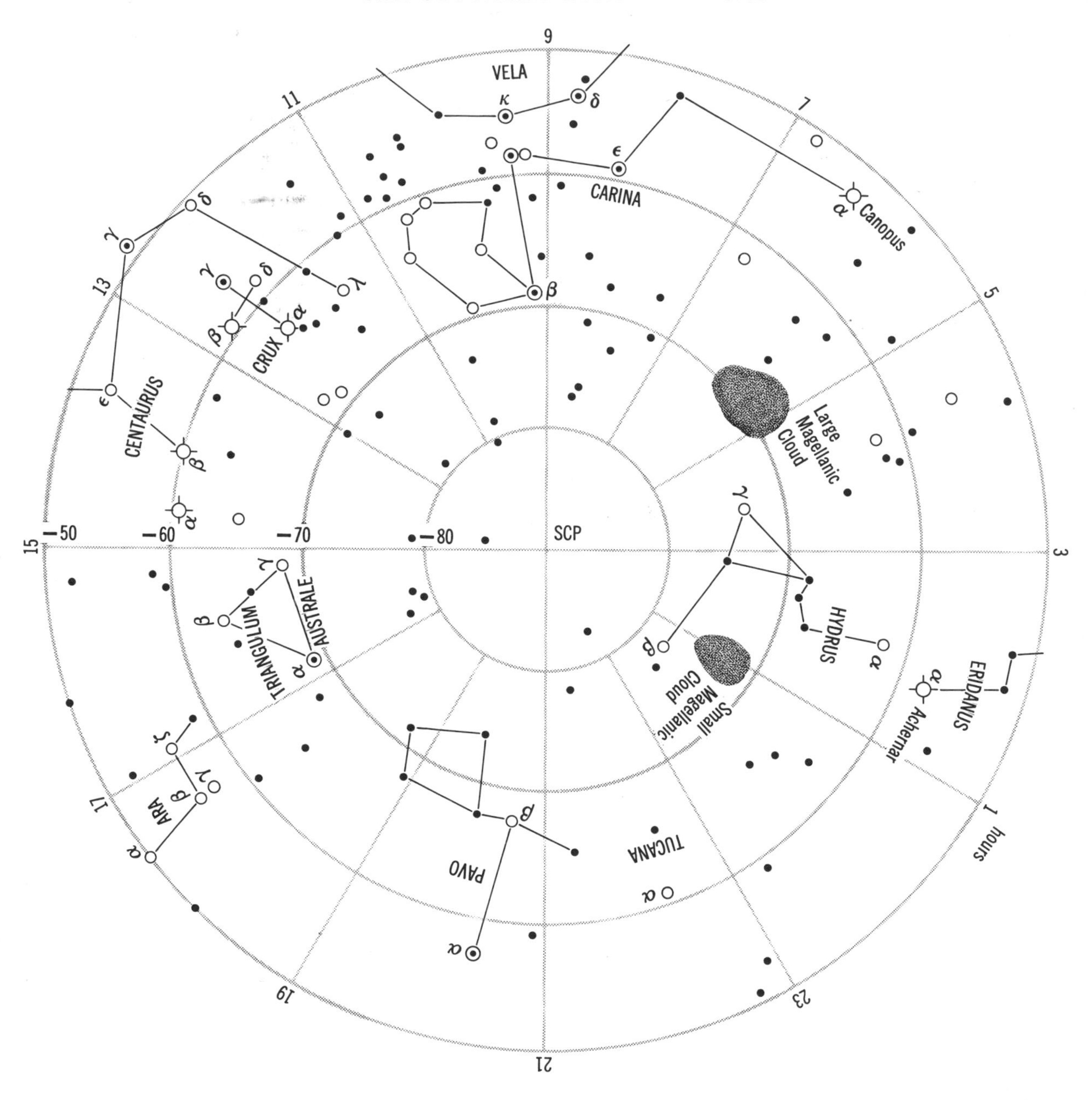

brighter than magnitude 1.5

between magnitudes 1.5 and 2.5

between magnitudes 2.5 and 3.5

fainter than magnitude 3.5